SpringerWienNewYork

Mediatecture

The Design of Medially Augmented Spaces

Christoph Kronhagel (Ed.)

SpringerWienNewYork

Christoph Kronhagel (Ed.) | Bonn | www.kronhagel-mediatecture.de

With contributions by:

- Bruno Schindler, Aachen
- Prof. Erkki Huhtamo, University of California
- Harald Singer, ag4
- Dr. Dirk Pörschmann, Zero Foundation
- Bernhard Pörksen und Heinz von Foerster
- Jan Edler, Realities United
- Manuel Abendroth, Lab(au)
- Wolfram Lusche, Bayer
- Andree Martens, Manager Seminare
- Prof. Lev Manovich, San Diego
- Prof. Josef Lukas, Martin-Luther-University Halle-Wittenberg
- Klaus Wassermann und Dr. Vera Bühlmannn, ETH Zurich
- Prof. Axel Thallemer, Industrial Design Education, Austria, Linz
- Wolf Lieser, DAM – Digital Art Museum, Berlin
- Dr. Christoph Köller, Cologne
- Marius Watz, New York
- Christoph Kronhagel, Bonn

Interviews with:

- Andree Verleger, Düsseldorf
- Harald Fuchs, Cologne
- Prof. Alexander Nützenadel, Humboldt-University Berlin
- Prof. Albert Speer, Frankfurt
- Prof. Bazon Brock, Wuppertal
- Prof. Frank T. Piller, TH Aachen
- Larry Weber, Boston

Layout and Cover Design: Christoph Kronhagel; Sebastian Völkel;
Image processing: Jens Kamin, Arne Hofmann;
Translation: Rupert Hebblethwaite; Peter Blakeney & Christine Schöffler; Harald Weiler; Reinhart R. Fischer, Nyon - CH;

Translations by Peter Blakeney & Christine Schöffler:
p. 180 / Indoor City
p. 212 / The Medial Projection of a Corporate Identity
p. 268 / Making Corporate Values Visible
p. 324 / How Media Create Reality in Our Heads
p. 364 / The Influence of "Mass Customization" on the Work of Mediatects
p. 398 / Digital Art for Media Façades
p. 440 / Perspectives for Mediatecture

Translations by Harald Weiler:
p. 98 / Second Order Cybernetics – A Paradigm for Mediatecture?
p. 368 / How Bionic Innovations Redefine Design Tasks

Translation by Reinhart R. Fischer:
p 334 / Streaming Spaces

All other translations by Rupert Hebblethwaite

Printing and binding: Holzhausen Druck GmbH, A–Vienna

Printed on acid-free and chlorine-free bleached paper

SPIN: 12667765
Library of Congress Control Number: 2010931689

With numerous colored figures

ISBN 978-3-7091-0299-2
SpringerWienNewYork

Note: *In order to be able to read the QR codes, you will need a compatible device (e.g. a smartphone) with a suitable scanner program. The legibility of the codes is dependent on your device's camera resolution and imaging software.*

The QR codes provide access to additional, interactive content.
All content can be accessed at www.mediatecture.info.

Foreword

Every building tells a story. Whether it swaggers with pomp and circumstance to demonstrate the pride of its owner, whether its complex mix of building styles bear witness to its dramatic history or whether its pragmatic appearance is a signal of simple functionality, a building always reflects the beliefs or at least the status of its current occupants. Even when façades recede into apparent inexpressiveness or appear to have been dropped at random into their urban environment they are telling us something.

The people behind these façades are being automatically labeled, assigned and classified. Some can influence this labeling process in that they are still in a position to choose - or at least influence - their architectural environment, but some cannot. But whether they can exercise this influence or not, this process is reasonably stable. The user of a piece of architecture can hardly escape from his label and this, in turn, influences both how he sees himself and how he is seen in his social context.

The digital revolution has given this process of labeling – now we can better call it encoding – a central significance. Everything is encoded; everything can be described by a binary sequence. And each of these digital "descriptions" can, at any time, be overwritten and adapted to a new set of circumstances. This digital flexibility can be seen in many areas of life. But does this represent progress? This is a subject of intense debate - and who, in any case, is in a position to define progress in this context? Beyond dispute however remains the fact that this encoding is the basis for a range of developments which (if one accepts that success is defined by levels of acceptance and proliferation) have been extremely successful.

And now we have the opportunity to encode our urban structures in ways which are flexible and can be adapted at short notice. This opportunity is offered to us by mediatecture.

The digital revolution is overwhelming most areas of life like a huge wave. Basic patterns of behavior are being transformed and people are becoming dizzy: to what extent are the new media a passing phenomenon and to what extent must we transform the way we think and act?

As I have now been engaged in mediatectonic projects for almost twenty years it was clearly time to take stock. Describing mediatecture was for me a delicate balancing act. I had to engage with topics, some highly technical in nature, which had little to do with my own work as an architect. If mediatecture makes it possible to develop new ways of encoding spatial structures then this automatically leads to a whole new set of relationships. For example – the spontaneous encoding of a façade immediately touches on the area of social processes which, in turn, raises questions of reception and communication. The act of taking stock of mediatecture can thus not be restricted to an examination of its technical aspects; it is not enough to merely describe – or explain – the use of media technology in architectural settings. The media façade may indeed enjoy a particularly prominent position in the world of mediatecture because it gives the subject such visual prominence. But there are also many other aspects of mediatecture which have a great influence over the design of our built environment yet which are not so directly part of the visible media landscape.

Fortunately I was able to persuade a number of mediatects and other experts to take part in this experiment – from the world of architecture and also from some very different fields. Without their contributions and the conversations which I was able to enjoy with them, I would have soon lost my balance. The book shows that a precise definition of mediatecture is not easy – and that is a good thing. It is precisely the complexity of this new area which is so alluring.

C. Kronhagel

Bonn, July 2010

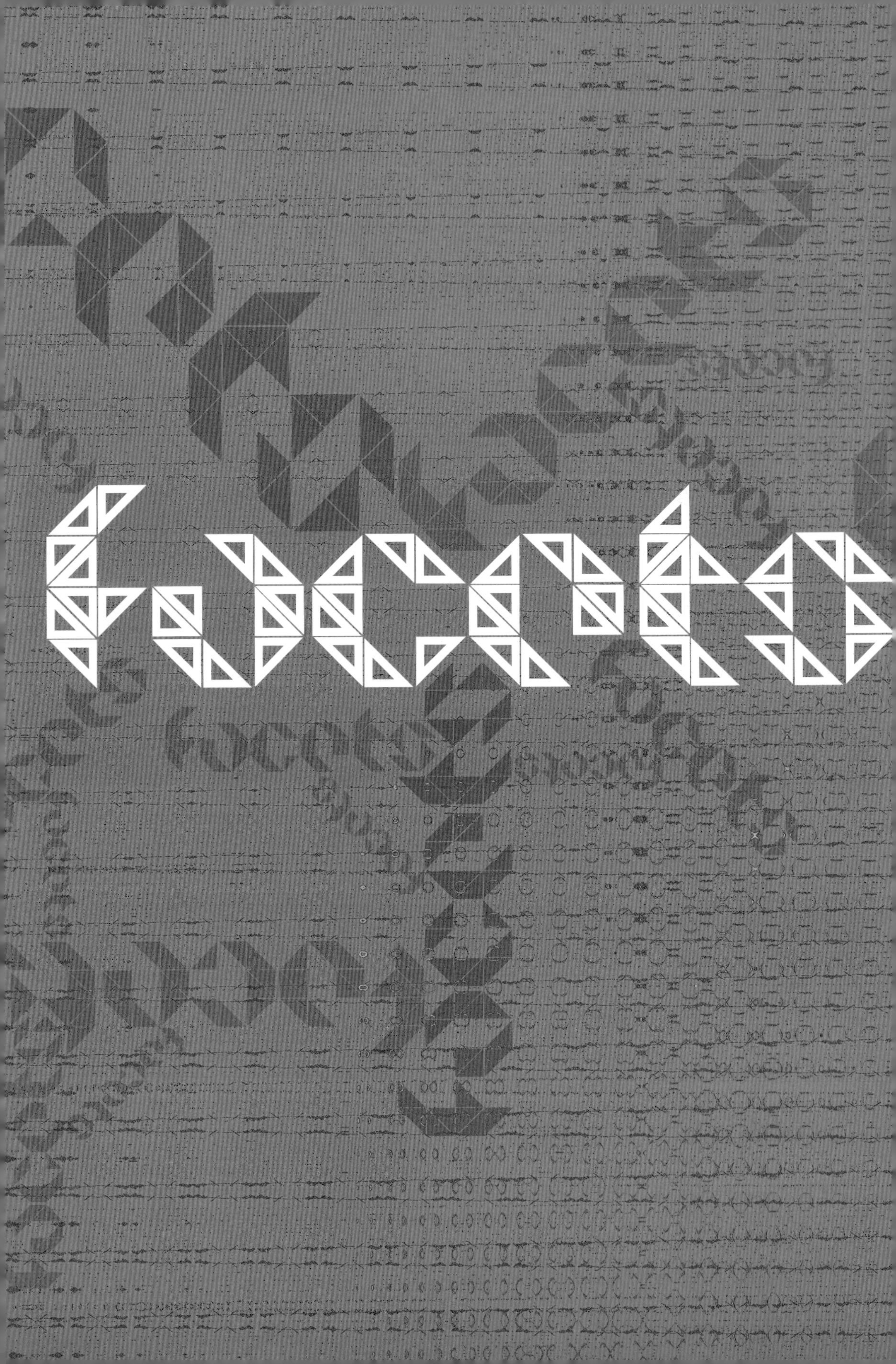

Aspects of Mediatecture

The design potential of mediatecture is currently being tested in very different sorts of public space. Any attempt in this book to describe some settled methodology could both interfere with this ongoing process of evolution and make mediatecture less accessible. From today's perspective it is not even clear if we are dealing here with a very new or a very old phenomenon. For this reason, this chapter is intended as an introduction which presents "aspects" of the mediatectonic way of working or, better, the mediatectonic way of thinking. This list of aspects is certainly not exhaustive, but hopefully it is at least the start of an attempt to create a proper place for this discipline which we know as "mediatecture".

The Principle of Sensual Orientation

I studied architecture at the Technical High School in Aachen between 1980 and 1988. But the individual who had the key influence on my later development came not from academia but was Bruno Schindler. Schindler is an architect who has spent his entire life dealing intensively with the question of what really lies at the heart of architecture and how architecture can be used to articulate key relationships. He is a brutally consistent lateral thinker who never compromises, which makes him quite unsuitable as a candidate for either an academic career or the realities of the architectural office. He lives in Aachen and has been involved for some time with the architecture magazine ARCH+. In 1987 he had the opportunity to produce a whole edition of the magazine and it was upon reading this that I came face-to-face with an idea that turned everything that the university had been trying to teach me on its head. This idea formed the basis of my understanding of mediatecture.

As I made my first attempts to understand architecture through the act of writing I contacted Bruno Schindler and this resulted in me being supervised by him for the rest of my studies. He has spent his whole life quietly surrounded by a small group of loyal students and he is very proud of the fact that most of these students have gone on to really achieve something. Writing this book gave me an opportunity to revive this contact and this renewed dialogue with Bruno Schindler allowed me to both sharpen my own position and improve the focus of the book which, up to that point had become somewhat indiscriminate in its attempt to cover too much ground.

In my attempt to present aspects of mediatecture in this book it appears to me very important that I also include the text that sparked my initial interest in the subject. Bruno Schindler has kindly granted us permission to reprint here his central article from ARCH+. It is short but to the point – almost, indeed, a manifesto. I will show the same images in the same sequence – because the sequential way in which images are used by Bruno Schindler is almost a language in itself. Which is hardly surprising considering that his central message was that architecture should offer us a form of sensitive orientation.

As I went out into the real world of architectural practice I gained a sense of comfort from knowing that I was carrying this idea with me.

A sense of comfort which was misplaced perhaps, because the real world of architectural practice looked very different: But firstly, here is the article:

Theaters of Power

Bruno Schindler

Book of Hours of the Duc de Berry by the Limbourg brothers from around 1415

Architecture is that part of building which seeks to demonstrate and legitimize power. Pyramids are architecture but farmhouses are not. The power of the Pharaoh stretches far beyond the horizon while the influence of the farmer ends at the bottom of the field. Architectural monuments are theaters of power. They are like the mighty flanks of warships bristling with cannons below and masts above, sails billowing in the wind and colorful flags proclaiming dominance and promising energy, profit and the inevitable victory celebrations. But architecture, on the other hand, has very little to do with the ebb and flow of the needy, the churning sea of houses, workshops, shops, huts and hovels; with – putting it another way – the needs of those without power. And hence books about architecture are full of temples, palaces and castles – as well as stations, arcades and skyscrapers – as is well demonstrated by this old plan of Rome.

Architecture has, from the very beginning, been concerned with power. Constantly surrounded by competitors and opponents, the powerful recognized early on that architecture has the ability to convince the public that it is better off with the leaders that it knows. Architectural progress became driven by the relationship between the purposes of power and the virtuosity of buildings – and this is a relationship in which the agenda of the powerful invariably took precedence over the skill of the builders. This explains why the baths and the fora of Rome remain known by the names of the emperors who built them despite the fact that they have lain in ruins for centuries. And why Versailles is, quite correctly, known as the palace of Louis XIV - and NASA's launch pad in Florida is named after an American president. From the very beginning, many works of architecture have been seen principally in terms of their relationship with the powerful figures in a city, a region or an entire country – and some have even come to represent the fortunes of an entire people.

This brief reference to the relationship between power and architecture intends to show that there is hardly any area of human activity which is less able to be an end in itself. And should a building have the temerity to attempt to remain autonomous – while at the same time avoiding the danger of drifting into insignificant formality – then its only chance of doing so successfully is by turning one blind eye to authority while the other eye squints warily at everything that the building must not be. In this connection it is highly instructive to look at the arms industry (and its civilian counterparts), to see the buildings which result from a constantly shifting balance of power – buildings which would hardly be worth mentioning were it not for the fact that the last two hundred years of European history have been marked by such a fatal complicity between the concealed exercise of power, autonomous architecture and meaning-based interpretation.

In the 18th century the powerful began to hide themselves increasingly behind philosophical speculation and formal institutions, which reduced their reliance on architecture alone as a means of intimidation and persuasion. Castles were abandoned and replaced with museums while nature, art and science each gained a virtuous aura as they became part of the secularized "better world" under cover of which the concentration and extension of power could continue much as before. Architects for their part were forced to work out the new rules of the game and they discovered these in the airy reaches of artistic theory, with the natural result that they were able to extend their influence over all those activities which could be interpreted as belonging to that "better world"; In short, architecture came to dominate virtually all building. Only those objects (which, ironically, were the true sources of power) such as factories, railways and bridges which it was difficult to envisage as part of the better world were left to the (new) engineering profession, the success of which was based on completely different, that is to say scientific, principles.

This completely speculative "better world" had, of course, the one severe drawback that it could not be judged in terms of the five senses alone: rather, it depended on continuous interpretation. The result was the growth of a new profession of "givers of meaning", who did not tire of explaining that progress, having started with the monkeys, was now in the hands of the English Gentleman - as a result of which the future looked extremely rosy.

Concealed power structures, autonomous architecture and meaning-based interpretation became increasingly intertwined and involved in ever more areas of life to the point at which even the humblest tenement building was given the architectural treatment, which made it ever more difficult to guess, on the basis of simple observation, which façade concealed a seat of power and which did not. It was precisely in those buildings where architecture should have been able to make a social statement that such architecture was missing – whereas it was precisely in those buildings where architecture logically appeared not to belong that it flourished. The act of building became a dark exercise in social deception and only with the images of the blue planet earth rising above the deserts of the moon and the writings of the Club of Rome did this situation change. Here, finally, autonomous conjecture independent of all sensual perception was exposed as a dangerous imperative of the present rather than a sure way towards a better future. Since this moment, powerful institutions have recognized that the wider public demands a more sensitive approach, as a result of which they now prefer to

clearly demonstrate how they intend to use architecture as a means of presenting and legitimizing their power. This is a very old approach – and, perhaps for this very reason, also completely new. (Corporate Style).

Every sort of power should arouse suspicion, but it is at least infinitely preferable that power structures are represented in – rather than hidden from – a society. Not so much because it is against the interests of the powerful to see support eroded by an atmosphere of fear but rather because it is to be hoped that it is better for their subjects to live in awe of the visible rather than in dread of the inscrutable. It is clear that architecture has the role of creating social orientation and showing everyone exactly what their position is – indeed, this seems the only obvious reason why architecture must be comprehensible at all. This is clearly demonstrated by castles and cathedrals – but it is also true of railway stations and stadia. These are all building forms of which not only every child has a clear image but which are informed by a simple basic idea.

In addition to this, any architecture must be based on principles and common beliefs which are shared, unquestioningly, by an entire society – regardless of political and social division. But this question of belief is of course not so easy; take the example of the prudent woman who considers the alternatives when deciding whether she is to baptize – or inoculate – her new-born baby….Or of you yourself when deciding in which bank you wish to deposit your money: the bank with the Doric columns or the one plastered with security equipment. Of course a bank can have both – classical columns and state-of-the-art security apparatus but the concern remains that the two together create a ludicrous impression.

The old belief that further scientific and technological progress are the indisputable guarantors of a better future has disappeared over the past 40 years like sand slipping between trembling figures; the certainty which drove the heroic modern, despite the lessons of Verdun, has been driven away like smoke in a strong wind. This has left us with a naked reliance on science and technology and due to the fact that, ever since the writings of the Club of Rome, there are no longer any winners who are in a position to point to past victories or promise a better future, the only option left to the powerful is to hunt out and then architectonically present all possible imperatives. This was true after the Black Death (14th century) and the Wars of Religion (17th century) and it is equally true today.

Bruno Schindler's conclusion, that architecture has no option other than to methodically surrender to the balance of power has unfortunately not been borne out. Architecture remains too often in awe of the investors who are very reluctant to accept any sort of meaningful content because this would limit their ability to let or sell a property to any possible end-user: real estate objects should remain as socially and commercially mobile as possible. And while this state of affairs can be frustrating, such is the interchangeability of international architecture that it can hardly be challenged.

And, as for those exceptional projects which are not directly the responsibility of investors, these are generally shared out amongst an elite group of star architects in the hope that a little of their stardust will rub off. This is a question of marketing.

Of course this process often results in extraordinary buildings, but these are seldom convincing examples of an architecture based on sensual orientation. This indeed is not the goal of the major competitions involving star architects, and for this reason competition briefs are not couched in these terms. It is much easier to allow oneself to be surprised by the complex language of shapes spoken by the stars and rely on the autonomy of architectural genius than to deal directly with the more subjective, internal complexities of a project. (This drifting apart of internal meaning and external appearance is very clearly demonstrated in the process which led to the decision to rebuild the Stadtschloss in Berlin.)

Fortunately I was still young when I learned this difficult lesson and my reaction was the concept of mediatecture. If a building is not in a position to react to the communicative requirements of sensual orientation then at least its programmable surfaces are. Programming is so variable and flexible that it can react to any requirements, which explains why the concept is, in principle, so well received by investors – although one can naturally not expect support from these same investors when it comes to the deeper question of renewing the paradigm of the sensual effect of architecture in the public space. There is much to do – but that is of course what this book is about.

The question of whether such an approach will allow the balance of power to be rendered sensually communicable in the way intended by Bruno Schindler becomes a dialectic: in many cases power structures search for ways in which they can present themselves in a more positive light – to which end a media façade is very well suited. And this also means that the very existence of a media façade is often evidence of the power structure which lies behind it – "the medium is the message". The ways in which content is then presented on this media façade are then highly variable. But even if it is purely used for the display of a program of media art (see Vattenfall's media façade in Berlin) this does nothing to diminish its importance to the power that lies behind it.

And despite all the critical connotations of what here may sound like a voice from 1968: this is good thing! We work "creatively" in order to make power visible. And when this also results in the production of art, this can only be a good thing. But my point is that it is just as easy for me as a creator to articulate the communicative interests of a power structure - and to "create social orientation and show everyone exactly what their position is" - without requiring the fig leaf of art. The same incidentally applies to advertising, because advertising is more or less the oldest way of communicating power. Bruno Schindler always quotes the example of the blossom which uses its conspicuous appearance to exercise power over and to attract the bee. And here is another story which Bruno Schindler told me:

"Everybody knows Mies van der Rohe's 1921 design for an office tower in the Friedrichstraße in Berlin. Mies was naturally widely attacked for such a design. Taking the elegant façades of the existing buildings on the Friedrichstraße as their starting point, critics claimed that the pure glass façade offered nothing for the eye to hold on to. Mies answered by referring to the growth in advertising on façades and pointed out that, especially in the Friedrichstraße, there would soon be very little to see of these so loved historical façades. For exactly this same reason, he added, no one should be concerned about his building because its façade too would soon enough be covered with advertising and, hence, would become just another integral part of the Friedrichstraße ensemble."

Similarly applicable in this context would be the system theory of Niklas Luhmann and the argument that power is in any case nothing other than a means of communication – but to digress further in this direction would soon take me outside the remit of this book.

And this brings us full circle. I wrote above of how the idea of mediatecture began from my reading of the article by Bruno Schindler and my hope is that this book is also a beginning. Because even when we have already come so far that this book contains some impressive results, the idea of mediatecture must still properly assert itself. It is a new idea – and an unusual one in a culture so dominated by the mechanical results of industrial production or a culture which generally understands the term "media" in terms of television or the printed press. But this is changing: the social networks which, in turn, are activated by "social media" are providing many people with a new way of understanding media.

And this changes the game.

Erkki Huhtamo *was born in Helsinki, Finland in 1958, and holds a Ph.D. in Cultural History. He works as Professor of Media History and Theory at the University of California Los Angeles (UCLA), Department of Design | Media Arts. His work has explored many aspects of media history and the media arts. Two new books, Illusions in Motion, a history of the moving panorama, and Media Archaeologies, a collection of writings (edited with Dr. Jussi Parikka), are forthcoming from the University of California Press.*

Pre-envisioning Mediatecture: A Media-archaeological Perspective

As a hybridization of architecture, urban design, media and new technology, mediatecture is rapidly gaining ground in the urban environment. Buildings are becoming "media machines" for living in, and they often display themselves as such. Not only are they increasingly conceived to carry display screens as a form of electronic windows, but matrixes of super-bright LEDs can also cover them entirely as a "skin" that responds to stimuli from the outside. All this seems a natural consequence of the desires of the media-obsessed society of late modernity. We only feel comfortable when we know that media surround us. The devices we carry in our bags and pockets give us a sense of safety and assurance; directly or indirectly, they resonate with the media-saturated surroundings we live and work in.

For most observers, this cultural condition seems very recent. Yet, as media archaeology can reveal, it has a much longer history - and this history deserves to be excavated. In the 1960s, groups of architects and urban designers such as Archigram purported to introduce new dynamic elements into the built environment, emphasizing increasing mergers between and crossovers of media, technology and architecture. Similar ideas were reflected upon by early postmodern architectural theory, in particular by Venturi, Scott Brown and Izenour in their classic *Learning from Las Vegas*.[1] Even earlier, Nicolas Schöffer envisioned in his techno-utopian writings and artworks the cybernetic city as a place where a dynamic interplay would take place between natural elements (winds, temperature, etc.), cybernetic buildings and monuments, and citizens. A giant cybernetic tower was to be erected in Paris, containing thousands of electric lights and projectors. Its parts were to be physically rotated by motors and all operations controlled by a central computer. Its multiple functions would have extended from the practical to the symbolic.[2]

Yet, in the first half of the twentieth century, avant-garde visionaries like László Moholy-Nagy, Zdenek Pesanek and Thomas Wilfred had already imagined - and in a few cases, realized - light spectacles and other technological enhancements to the emerging modern city. They longed for a new unity that Moholy-Nagy called "total work" (*Gesamtwerk*), obviously referring to Richard Wagner's *Gesamtkunstwerk*, but dropping the word "art" - the distinction between "art" and "design" was becoming superfluous - and Moholy-Nagy's concept embraced nothing less than life itself.[3] The avant-gardists drew inspiration from the billboards, electrified signs, sky projections and other media spectacles that had already been realized for commercial and civic purposes. They wanted to raise these to a new level in terms of both their effect and their cultural and social relevance. Their dreams were rarely realized, and the fact that Albert Speer's architecture of light was so effectively used in the service of Nazi propaganda cast an ambiguous and ominous shadow over their plans.

That is, however, another story. This article will look even further back in time, searching for seeds of urban mediatecture in contexts that have rarely been explored before. The nineteenth century witnessed the development of the urban "adscape", where posters and billboards not only grew in size, but were also enhanced by new technology. Even earlier, spectacular light effects had already been introduced into the urban environment by fireworks displays and *son et lumière* presentations. Magic lanterns were used for public projections on walls or on screens erected on rooftops. Electricity was harnessed by advertisers to create kinetic billboards and "sky-signs". And even the sky itself was integrated into the mediated urban space by giant searchlights. Considering mediatecture solely as a contemporary issue and innovation would trivialize it. We can gain new insights by re-visiting phenomena buried in the thick of time.

The Origins of Public Displays

Unlike the cinema spectator, the person experiencing outdoor media is often a casual passer-by. Whether a pedestrian or a driver behind a steering wheel, s/he is merely traversing a cityscape in which multiple features compete for attention. Media displays posted in the urban space are one element in a mosaic of attractions. The origins of such a 'logic' go far back in time.[4] Painted or carved wall inscriptions were already known in ancient Rome. Romans also used signboards to identify craftsmen's workshops and various services. Similar practices existed in Medieval Europe. Metal emblems with symbolic objects or coats-of-arms suspended from a rod protruding from the front of a house were widely used as means of identification; before the adoption of house numbering, they also served as address indicators.

The habit of posting printed bills on doors and walls made more varied linguistic messages possible. It began in the late fifteenth century in the wake of the Gutenbergian revolution. As communities grew and the hold of capitalism became stronger, the role of public advertising gained ever more importance. Signboards also developed into vehicles for distinguishing between *similar* competing products and services. Strategies of persuasion began to overshadow the seemingly neutral notices that announced services. Fairs, carnivals and other similar public gatherings played a role in this development. Market stalls for touring theater troupes, circuses and other entertainments often used large painted banners as attractors and previews of the show's content. The successful itinerant theatrical entrepreneur John Richardson (1766-1836), who began his career as a humble penny showman, used the services of the most famous scene painters from London's royal patent theatres.[5]

Printed outdoor notices became widespread in the nineteenth century. In London, printers and booksellers began to display satirical political engravings and novelty prints in their shop windows, turning these into free street galleries for media imagery.[6] Advertising broadsides promoting products and popular spectacles began to be used in great numbers on any available surfaces from fences to walls and even doors. The first half of the nineteenth century was the wild period.[7] Bill-posters obeyed no rules, using any available surfaces. Layer upon layer they posted their broadsides on walls that were often already covered. They competed and even physically fought with each other, paying little attention to the official edicts meant to control the situation. The cityscape turned into a constantly metamorphosing, tension-filled patchwork of overlapping messages.[8] Texts and imagery came to envelop the lives of citizens at every step.

Claiming that randomly posted broadsides already functioned as *proto-screens* would however be exaggerated. As an information interface, the screen should incorporate a separation between hardware and software. It should function both as a frame and a gateway through which messages are transmitted and retrieved. Such conditions began appearing when, after decades of bill-posting anarchy, some enterprises began acquiring legal rights to use certain surfaces and rent them out to advertisers. In England this practice came to be known as "placard advertising." Advertising space was rationally divided into framed lots (much like farmland), and these were "cultivated" by companies who had bought the rights to use them. Railroad stations were early examples of public places where the newly organized adscape was realized – the constant flow of passers-by made them attractive, and passengers could see the placards from the windows of the train as well.

In the United States the word "billboard" was adopted, reflecting a similar institutional and commercial development.[9] Catherine Gudis has aptly summarized its significance: "Like the buildings rising in growing metropolises, billboards contributed to the accretion of commercial centers and formalized the incursion of pictures and texts to the public sphere."[10] Although billboards represented an effort to tame (and, quite concretely, frame) the wild advertising, they did not manage to dampen criticism of its harmful effects. On the contrary, a heated and long-lasting public debate ensued. This manifested itself in the formation of trade organizations to defend the business, including the International Bill Posters' Association of North America (1872) and the Associated Bill Posters' Association of the US and Canada (1891), which eventually developed into today's Outdoor Advertising Association of America (OAAA). Citizens' watch groups were also formed, such as the British S.C.A.P.A. (Society for the Checking of Abuses in Public Advertising, 1893).

Like similar organizations elsewhere, the S.C.A.P.A. encouraged its members to document the misuses and excesses of public advertising and fought for the removal of misplaced and improper billboards. The tension became even more intense; the introduction of the automobile encouraged advertisers to erect billboards not just in the cities, but by the roadsides as well. The billboards were frequently accused of blocking access to scenic landscapes. The issue did not escape the attention of satirical cartoonists. One cartoon, titled "Go prepared if you wish to enjoy American scenery" (1925), depicts a car parked by the roadside that is flanked by billboards.[11] The passengers are seen enjoying the scenic view from the top of long ladders leaning against them. The fight against "rubbish, weeds, and billboards" was declared a "crusade."[12] Taking this literally, the pioneer feminist Frances Power Cobbe attacked billboards with a "pot of paint and a long-handled brush on her rural drives in order to deface defacements," setting an example for the 'adbusters' and street artists of later times.[13]

The Gigantic meets the Electric

Early signboards and printed broadsides had been relatively small. Their scale could be characterized as anthropomorphic, matched with the dimensions of the human body and the environments in which most people lived. Elements that did not conform to this principle usually had to do with power. Cathedrals, city walls, castles and town halls were meant to impress the common people with their size. Gothic cathedrals had enormous rose windows made of thousands of pieces of stained glass. Standing under Bernini's immense cupola in the new St. Peter's in Rome was meant to convince the visitor of the might of the Catholic Church. And yet, even extraordinary public sights, such as the astronomical clocks built into the walls of churches or town halls (sometimes on the outside), often consisted of relatively small elements. Their clockwork-operated *Jaquemarts* performed at regular intervals, but the mechanical moving figures were not necessarily any larger than the automata demonstrated at fairs by itinerant showmen.

During the nineteenth century the situation changed dramatically, partly due to economic developments and changes in the urban environment and partly because of improvements in printing techniques. It became possible to produce very large, chromolithographic posters, even in multiple colors. Graphic designers learned to deal with large size, concentrating on elements that could raise interest and be detected from a distance. They simplified the textual part of the message, focusing on trademark and branding. Advertisers also began to take into consideration the placement of the billboard within the adscape, playing with issues of scale and perspective. Quite understandably, billboards became a target for modernist architectural critics, who called for a city cleaned of ornamentation, historicist references and non-functional(ist) spectacles.
But there were other kinds of voices as well. True to his non-conformist stance, Oscar Wilde praised street advertising for bringing "colour into the drab monotony of the English streets."[14]

The traditional billboard could suggest a narrative, but it wasn't a medium for sequential presentations. No matter how gigantic, it was a frozen printed image; it neither moved nor evolved. But the situation began to change in the late nineteenth century and the technical *primum mobile* was electricity.[15] Soon after the incandescent electric bulb had been introduced in the late 1870s, it was applied to advertising. In New York, the Broadway came to be known as the "Great White Way," which referred to the electrified advertisements and illuminated shop windows that turned the street into a luminous attraction after dark. "The transparent posters on which electricity wrote advertising texts with letters of fire" (Jules Verne) had qualities lacking from normal billboards.[16] Not only did they lengthen the time for which passers-by were exposed to their messages but animations could also be produced by rhythmically switching the illuminated parts on and off. A sign depicting an illuminated Roman chariot race on the roof of the Hotel Normandie in New York was seven stories high and required twenty thousand light bulbs.

Electricity made spectacular light effects familiar, but these were not without predecessors. Fireworks had already been used much earlier to illuminate architectural structures to celebrate royal births and weddings, or war victories.[17] "Machines" (or "temples") had been used for this purpose already as early as the seventeenth century.[18] These were "elaborate ornamental structures, usually in the form of buildings, which were decorated with paintings, usually of allegorical figures, flowers, and lamps which were cut out in silhouette to glow from behind."[19] The "machines" often resembled fountains, palaces or boats. They were displayed on floats on a river, bridges or open squares, producing pictorial scenes which were "animated" by explosives and finally consumed by fire. Carolyn Marvin has acknowledged the relationship between such earlier extravaganzas and the late nineteenth-century electric illumination of bridges, buildings, statues and other elements of the urban environment.[20] The transformation of traditional effects from the era of fire to that of electricity was "very gradual."[21] Many classical motives were simply translated into the "vocabulary of electric light effects."[22]

The Advent of Large-scale Public Projections

Another way of introducing large-scale dynamic visuals into the public space was projection. This meant turning to another piece of old technology, the magic lantern, which, until then, had only been considered suitable for darkened interior spaces.[23] This is understandable, because the available light sources had been weak, and the projected images dim. But improvements in lighting technology, in particular the introduction of the oxy-hydrogen limelight ('calcium light') and the electric arc-light made projections in outdoor public spaces possible.[24] According to Schievelbusch, powerful arc-lights were experimented with as early as the 1840s for the illumination of public monuments in Paris.[25] The idea of projecting not just a beam of light, but pictures and texts as well, was a logical next step.

Particularly in the United States (where the magic lantern was known as the "stereopticon"), slides were projected outdoors on screens, blank walls and even public monuments from the 1860s.[26] An enterprise based in Boston claimed that its "Automatic Stereopticon Advertiser Works All Night," displaying "your Advertisement to wondering crowds."[27] The messages were projected on a large screen erected on a horse-drawn cart. Commercial outdoor projections became a well-established tradition. Later in the century the American soap manufacturer Benjamin T. Babbitt advertised his free touring "Magnificent Stereopticon Exhibition and Musical Entertainment using the Oxyhydrogen or Calcium Light," shown in city centers after dark. [28] The views were said to be projected "to a greater size than the largest Panorama," and the "Magnificent Horses and Wagons used for transportation" provided additional "pleasure to thousands."

In its 1904 catalogue the magic lantern manufacturer T.H. McAllister described five ways of projecting slides in outdoors spaces with its "advertising stereopticons": on walls, shop windows, screens mounted on moving horse-drawn carts and screens erected on the roof (using either front or rear projection).[29] The roof projections were recommended for displaying "latest news bulletins" or "election returns." The last mentioned had been used for decades. On November 24, 1866, *Harper's Weekly* published a full page illustration showing the presentation of "election returns" by means of a magic lantern outside the *New York Tribune's* office building.[30] If the illustration can be trusted, a huge crowd was present. Similar reports were published frequently over the years.[31] The illustrations indicate that the projected slides contained handwritten statistical data about the ballot count, as well as portraits of the candidates. The text slides had been scribbled on the spot by a needle on blank coated magic lantern slides, based on data received by telegraph (or, later, telephone).[32]

Sometimes more than one magic lantern and screen were used, and the announcement of the results could also be accompanied by fireworks, evoking a much older tradition for public celebrations. Although a short time-lag was unavoidable, the use of visual media for this purpose anticipated today's election-night broadcasts on television. The election night in the city space in front of the large projection seems to have been a festive occasion. And on at least one occasion the spectators were encouraged to turn from passive bystanders into active participants. *The Century Magazine* reported in 1896: "When he has nothing to report the operator displays a portrait of a candidate, or an impromptu cartoon, exhibiting in comical allegory the success of his man, or his side, and the disconfiture of the other fellow. Of late a favorite bit of fun has been to throw upon the screen a question like this: "What's the matter with Cleveland?" Promptly comes the answer from ten thousand throats: "He's all right!" Then shines out: "Who's all right?" And the windows rattle with the acclamation: "C-l-e-v-e-l-a-n-d!"[33]

Envisioning the Sky as a Display

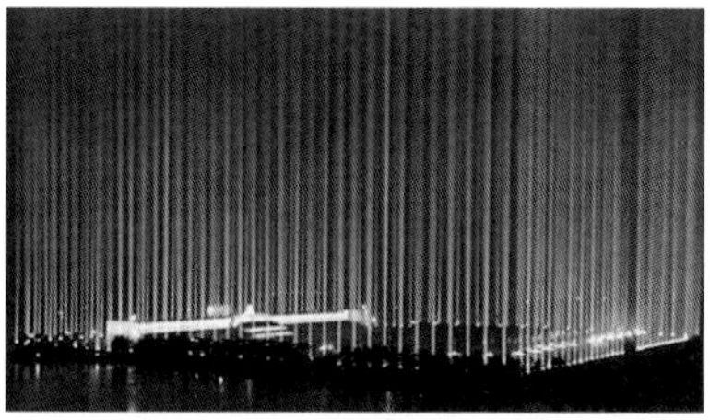

The cathedral of light by Albert Speer, Nürnberg 1936

Another element of the developing urban adscape was the "sky-sign." Sky-signs were defined by a contemporary as "advertisements, whose letters, standing clear of the structure's top, show against the sky."[34] Depending on the time of day, they could be seen from the street below as if written on the sky. In the evening they were illuminated, which further enhanced their effect. Depending on the viewing angle, it might even have been possible to detect the moon behind the letters. The appearance of such translucent rooftop signs could be read as an expression of the era's haste to move upwards.[35] The urban environment was getting cluttered, and catching the eye becoming more and more difficult. And although buildings were getting taller, their walls still weren't high enough for the advertisers.[36]

From the rooftop sky-sign it was a logical step to define the sky itself as a display. From primordial times, humans have converted the skies into cultural material by means of mental projections, attributing abstract and symbolic meanings to celestial phenomena. The sun, the moon and the stars have been interpreted as anthropomorphic beings with faces; the fantastic designs created by Georges Méliès for his silent film *Voyage au lune* (1902) were a manifestation of an important tradition.[37] The constellations have been represented as mythological beings and scenes; cosmic "plays" are believed to have secret correspondences with events on earth. Last but not least, the sky has functioned as a screen, where a divine intelligence has manifested itself.
The cross and the words *in hoc signo vinces* which Constantine the Great is said to have seen in the sky before the Battle of Milvian Bridge (312 AD) is a well-known example, but not the only one.

As often happens in the history of technology, in the nineteenth century sky projections were first realized in the imagination. In 1873 the French writer Auguste de Villiers de L'Isle-Adam published a satirical short story that fantasized about this possibility.[38] Its central character was an engineer named Grave, who, inspired by the electrical experiments of Benjamin Franklin and the "enormous lenses and gigantic reflectors of the American engineers," manages to perfect a system for "an absolute Publicity." With "powerful bursts of magnesium or electric light, magnified one hundred thousand times," spectacles could be beamed up to the sky "from the summit of some flowery hillock, the paradise of young couples." Villiers presents a veritable firework of hilarious and sarcastic suggestions for ads and other messages that could be beamed among the constellations. Between "the sublime paws" of the Great Bear there would "burst forth this disturbing announcement: DO YOU NEED CORSETS? YES – OR NO?" Even the moon would be inscribed with a slogan. The "sterile spaces" of the sky could therefore be "converted into really and fruitfully instructive spectacles" in the name of progress.

Villiers suggests projecting photographic lantern slides of "absconding bankers" and notorious criminals to the skies. Sky projections could also be used for political purposes. Portraits of the candidates would be beamed up "during the evening preceding the poll." With the "aid of a little wheel," the physiognomies of their faces would be manipulated, so that they would "smile at the Future, shed tears for our disappointments, open the mouth, wrinkle the brow, swell the nostrils in anger, assume an air of dignity."[39] Here the tone of Villiers' discourse is at its most sarcastic: the involuntary expressions of the faces (derived from the rhetoric repertory of public speech) would have the advantage of giving a 'dignified' idea of the candidate to the elector, who would hence no longer have to buy "a pig in a poke." The well-known idea of the politician as a media product is already expressed here.

The first more or less successful experiments in projecting texts and images onto clouds date from the early 1890s.[40] As A.E. Dolbear, a well-known specialist in optical projections, explained in 1893, "[b]y employing large lenses of proper focal length it has been found possible to project pictures upon the clouds as upon a great screen. The pictures to be thus projected may be prepared in stencil on sheets of tin or iron. Advertisements prepared in this way may be read a mile or more away, as the letters may be more than a hundred feet long. So the cloudy sky may soon be made an advertising sheet!"[41] The First World War intensified the development of mobile searchlights for marine search and destroy as well as air defence purposes. New civilian uses had to be found after the fighting had ended, and so searchlights became a fixture of movie premieres, exhibitions, and the inaugurations of skyscrapers, particularly in the United States, a country that hadn't experienced the horrors of the war at first-hand.[42]

Including sky projections in a media archaeological exploration of mediatecture may seem far-fetched, but it makes sense. Architectural discourse now routinely deals with both material and immaterial forms, including digital "liquid architecture." Likewise, any form of mediatecture contains an immaterial component contributed by the media component. Sky projections were merely an effort to extend this problematic in space, as Albert Speer well understood when he designed his "Lichtdoms" around the Zeppelinfeld in Nuremberg in order to accommodate and "frame" the Nazi mass rallies of the 1930s. For further evidence one could also refer to the Relational Architecture series by the Mexican-Canadian artist Rafael Lozano-Hemmer. Works like *Vectorial Elevation* (1999-2000) and *Pulse Front* (2007) extended the built urban environment skyward by means of massive concentrations of interactively operated batteries of computer-controlled searchlights, beaming shifting and / or pulsating matrixes of light into the night sky.

1 Robert Venturi, Denise Scott Brown and Steven Izenour, Learning from Las Vegas (Cambridge, Mass.: The MIT Press, 1993, orig. 1972).
2 Nicolas Schöffer, La ville cybernetique (Paris: Denoël-Gonthier, 1972), pp. 179-185. See also: Nicolas Schöffer, ed. Maude Ligier (Paris: Les presses du réel, 2004).
3 Krisztina Passuth, Moholy-Nagy (Dresden: VEB Verlag der Kunst, 1982), p. 49.
4 See Jacob Larwood [Hermann Schesoichaven] and John Camden Hotten, The History of Signboards From the Earliest Times to the Present Day, Eighth Edition (London: Chatto & Windus, 1875, orig. 1866).
5 See Richardson's obituary in The Gentleman's Magazine (London: William Pickering; John Bowyer Nichols and Son), Vol. VII, New Series (Jan-June 1837), pp. 326-327.
6 Diana Donald, The Age of Caricature. Satirical Prints in the Reign of George III (NewHaven and London: Yale University Press, 1996), pp. 3, 7, 28, 33, passim. W. Weir paid attention to this in 1851 when he wrote that the "windows of the print-shops – especially of those in which caricatures are exhibited – have great attractions [...]," although they could not rival (anymore) the placards at the stations. See W. Weir, "Advertisements," in London, ed. Charles Knight (London: Henry G. Bohn, 1851), Vol.V, p. 33.
7 See Blanche B. Elliott, A History of English Advertising (London: Business Publications Limited in association with B.T.Batsford Limited, 1962), pp. 164-167; Philippe Schuwer, Geschichte der Werbung (Vebey: Editions Rencontre, n.d.), pp.62-63; see also Henry Sampson, A History of Advertising from the Earliest Times (London: Chatto & Windus, 1875).
8 Sally Henderson and Robert Landau, Billboard Art, ed. Michelle Feldman (San Francisco: Chronicle Books, n.d.), p. 11.
9 This probably happened in the 1870s, although the Oxford English Dictionary mentions a British example from 1851: "The bill-boards of the Park [...] still continued to style the Park 'The Theatre'." This doesn't seem to be directly related to the later American usage.
10 Catherine Gudis, Buyways. Billboards, Automobiles, and the American Landscape (New York and London: Routledge, 2004), p. 19.
11 Repr. in Gudis, Buyways, p. 186.
12 Clinton Rogers Woodruff, "The Crusade Against Billboards," The American Review of Reviews, ed. Albert Shaw (New York: The Review of Reviews Company), Vol. 36 (July-December 1907), pp. 345-347, quot. p.345. Woodruff was the first President of the American Civic Association.
13 E[rnest] S. Turner, The Shocking History of Advertising (New York: Ballantine Books, 1953), p.124. Cobbe does not mention this episode in her autobiography Life of Frances Power Cobbe as Told by Herself (London: Swan Sonnenschein & Co., 1904).
14 Quot. Elliott, A History of English Advertising, p. 165. No source is given.
15 David E. Nye, Electrifying America. Social Meanings of a New Technology (Cambridge, Mass.: The MIT Press, 1992, orig. 1990), p. 32.
16 Jules Verne, Paris au XXe Siècle (Paris: Hachette, 1994, written in 1863), p. 197 (author's translation).
17 Wolfgang Schivelbusch sees their origin in medieval bonfires. Already in the seventeenth century they were developed into an 'artform' executed by fireworks masters with strict rules and their own aesthetics.. See Schivelbusch, Disenchanted Night. The Industrialization of Light in the Nineteenth Century, trans. Angela Davies (Berkeley & Los Angeles: University of California Press, 1995, orig.1983), pp. 137-139.
18 George Plimpton, Fireworks (Garden City, New York: Doubleday & Company, 1984).
19 Plimpton, Fireworks, pp. 34-35.
20 Carolyn Marvin, When Old Technologies Were New. Thinking About Electric Communication in the Late Nineteenth Century (New York and Oxford: Oxford University Press, 1988).
21 Marvin, When Old Technologies Were New, p. 153.
22 Marvin, When Old Technologies Were New, p. 164, 167.
23 For the early history of the magic lantern, see Deac Rossell, Laterna Magica - Magic Lantern, Band 1 / Vol.1 , German trans. Marita Kuhn (Stuttgart: Fuesslin Verlag, 2008).
24 The history of light sources is complex. Both reached brightness levels at which they could be used for practical purposes towards the mid 19th century. See Patrice Guerin, Du soleil au xenon. Les techniques d'eclairage à travers deux siècles de projection (Paris: Prodiex, 1995); Terence Rees, Theatre Lighting in the Age of Gas (London: The Society for Theatre Research, 1978).

The idea of projecting advertisements onto the moon appeared again recently, as if spontaneously, in the "moonvertising" campaign developed by the brewing company Rolling Rock. In the spring of 2008, billboards were placed in American cities asking passers-by to have a look at the next full moon; purportedly they would see the Rolling Rock logo projected on its surface. Although it was a hoax, the discussion spread virally on the Internet, which was, of course, the intention.[43] This example demonstrates something that every media archaeologist knows: ideas of the past are constantly recycled within contemporary media culture and given new meanings in ever-changing contexts. Although mediatecture is, in a technical sense, an outcome of recent advances in lighting, programming and controlling technologies, it also re-connects with ideas that were tried decades and perhaps hundreds of years ago. Understanding this process will give us a better idea of its true prospects.

25 Schievelbusch, Disenchanted Night, pp. 54-55. Later, during the colonial wars in Africa, the French used them to scare away the enemy, thus proving the ideological and military benefits of a 'blinding' technology (p.57).

26 In The Shocking History of Advertising, Turner records a commercial magic lantern projection at Trafalgar Square in London on Trafalgar Day, 1894. Ads for "pills, blacking, and watches" were projected on the side of Nelson's Column and the pillars of the National Gallery. In The Times a reader "suggested 'jamming' the advertisements by a more powerful beam." (pp. 126-127).

27 Tradecard, undated (c.1860s), author's collection.

28 Undated broadside (c.1880s) in the author's collection. The exhibition in question was at the corner of Broad and Middle Streets (city unknown) at 8 o'clock. The broadside mentions earlier presentations in Charleston (South Carolina); Augusta, (Georgia), and Hinesville (Georgia). Babbitt's soap works was based in New York City.

29 T.H. McAllister, Catalogue of Stereopticons, Dissolving Apparatus, Magic Lanterns, Moving Picture Machines and Accessories (New York: T.H. McAllister, April 1904), p.35 (author's collection).

30 P. 744. The event had taken place around midnight, November 6-7, 1866.

31 For example, Frank Leslie's Illustrated Newspaper, Nov.23, 1872 (cover); Oct.25, 1884 (cover); Nov.17, 1888, p.223-224; Nov.15, 1890, p.262; Harper's Weekly, Nov.17, 1888, p.877; Collier's Magazine, Vol.XXXIV, No.4 (Oct.22, 1904), cover. I found these in the archives of the Magic Lantern Castle Museum, San Antonio, Texas (thanks to Jack Judson).

32 This is confirmed by the T.H. McAllister catalogue mentioned above: "LANTERN ADVERTISEMENTS for temporary use – Election Returns, etc. – can be easily made by writing or painting them on glass, with India Ink, or with the 'opaque' used by Photographers." (McAllister, Catalogue of Stereopticons [...], p.35).

33 "Election Day in New York", The Century Magazine, Vol. LIII, No. 1 (Nov. 1896), p. 12.

34 Charles Mulford Robinson, "Making the City Beautiful," Current Literature. A Magazine of Record and Review (New York: The Current Literature Publishing Co.), Vol. 31 (July-December, 1901), pp.139-143, quot. p.139.

35 Neon tubes eventually replaced the bulbs. It should be noted that Billboards sometimes contain cut-out elements that 'exceed the frame.' In a way this feature resembles sky-signs, because it creates a silhouette that can be easily perceived from a distance.

36 Gudis has noted the late 19th century tendency for billboards to "grow upward with double- and even triple-decker panels paralleling the multiple stories of adjacent structures." This led to the "rooftop signs, now illuminated with electricity, creating a new spectacle and skyline." (Buyways, p. 19).

37 See Scott L. Montgomery, The Moon and the Western Imagination (Tucson: University of Arizona Press, 1999).

38 «La Découverte de M. Grave,» La Renaissance littéraire et artistique, November 30, 1873, later included as "L'Affichage céleste" in the collection Contes cruels (1883). Translation used here: "Celestial Advertising," trans. Hamish Miles, online at http://gaslight.mtroyal.ca/villier2.htm .

39 Villiers is obviously referring here to the mechanical magic lantern slides that were widely used at the time.

40 At the Chicago World's Fair (1893) a cloud projector situated on the roof of the Manufacturers Building is said to have announced the daily number of visitors on the sky (Turner, The Shocking History of Advertising, p. 247).

41 A.E. Dolbear, "The Electric Searchlight," The Cosmopolitan, a Monthly Illustrated Magazine, Vol.16, No.2 (December 1893), p.254.

42 The creation of an artificial sunset for the Panama Pacific Exhibition (San Francisco, 1915) by a battery of General Electric's sky projectors was a task for which the technology was suited.Nye, Electrifying America, pp. 63-65.

43 For an idea of the project and the discussion it triggered, make a Google search for "moonvertising." The idea of moon projection also appeared in the recent Hollywood film Hancock (2008).

Andree Verleger *is a media artist and show producer in Düsseldorf who works with visual concepts in which the real and virtual worlds coalesce.*
He began his career in the music branch, concentrating on marketing events and developing his own form of multimedia shows, in which live acts were synchronized with complex video projections. Since 2004 he has been responsible for marketing productions and spectacular opening parties worldwide. In 2008 he played an important role in the opening ceremony of the Olympic Games in Beijing.

"The concept itself emerges from the urge to discover an emotionality which surprises even me"

I would like to introduce you to the readers as a mediatectonic artist. Can you please tell us which technical components you directly handle yourself and which ones you delegate, while remaining in complete control of the way in which they are implemented.

I create the pictorial language, but before doing so I have to determine an emotional starting point. This is the point in which I differentiate my work from what one usually calls "themed environments". My objective is to create a certain atmosphere. And this atmosphere will unquestionably be strongly influenced by the location for which the work is to be created. My first task is to absorb this location or, one could say, to get the taste of it.

After this, music is very important for me because music is where I originally come from. Music is the basis of our emotions and without it any sort of intervention is, for me, quite impossible. At this point the first ideas for the installation emerge and, as they do so, I am very careful to allow myself not to be limited by the technical possibilities. Instead, I always try to widen the technical spectrum in line with the content of my emotional idea. I am no technician myself but I understand quite a lot about what is implementable. The same is true with the pictorial language. Naturally I have a team that generates the images – but before they can do so I must have developed them very concretely on the computer.

The composer Olivier Messiaen was a synesthete. He could see the colors of sounds and vice versa. Could it be that mediatectonic work generates similar effects? That one translates the image that one sees into a spatial order or that, upon experiencing a space; one automatically imagines a pictorial sequence?

Well, that's how it works with me in the case of, for example, music. I can only develop a spatial idea when I hear the appropriate music. In the work for the opening of the Olympic Games in Beijing I only received the music very late which meant that I first had to produce my own music in order to be able to generate emotional images.

Another example of a synesthetic situation is when I see a tree and ask myself what it actually is. I start by seeing many irregular polygons which, in principle, could be projection surfaces. The next stage is that I construct an installation with many leaves which are both kinetically charged and able to transform wind energy into electricity which will then be used for the projection onto this tree at night. The interesting thing is that, when one projects, for instance, a face onto the tree, this is only recognizable when one is driving past because it is essential that one is looking in the same direction as the projector. If you are not in the line of the image then the tree is merely a secretive play with light – a process which one can perhaps describe

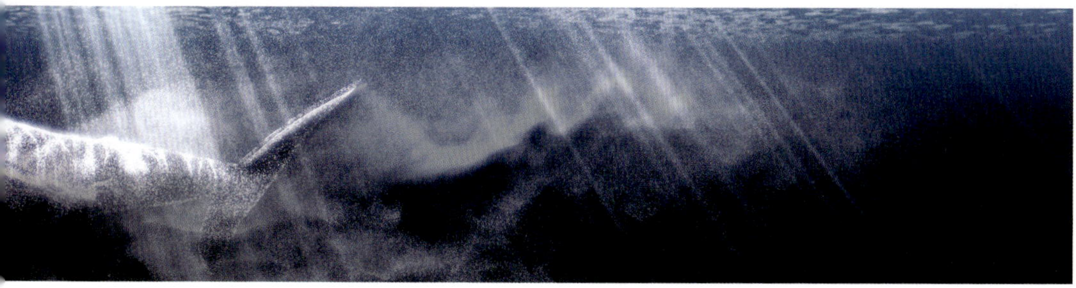

as a sort of natural arch. What interests me here is the relationship between order and chaos. There is chaos which the observer considers as beautiful. Or there is order which one very consciously destroys – such as the representation of a cigarette which is rectangular rather than cylindrical. This immediately grabs the attention. It is always a question of perception.

How can one develop an overall conceptual structure for such a complex design approach? Where does the concept itself come from?

The concept itself emerges from the urge to discover an emotionality which surprises even me. I have no preconception about what is going to come and simply try things out until something strikes a chord with me. In doing so I always leave chartered territory behind and let things collide and find their own existence.

This way of working reminds me of the architect Günther Behnisch, who often did not decide about the next stage of his buildings until he, for instance, was able to see and react to the raw concrete of the spatial structure. But even such an apparently spontaneous approach needs some rules if the end product is to work. What are your rules?

There are always questions about the sensual effects of installations. If I want to get a hold on the space in a huge stadium then the ring form is always crucial. And my productions always draw people in to become part of the performance. It is precisely the relationship between this combination (life-performance-people) and the virtual images that can reveal the desired emotionality.

The work for the opening of the Olympic Games in Beijing reminds one of a clockwork whose complex mechanism produces effects which are barely comprehensible – which is also what creates the magic.

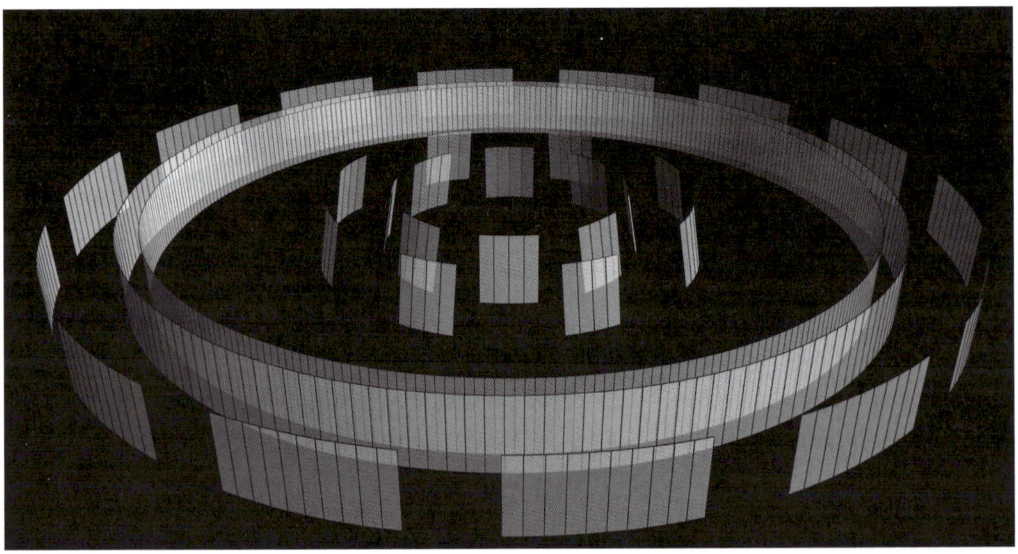

A clock was also the basic idea. People carried screens constructed of gauze stretched over a framework which were then arranged in a series of rows. The projection was then set up in such a way that the highly varying pictorial content was projected with perspectival precision onto these surfaces. The additional complexity created by the fact that each projection consisted of three images meant that I was always close to failure. Rather than the order of the screens being stable, the people carrying the screens had to change position – in a sequence which was naturally perfectly coordinated with the video images. You can imagine how difficult it was even to determine the correct size of the screens.

What drives you to create such complicated pictorial constructions? Could it be that this conscious fragmentation of the image in the mind of the viewer allows him to generate his own image?

Certainly; and in the case of my installations there is also the fact that I give depth to the arrangement of images. A simple two-dimensional image can create no sense of reality, especially when it portrays an actual object. But if the object is only partially presented then the brain is encouraged to think for itself and this leads to its own reality. It is for example false to imagine that real 3D cinema images bring one closer to reality because so much input is simply missing. The emotional intensity of my installations comes precisely from the fact that what one is seeing and hearing has, at exactly that moment, a concrete relationship with the space in which one finds oneself. The result is that one perceives a reality – but one that is, at the same time, magic.

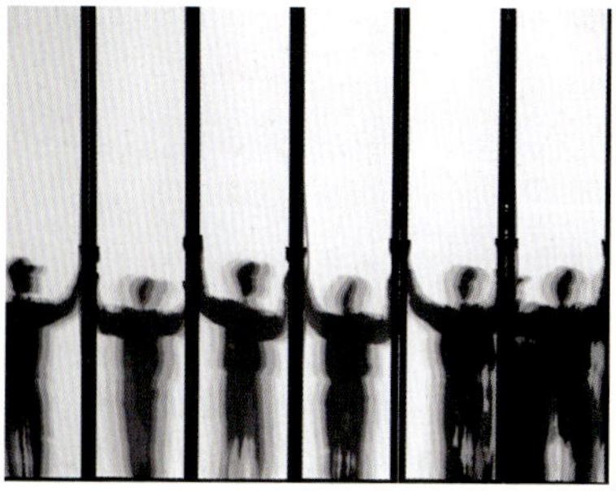

That reminds me that at the end of the 1990s everybody was talking about the design of three-dimensional websites. Special 3D tools were even developed because everyone wanted to talk about virtual space. But it failed to catch on.

Precisely – the three dimensional image of a space delivers no really authentic sensation. But the organization of images in space – whether superimposed, fragmented or otherwise arranged – forces our sense of perception to think for itself and this creates reality. This is the way in which we humans are conditioned.

The creative power of the viewer is much more challenged by a non-real image as by an image of a reality – this is my guiding principle.

A wonderful principle! I would like to consider it directly in my experience of working with media façades. Clients are often eager to combine their building with clear, direct images and videos of televisual quality – with football, incidentally, as the preferred content! But this creates a dead façade; a closed space. And this in turn leads to the justified criticism of media façades, that they simply overload urban spaces and architecture with virtual images. This sense of overloading occurs because – as you say – the depth of the space cannot be explored.

These thoughts also went through my head as I developed the concept for the opening of the media façade for Bayer. Firstly I thought about the many media façades in Asia – and this made me uneasy because such advertising surfaces are something of an epidemic.

But the interest thing about the Bayer project is the way in which it was arranged in many levels. The overlaying of the media skin with the green core and the bare construction of the building creates precisely the sense of depth which allows the brain to create its own images.

But this complexity was of course deliberate. And it is precisely this complexity which fascinates me about your work – and the fact that it also incorporates the human dimension.

That is very important. As a human I cannot identify myself with objects and the fact that humans are directly involved in my installations means that the viewer is also involved. It is this combination of the human system with the system of virtual images that generates the special tension which is so exciting.

If we two were to develop a media façade together then perhaps a glass bridge between two buildings would be a suitable location. The two glass façades – one in front of the other - could be given a mediatectonic treatment and people would be passing between and interacting with these media images.

Yes – but I would have to be in control of the movement of the people because the people will start no interesting action of their own accord which I can coordinate with my images.

Perhaps, but I am interested in the idea of "anyhow". Precisely this unplanned combination of two systems can lead to very stimulating surprises. I am always searching for open and organically growing systems for public space.But it would not be a problem to set aside time for a performance or, let us say "a performation": a mixture of performance and installation.

Of course: have you already created an installation in which you did not have control over the human content?

No, but one could think about this. A ring similar to that which we created in Beijing could be set up as a permanent installation on a public square in a city as an opportunity for the citizens to trigger their own pictorial effects.

Naturally one would first have to find the suitable technology for such an installation. Of one works with projection it would be necessary for the installation by day to have the artistic power of, say, a sculpture by Richard Serra. And at night one could use a technology such as camera recognition to mediate between the behavior of visitors and the video performance.

Or you could have an external image viewer or referee who logs in in the evening to control the performance. Then one could set up the same installation in various places around the world and synchronize or exchange images from, say, Tokyo and Cologne – or use the people in Cologne to trigger images in Tokyo. This could become a sort of virtual modern Stonehenge.

A very good concept: meaning that this would be a sort of global spatial archetype?

Precisely: a physical installation in which people from all over the globe meet and make contact by means of media-based animation.

But there are problems with our use of the classical notion of the sculpture in the public space. A lover of Serra would probably completely reject our idea because the special quality of a Serra sculpture is that it triggers something in the imagination of the viewer precisely because it is so purist and reduced. As mediatects our task is to create a media-based language which, as you say, rather than delivering finished images uses images to encourage contemplation.
Exactly, the difference between such an installation and a two-dimensional medium is simply the depth of the space and it is this which makes it possible to stimulate new spaces in the mind.
At the end of our discussion I gained an insight into the boundless nature of Andree Verleger's creative imagination as he talked about what would happen if we could cut up light; if we could take slices from a beam of light and then freely distribute them as a pixel structure in space.
Just as we architects dream of creating buildings which float in defiance of gravity and, hence, require no columns, Andree Verleger dreams of images consisting of points of light which can be delivered into a space with no need for a technical infrastructure.

If only ...

www.mediatecture.info/01

Harald Singer *describes the development of mediatectural ways of thinking and working and the emergence of the term "mediatecture" in ag4 (Arbeitsgemeinschaft für 4-dimensionales Bauen - working group for four-dimensional building). The group was founded in 1991 in Cologne by Harald Singer (director, still a member today), Reinhard Lepel (architect, member until 1993) and Christoph Kronhagel (architect). ag4 turned into an interdisciplinary group with members from a wide range of specialist areas, with Ralf Müller (economist) joining in 1997. Under the name ag4 mediatecture company, the members of the group shared the commercial responsibility for implementing projects. Between 2003 and 2008 ag4 was managed by Ralf Müller and Christoph Kronhagel. Since 2009 Ralf Müller has been the majority owner.*
Harald Singer studied photographic engineering in Cologne. Alongside his work with ag4 he runs the media production companies EXPONENT 3 (advertising, media installations and documentary films) and LivinGlobe (immersive cinema, fulldome installations) in Munich.

Origins of Mediatecture in ag4

Our original idea, based on the experience that architects and film-directors have very similar approaches, was to create a simple cooperative office. It soon became very clear, however, that, in addition to such similar working methods, a huge potential for contemporary communication and presentation is hidden at the interface between the two disciplines. We had the sense of something new - but first of all we had to find out exactly what this was.

A fascinating period ensued in which, in intense discussions and creative projects covering questions of how architecture and media could work together, we developed the basis of what we would later call „mediatecture“. The things which we developed in our concepts and projects could no longer be described using standard terms - this was no longer just film or pure architecture, design or communication. This was something else.

In 1993 the members of ag4 finally found the ideal term, „mediatecture“, a made-up word combining media and architecture. But the result, rather than being a simple combination of existing things, is something new. We have used the term ever since to describe our work and it also became part of the name of the group: ag4-Gesellschaft für Mediatektur and, finally, ag4 | mediatecture company. If you google the word „mediatecture“ today (2010) you find over 16,000 entries in German and 35,000 in English. At the beginning, we very deliberately decided to leave the term „mediatecture“ open, in order to allow it to become an accepted term for a new genre - and this obviously happened.

Photo © Arne Hofmann

1st project by ag4: Sony trade fair booth at the Internationale Funkausstellung Berlin 1991 (consumer electronics trade fair)

Mediatecture as Built Communication in a Contemporary Language

If one wants to understand mediatecture one also has to consider the period in which the idea arose. It seems to me to be no coincidence that this development occurred at the beginning of a new decade and as a millennium was coming to a close. The nineties at the end of the last millennium was a very energetic decade in which everything not only both appeared possible but was actually achieved. It was a time in which such notions as virtuality and networks also began to be heard in the public realm and it was certainly the last decade in which Marshall McLuhan's famous statement that "the medium is the message" was valid. Perhaps one could even risk the opinion that the developments of the technologically driven 20th century cumulated in mediatecture – for it is mediatecture which renders visible the communication and information society.

In this context it is very interesting to note that, at that time, network space was no more than a theoretical idea, with the internet not becoming publicly accessible until the introduction of the World Wide Web in 1991 – the year in which ag4 was founded. Mobile telephony and the possibility of omni-present communication for all then followed in Germany in 1992 with the founding of the D-Net. The world was becoming virtual – and the virtual was becoming real - via a series of huge steps. The digital revolution was underway with a vengeance.

It was no longer possible to look at a society as a uniform whole. The present was becoming complex – an interface between a number of truths and realities. The world could no longer be taken in with a single glance and the consequence was the "Fragmentation of the View" (Peter Weibel) or, as we say, a "layered work" or a "layering of stories".

But how could we formulate a relevant statement about this age and reflect the complexity of society? Because this was always our aim: mediatecture is not only decoration, design and form – mediatecture is built communication in an up-to-date language. We were engaged at that point with two main currents of architectural discourse: deconstructivism and building with glass. Deconstructivist architecture turns the simple form into a building volume that "explodes" as a means of temporalizing it. The static building becomes a dynamic sculpture. Time enters the timelessness of architecture, and the "immobile" becomes "mobile". But we found, that this was only a vision of dynamics - deconstructivist architecture itself remained static, which meant that it was not really the ideal way of representing our time

But the idea was born: the relationship between space, dynamics and time. This was already the inspiration for the logo of ag4: a runner, based on the studies of movement by Eadweard Muybridge. This pioneer of film positioned several cameras along a runner's track which he then triggered, one after the other, as the runner passed. The movement in space was hence transformed into a temporal sequence of pictures. It was, after all, this sequential photography which was the basis of film, which we were now seeking to bring back into the space.

The second key architectural trend of that period was – and remains today – building with glass. No longer should an architectural building be a safe, defendable and closed place – rather it should become open, allowing views in and out. Whereas architecture is loud and dominant in deconstructivism – in the case of glass buildings it dissolves into nothingness. Architecture becomes transparent and the inside becomes visible.

Whereas architecture requires glass, in mediatecture we use such media as monitors, projections or LED panels. Such surfaces are more transparent than glass can ever be because they can permit views into a real world - but also into the virtual content space of the user of the building. Concrete becomes transparent and the displays show the background. The interior world of a building appears to the outside and the façade has become an interface.

It was clear to us that we were actually addressing a very old architectural subject using contemporary methods, techniques and themes. Orchestrated interiors, sculptural architecture, façade design, ornament, painting and glass windows are, just like media façades, medialized and interactive methods, which give space a meaning and force it to speak as a communication-sculpture for the client. In doing so, mediatecture does not only use media but is also a medium itself – a mediator between the worlds of built and physical realities on the one hand and of imagined identities and visions on the other.

Interdisciplinary Working

Although ag4 was always a meeting point for extensive artistic potential, we never saw ourselves as a pure group of artists – and even sometimes consciously rejected the use of the word "art" in connection with our work. We didn't want mediatecture to be a subject for the ivory tower – not that this prevented the occasional emergence of works of art.

Mediatecture is a result of interdisciplinary work – of the cooperation between diverse professionals driven by the curious and open desire to "look beyond the horizon". The media designer and architect merge into someone new – a "mediatect". This interdisciplinary profession did not exist at that time. Therefore, the members and owners of ag4 belonged to a wide range of professions. The partners of ag4 and its various subsidiaries over the years included: Christoph Kronhagel, architect; Harald Singer, film director; Reinhard Lepel, architect; Professor

Ralf Sommer, designer; Lynn Spiegl, filmmaker; Johannes Keh, cameraman; Ralf Müller, economist, Thorsten Hofen, designer; Martin Bauer, technical manager; Wolfgang Rühl, film director; Joachim Heinz, web designer; Martin Esser, economist and Stefan Mosblech, journalist. And in addition to these, the group always had a range of valuable, key employees; architects, interior designers, designers, media designers, engineers and technicians, without whom our projects, in all their complexity, could never have been realized. In most of our projects the development of new technology was a key aspect of the design. We always sought to go right to the limit of what was considered achievable - and often beyond.

Space as Temporalized Sculpture

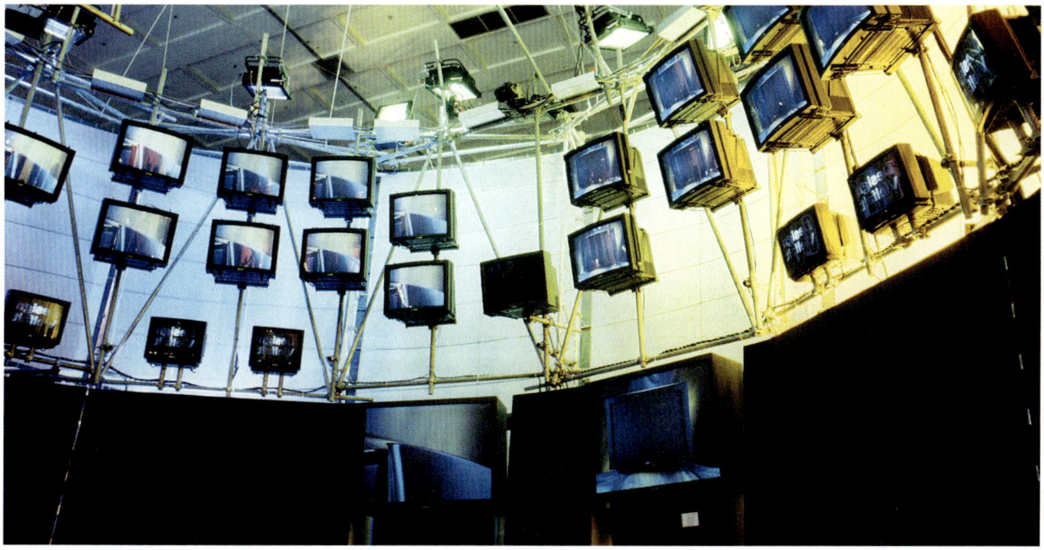

Fairs and exhibitions became one of the first areas in which mediatecture became feasible. In 1991 we were commissioned by Sony Germany to design their booth at IFA – the electronic trade show – in Berlin. The main idea of the installation "Worlds of Sony" was an open, media-sculpture – a polymorphous structure. Dozens of monitors surrounded the viewers who stood in an open viewing space. The building itself was as immaterial as it could be – not much more than a simple scaffolding construction spanned with transparent gauze. When illuminated from within, the gauze is opaque so that, for the viewer inside the space, the construction has become closed, without any constructional movement being necessary. If this internal light is reduced, the gauze becomes transparent again. With this simple stage trick we created a medial and almost immaterial, "breathing" body. The lighting of the gauze and, hence, the transparency of the volume are synchronized with further media. The object opens and closes – which is somehow appropriate to the communicative qualities of the content.

Inside the installation, 40 monitors displayed sophisticatedly produced media which use Sony products to show Sony products. The space could be dynamically speeded up and slowed down. Sometimes one saw a mosaic of individual images and sometimes a movement being carried through all monitors.

Mediatecture is the orchestration and temporalizing of space, the filling of space with meaning and the creation of a sphere of communication.

The First Concepts for Media Façades

We developed the CTC (Cologne Trade Center) vision for the **Architectural Competition "Breslauer Platz in Cologne"** in 1992, interpreting the space in front of the cathedral as a stage. In our concept a trade center became the pulsating center of the place, showing the activities inside at night on the outside.

Excerpts from the concept:

The CTC is a place of communication which communicates its own activities to the outside. The skin becomes a screen which can carry video images due to a grid of individually controllable and dimmable lighting elements with a wide pixels pitch. From the inside images are projected on solar sails in the internal space that will become visible at night from the outside.

By night many different images build a combined over-all image. By day the skin is transparent, allowing the view into the inside of the cell.

We required 12 years to develop the first transparent media façade for T-Mobile in Bonn. More than anything else we had to wait for LED technology. In 1992 the white and blue LEDs which are necessary for the creation of real color images did not exist.

As an intermediate step we created a first technical set-up for Deutsche Bahn in placing a huge transparent LED clock on the Sony Center, Potsdamer Platz, Berlin. Unfortunately the potential costs of such a piece of equipment exceeded the Deutsche Bahn budget.

Photo © Arne Hofmann

First animated renderings of an electronic media façade from 1992!

Concept for a media façade on Potsdamer Platz – 2001

The Mediatectural Film

As a film director and co-founder of the group I have intensively followed the sometimes difficult path from filmmaker to mediatect. Difficult? Because most of what defines a good filmmaker or director becomes suddenly worthless in the area of mediatecture. Mediatecture is not film. And yet – and I think that this was a particular feature of ag4 - our media and films were always of the highest quality (and, correspondingly, often received national and international awards in the category of "film"). We produced really complex films, such as "Window to the World" – a high quality film created for the **World Savings Banks Institute** at **EXPO 2000** which was filmed in 13 countries – and which was mediatecture too.

Mediatectural films have to be specially designed for the space in which they are to be presented. In this case we simulated the view from a window onto various global locations showing the different attitudes to money around the world. As if it was a real view from a window, the camera was completely static. It looked out onto these various places in which the contents were displayed as if on stage. This is certainly not classic film – with a story, editing and movement.

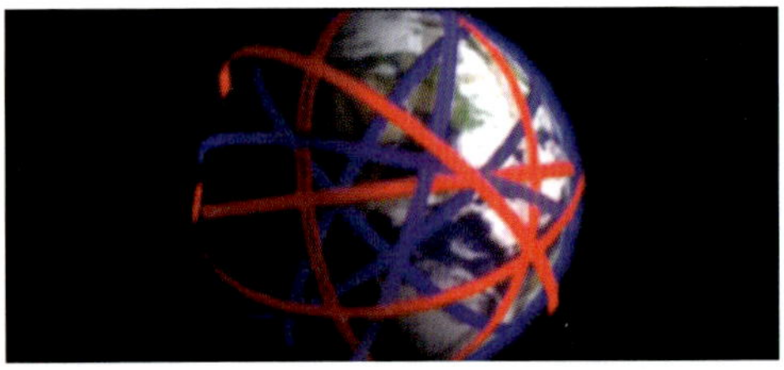

www.mediatecture.info/02

Non-linear Dramaturgy

A characteristic feature of mediatectural film is non-linear dramaturgy. Unlike conventional film – with its defined beginning and end – the mediatect cannot define when a viewer enters the space – and, hence, the "film". The films must thus be conceived in such a way that viewers can enter them at any time.

We achieved this by transforming the temporal organization of a film into a spatial organization. In conventional films, scenes follow each other in a chronological sequence in order to tell a story. In mediatectural film these individual components are either superimposed upon each other by means of multi-layering as a way of melting them into a new, complex picture or they are displayed in the form of a medial mosaic – alongside each other. The mediatectural film stretches itself across a room – filling and configuring it. This was exactly the case with the "Worlds of Sony" production of 1991 in which 40 monitors with 12 different monitors surrounded the viewers. In each scene of the film the different elements and aspects were simultaneously displayed alongside each other. The dramaturgy was non-chronological, non-linear and spatial.

The dramaturgy thus addresses not time but a space. Only when the viewer moves in the space does the non-linear, spatial dramaturgy become, once again, a linear, temporal dramaturgy. And the viewer becomes a director. From the wealth of contents on offer he "edits" his own "film" by simply looking around or moving through the installation. This gives media a sense of both space and body.

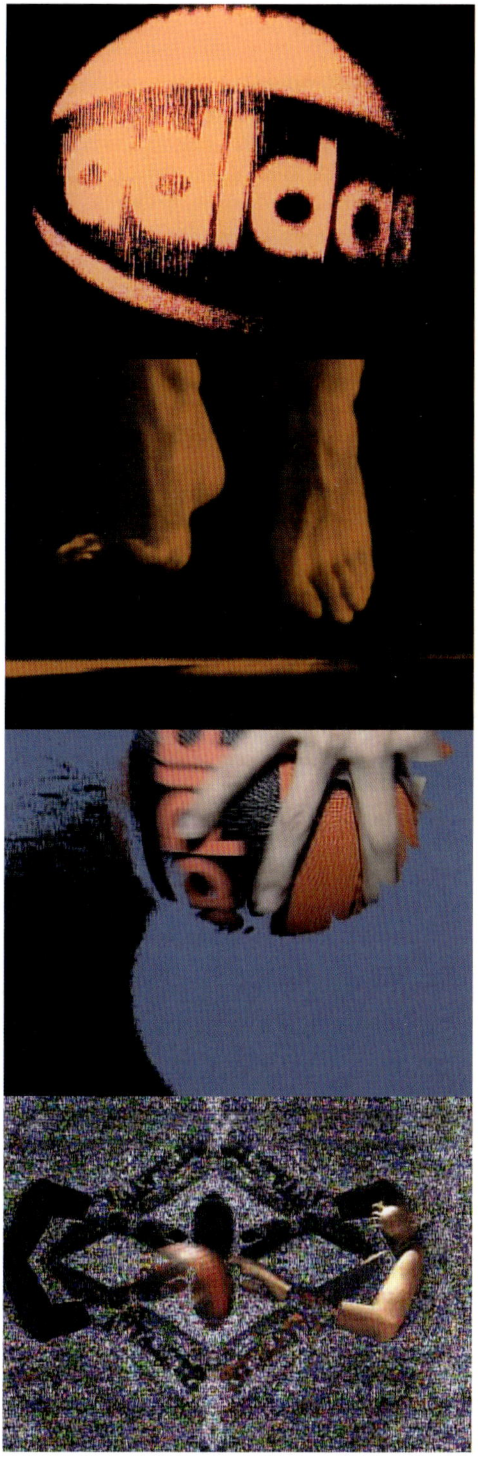

www.mediatecture.info/03

Ambient Videos

In contrast to the "Worlds of Sony" installation, where defined contents had to be communicated, many mediatectural films have the more specific role of creating a mood or atmosphere. In order to achieve this, the filmmaker has to use images and sounds as building material just like stone, timber and steel in order to create a specific atmosphere or mood for the space through media. What one needs here are not "films", but rather "ambient videos". With their restrained design, these define the mood of a space just as the warm timber of the floor or the cold steel of the architecture would define it.

Ambient videos either run for a long time, run as a loop or are generative – which is to say that they create themselves continuously on the basis of certain parameters. But at the beginning this was difficult to do elegantly because high definition film could only be displayed by means of analog tape machines or expensive laser discs. The solution for projecting ambient videos in those early days was hence to produce very long films (for the Hoechst project we even produced a run length of 14 hours). As a result of which one could be pretty sure that no visitor to the installation would notice a repeat of the content. For **adidas,** between 1993 and 1994, we realized an international shop (in-shop) system. Here we integrated ambient videos with a variety of themes - basketball, swimming or football - each of which lasted for more than an hour, creating long video sequences with music which was developed exclusively out of the sounds associated with the particular sport. The result was extremely intensive atmospheric spaces – at a time (1994) when the standard solution for such shops was the interchangeable streaming of MTV. The success of the concept can be seen in the fact that adidas has remained a customer of ag4 medien ever since.

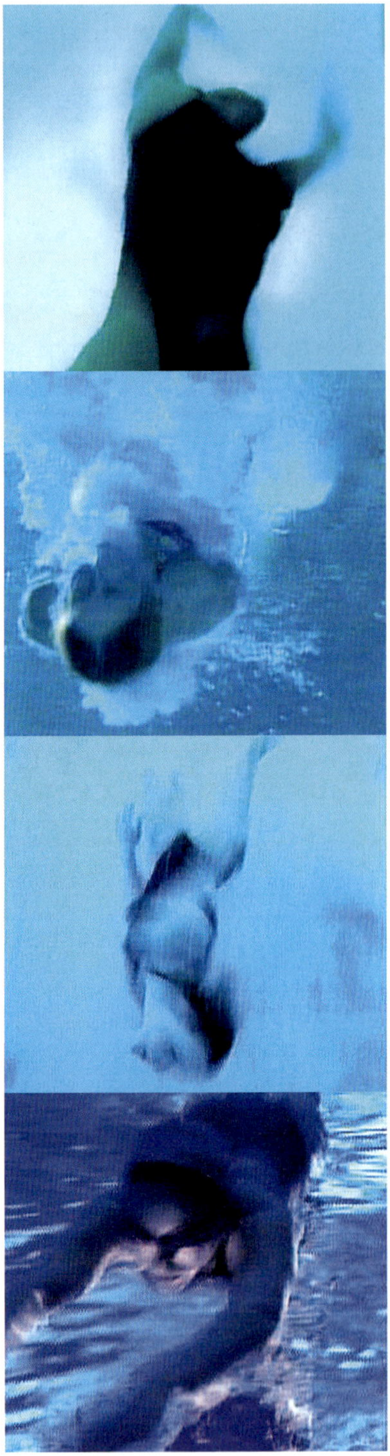

www.mediatecture.info/04

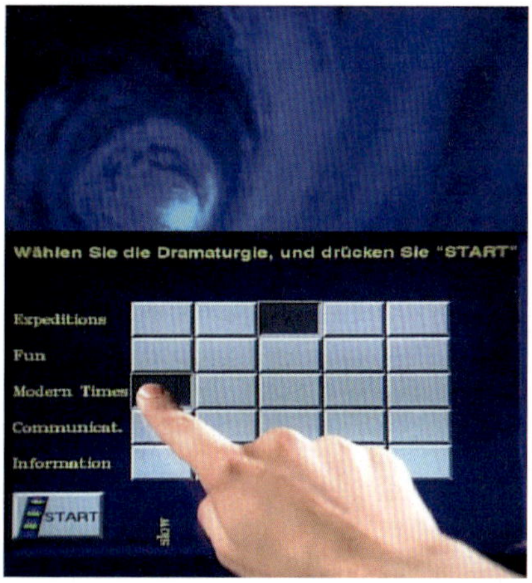

Process-related Media Control

Our greatest dream was always the self-generating film. But, unlike in the case of computer-generated film, real images cannot be generated – but just combined, manipulated and overlaid. Despite all the technical difficulties, we made our first important breakthrough here as early as 1993 by developing a form of interactive, process-related media design based on innovative, complex, program-based media control. In the installation for the **Electronic Forum in Frankfurt/Main** viewers were able to generate an entire media show. All available media and contents were categorized in a database in line with certain parameters. Visitors selected a dramaturgy in line with purely text-based criteria in the form of a matrix. The program behind the matrix calculated the contents which were to be shown. In case no visitor actively interacted with the system, we added such external parameters as weather, daylight and street-noise as inputs, in order to generate the program. The result was a generative, process-related media space – an organic, constantly changing show.

Out of these early ideas and first attempts at programming we developed the process-related media control which we use in diverse installations or in the programming of media façades to this day.

I can conclude by saying that, despite some pain at the loss of the cinematographic and narrative aspects of my earlier work, I have found a completely new sense of satisfaction from the transformation of a space into a truly atmospheric place of communication and from the combination of media and architecture to the point at which they can no longer survive without each other – from, that is to say, the creation of true mediatecture.

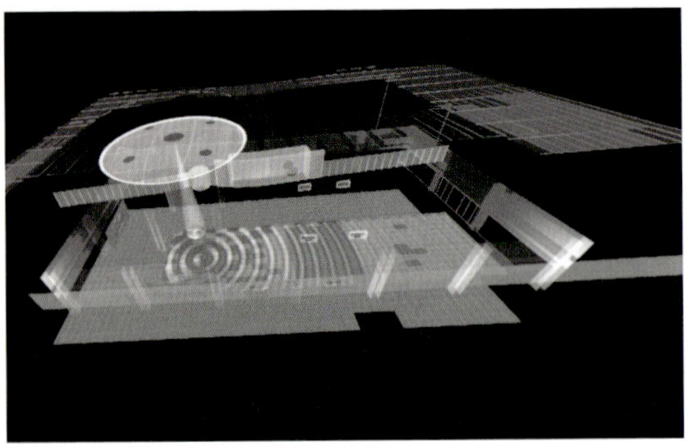

ag4 tasks: concept, planning and architectural design, planning of the media technique, media production for the stone, the animated wave in the floor and the world time clock

Mediatecture as a Supporter of Corporate Communications

For us, mediatecture has a social relevance in that it reflects a client and helps him in his communicative operations. The design of the **Corporate Forum Hoechst** in Frankfurt/Main was even an opportunity to reflect a piece of industrial history.

Hoechst was once the world's largest pharmaceuticals company having a peak of 170,000 employees before later breaking up into a series of smaller companies. Central to this transformation process was the massive program of restructuring and reorientation launched by the new Head of the Company Jürgen Dormann in 1994. Everything was in flux, subject to testing and change.

Our mediatecture seemed to find images which hit a nerve regarding this change process because Jürgen Dormann and the communications department confirmed to us that they often used the foyer to explain this restructuring program to visitors.

Above the lift which served the directors' offices we wrote out – in red neon – the words of the Greek philosopher Heraclites "panta rhei" – everything flows. Everything is moving: nothing stands still.

And it is quite true – the entire foyer was a dynamic sculpture: the 11,000 blue LEDs below the non-slip, frosted glass floor created waves which rippled out from the center across the entire space and back again. A metaphor for "communication": sending and receiving, speaking and listening. The source of the waves was a huge, white block of marble onto which artistically-enhanced video images in the form of waves were projected which referred to Hoechst's areas of activities. The entire room began to hum, to "breathe" – it had become dynamic. Whereas we had wondered if these moving images might unsettle visitors, the client took the opposite view and demanded more dynamism, requesting that we speed up the waves.

This installation also represented technical innovation. Blue LEDs were available in adequate numbers for the first time (and, back then, 11,000 blue LEDs cost a small fortune.)

A world clock opens up a global perspective. A satellite image of the world takes 24 hours to move across the video wall with the place at which it is midday always being right in the center. Frankfurt was no longer the center of the world but wandered across the space in line with the movement of the earth; Frankfurt, the old head office, Hoechst, wandering from the center until finally disappearing from view altogether.

With the phrase "panta rhei", the images of the waves spreading out across the world and the world clock with its disappearing Frankfurt we hit the nail right on the head. Such was the progress of the restructuring program that, by 1998, the parent company Hoechst AG consisted of just the corporate center: this directorate building with our installation and 200 employees. And in 1999 Hoechst finally merged with the French company Rhône-Poulenc to create Aventis. The entire management of the new group moved from Paris and Frankfurt to a newly created building in Strasbourg. Our role was to make the new identity and orientation of the company visible from the very start in the form of a new mediatectural foyer installation.

Aventis was now completely devoted to the "Life Sciences", a collective term for the modern areas of biotechnology and genetics, areas of activity which were already difficult and controversial for a company despite all the euphoria about constant scientific progress. Our serious mediatectural concept addressed the questions, insecurities, opportunities and visions related to these areas. We created no advertising and we showed, apart from the share price, no Aventis service or product, but we focused on providing materials which could inspire and encourage discussion and debate. And alongside all its attractive and successful design elements, this mediatectural installation became, above all, a space for intellectual discussion.

www.mediatecture.info/05

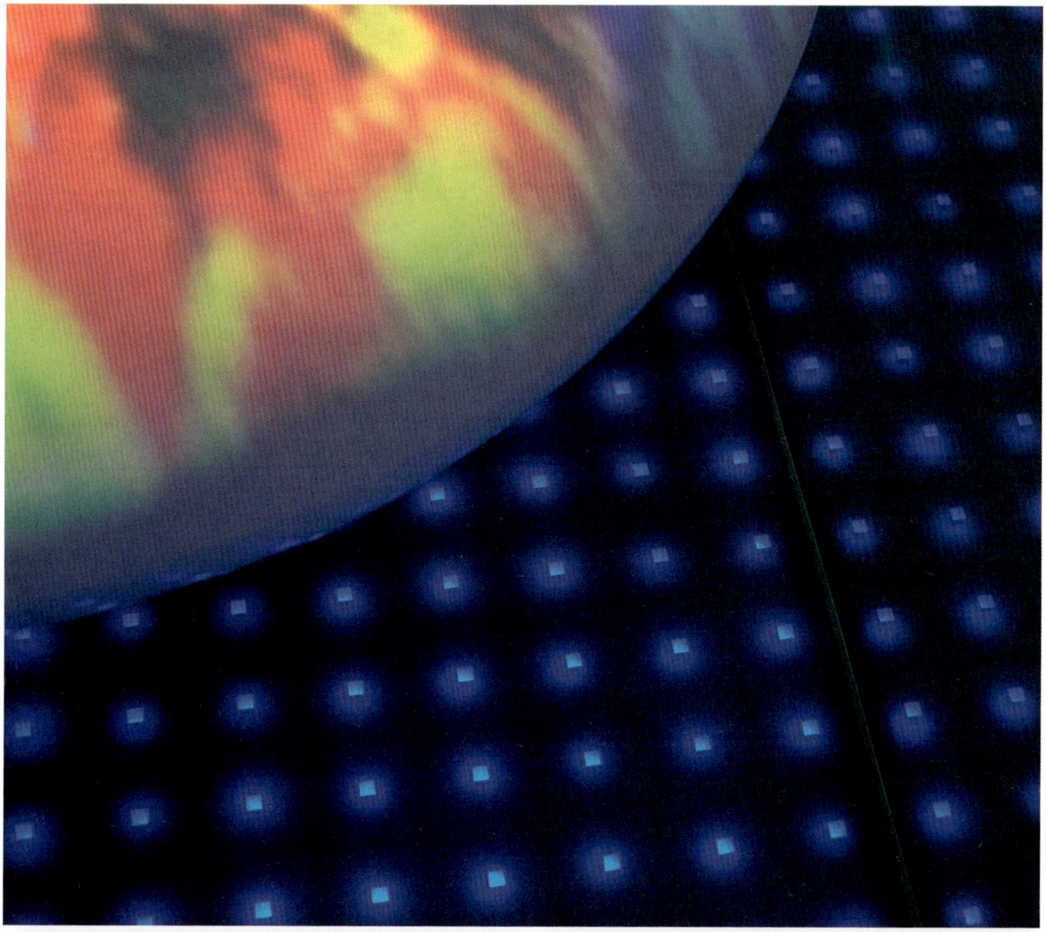

One of the waves that springs forth out of the video projection was synchronized with the animation in the floor.

The central image of the installation is an ancient allegory used by all cultures to denote life: a tree – the “tree of life”.

This is the cross-cultural symbol for insight: whether it is the historical Buddha finding wisdom sitting under a tree or Eva giving Adam the apple that she has plucked from the tree of wisdom. The wisdom that allows them to see that they are naked.

In the heart of the foyer a real tree was planted which, despite all doubts, has flourished. But this image alone would be too simple, too clichéd. Opposite the tree we placed a semi-transparent reflecting surface onto which a video could be projected from behind. The real tree, its reflection and the contents of the projected media program combine into a single entity. The video material consists of discussions on the themes of life sciences, biotechnology and genetics. The element of interaction which we have chosen is – what else – the apple of wisdom! Only by touching the apple can one navigate through the contents of the reflective wall – a sensual way of navigating through an intellectual discussion.

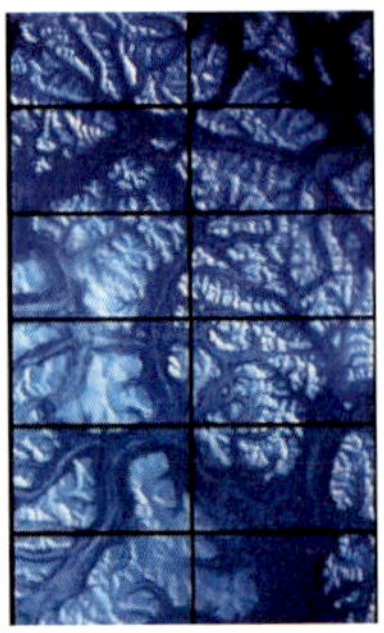
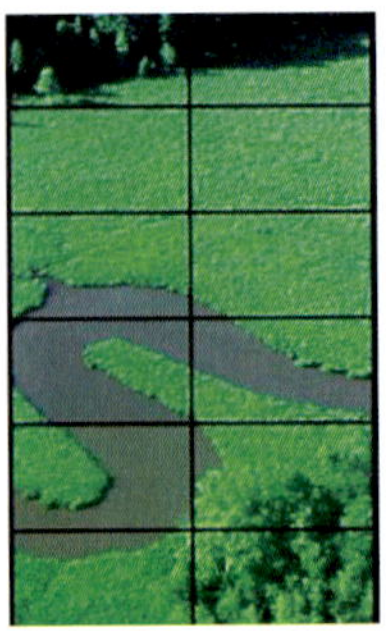

Photo © Arne Hofmann

The artistic media installation “Inner Space” uses 12 screens to project a complex image of the world as seen by Aventis which, as one of the leading life sciences companies, carries a large amount of responsibility. The view of the world becomes a view of the world within – of the objectives facing the company. Documentary scenes show the complexity of this challenge: the world’s peoples and markets, hands (which do both work and business) and feet (symbolizing life’s basics: the things which support us – whether bare feet or high heels).

A couple of years later Aventis merged with Sanofi-Synthelabo to create Sanofi-Aventis. Panta rhei.

www.mediatecture.info/06

An interactive video is projected on a holographic glass panel. Camera recognition technology enables menu navigation via hand movement on the apple.

The glass art at the reception displays a genetic code from leaves.

Mediatecture Creates Identities

We have always understood mediatecture as communicative sculpture, as built communication. In this way it can also create identities.

This is exemplified by the mediatecture which we developed in 2000 for **Start Amadeus**, one of the leading providers of IT solutions for the tourism and travel industry. In order to make the invisible service provided by the company visible and experienceable to both the employees of and visitors to the Start Amadeus headquarters in Bad Homburg we built an accessible interface in the foyer. The central interactive element was the "Virtual Globe": a huge glass "Trackball" with a diameter of 3m which could be used to interactively control not only the contents of the installation but also the entire building. An $11m^2$ LED video wall in the basement created an image which, with the help of over 100,000 fiber optic cables, was carried to the glass hemisphere in the foyer. This technology produced such projective clarity that it was even visible in daylight. With simple hand movements one used the trackball to navigate to specific global locations. The video wall then showed the corresponding country or city video (from the archive of VOXTOURS, the then leading German television travel program). And at the same time, the space was lit with colors appropriate to the corresponding time zone: interactive architecture was born.

The design principles developed by ag4 were then extended to the orientation and signage systems in the building as well as to the Amadeus Intranet and Training Centre. Hence, the interactive installation intended to help the company establish its corporate identity has truly become important for its visual identity, because the entire internally directed corporate identity is based on our basic design. Mediatecture creates identities.

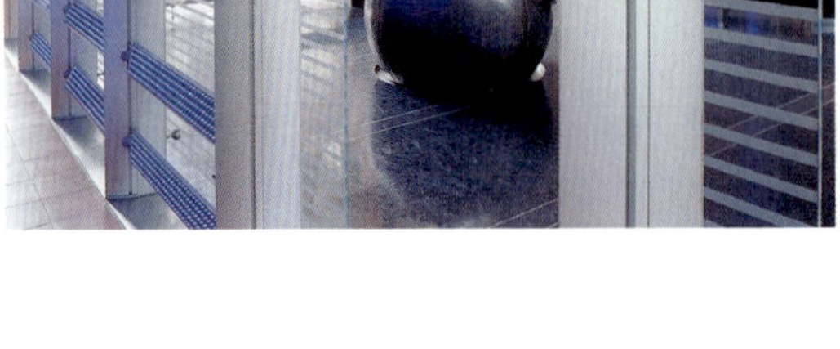

www.mediatecture.info/07

Light-emitting diodes project the image of a globe on a glass hemisphere. The globe can be rotated by hand via camera recognition technology.

The globe displays a climate map. Colored temperature information for the region in focus is presented. The white points on the map are interactive and trigger documentation about the respective place displayed on the monitor.

Mediatecture as Icon

Mediatecture has an iconographic function. An example is our installation for **Deutsche Bahn (German Railways)** at **EXPO 2000** in Hanover. The DB-Net, the world's largest railway network display panel, was installed above the visitors. The map showed all train movements on the DB network over a period of 24 hours compressed into 24 minutes. Watching these countless points and how they simultaneously and coordinatedly flow around Germany, each visitor gets a very sensual impression of the complexity of rail transport.

And thus mediatecture becomes an icon or a symbol for a company or a specific communication task. The more striking the icon, the quicker and more intuitively it is understood. In 2002 we developed such a symbol for sustainability for **BMW**. The automobile company was presenting itself at the **UN Summit on Sustainable Development** in Johannesburg, South Africa. The aim was to create a place for meeting, talking and debating – because, as the world's most sustainable automobile manufacturer, BMW had a lot to say on the topic. To this end we used the image of the medial globe as a mediatectural communications sculpture and created the "**BMW Earth Lounge**", an accessible hemisphere in the form of the earth with a diameter of 24m which served as a lounge, a plenum for daily discussions and presentations and as an immersive projection space.

The entire inside face of the sphere was filled with a video that showed, on the one hand, the people of the world (not just the rich who could afford BMWs but also those who probably had most to lose from a failure to develop the earth in a more sustainable fashion) and, on the other hand, a mirror image of the surface of the earth - the "world from within". This symbolized the fact that we must change our way of looking at the world if we really want to meet the challenges of the times.

The image worked. Most television reports from the summit were filmed at the Earth Lounge and BMW became linked with sustainability. This globe represented the link between BMW and sustainability for a number of years during which it was used for many follow-up events.

www.mediatecture.info/08

Seating landscape with a video projection onto a printed map of Germany

Challenges for the Future

The spirit of the times has changed. The world and its people are now truly networked while the challenges have become both more complex and more urgent. In its first ten years, ag4 developed tools and methods for mediatecture based on the spirit of the turn of the millennium. But the internet and digital media have, in the meanwhile, changed society so fundamentally that the scope of activities for mediatecture simply continues to grow.

All of which means that I think that my initial suspicions of 1991 have been proved right. Film seeks to immerse the viewer – to draw him in and surround him. Mediatecture is a potential means of doing this. 16:9 widescreen, high definition television (HD) and 3-D television and cinema are all leading in the same direction and confirm the thesis. The next step will most likely be the 360° x 180° film projected inside a sphere – so-called "Immersive Cinema". "R+J" (Romeo and Juliet), which I made in 2003, was the world's first feature film made in this new genre. I have always been personally interested in the development of the medium of film and can see that immersive cinema offers the possibility of combining my basic mediatectural interests with my passion as a filmmaker. This is the focus of my company LivinGlobe in Munich.

Moving pictures have started to really move out to conquer space: and the result is mediatecture.

ag4 has continued to develop the idea of mediatecture since 2001 and the more recent projects are included elsewhere in the different chapters of this book.

Low angle shot in the BMW Globe
A dome projection with laser technology

Harald Fuchs *(born 1954 in Rehau/Oberfranken) lives in Cologne and is an artist who works intensively with installations at the interface between virtual worlds and real places. He studied free graphics at the Stuttgart State Academy of Art and Design, was Professor of Visual Communication and Mixed Media at the University of Applied Sciences in Augsburg from1992 to 1995 and has been professor of the same subject in Düsseldorf since 1995.*
His 16 study trips to visit African religious sites and traditional medicine men have sharpened his view of the contrast between primitive civilizations and western culture, a view which he expresses in multimedia installations. Many museums at home and abroad offer him the opportunity to temporarily present his sensual productions in the form of spatial works.

What can I do as an artist in society? Which contracts can I fulfill?

I would like to introduce you to the readers as a mediatectonically working artist. What is the method with which you develop a project?

The first thing is the space in which I should create the installation. Such spaces can be extremely varied: a museum, some very stiff space - or an old factory. I have to come to terms with this space – must determine the extent to which I should work with wall-mounted images or with objects which one can move around. This is how the theme of the work develops. Theme and space then grow together.

And does the design of the images then immediately appear before you?

No, first of all I have to distinguish between the various projection surfaces in the space and decide where fixed images and where moving images are possible. The dimensions of the pictures must also be determined beforehand and this also has much to do with the room – as does the logical speed for the movement or development of the images.

In all it is essential to remember that the presence of people inside my installations is always very important to me and this is why my projections are always very knowingly organized in the space in a way that ensures that people are forced to move through the projecting beam and, hence, to become part of the work. This is all part of confronting the space in which the work should be situated and it is this confrontation which leads to the first images.

Artists are in the position to create connections between people and areas of life to which they normally do not have access. Which are the new worlds that your "visitors" can reach?

We humans tend to seek to understand the world as rationally as possible. But this is only works in part. There are always parallel worlds which cannot be directly understood. This is why I am so interested in scientific research,

"Die Hybrid-Kontroverse und die Trommeln von Malaga" (The Hybrid Controversy and the Drums from Malaga) Installation in the Fuhrwerkswaage art space, Cologne 2008

because this is also an activity which reaches areas which are normally inaccessible. The nuclear researcher, for example, considers elements which the senses can simply not experience – or calculated masses which cannot really exist at all. But the scientist believes precisely that he can deal with these things. Realities are created which work on paper but cannot be part of my sensual world. This interests me. On the one hand, the real, sensually experienceable world and on the other hand the surreal, fictional world. And here I also see the discrepancy in our world. There are things we experience and things we know. Experience and knowledge were for so long united, but now they have become separated. This creates a vacuum and it is this vacuum which I seek to occupy with my art.

Which a way of saying that you create interfaces between the truly ascertainable and the scientific worlds?

Yes. Art is apparently the only form which can do this – although whether this works every time is of course another question.

With reference to your work it is also important to mention another element that you incorporate very strongly in your work: your experiences of traveling among the primitive peoples of Africa. Can you describe these experiences?

My visits to medicine men in Africa offer me another opportunity to encounter a world that I cannot understand. On these visits my thinking and knowledge clash head on with magical worlds, whose inhabitants are just as convinced and convincing representatives of their background as I am of mine. And reality is something we can argue about. These are people for whom the spirits of the bush definitively exist – although we cannot experience them, but a character in a soap opera is, in reality, just as non-existent and yet, for us, somehow present.

This collision between western culture and primitive peoples creates another vacuum which I seek to fill with my art.

In your installations, one often simultaneously encounters both science and the origins of African culture.

This raises the obvious point that these themes – science and primitive peoples – are extremely different.

I don't agree. I have been documenting African medicine men for over 20 years and they have convinced me that they are as focused as we are on arriving at an understanding of the world. Their methods are simply different.

Your way of reaching these partly extremely archaic worlds is to use media technology. This is, of course, a very modern medium and your approach is to contrast it with, for example, fetishes, which you yourself introduce. Why are these high-tech methods so appropriate here?

First of all – the one does not exclude the other. And, in addition to this, an overhead projector is, for example, actually a relatively archaic medium. A fire (the light) is placed below an image or object which I lay on the projector and this creates an image on the wall. Essentially this is Plato's cave. And then I compose a digital world for this situation, a digital world influenced by our science. But, deep down, the content is unchanged. This act of projection – whether digital or archaic – is a way of expressing levels of life which we don't really understand – and hence in which we can only believe.

Your installations are always remarkable for the fact that you never use media in an obvious way. There is not a simple video wall onto which a film is projected. Your images appear fragmentarily in space – they, quite literally, tear it apart.

I employ my media in a space in such a way that it is the pictorial content and not the medium which stands in the foreground. And, rather than using a medium to generate a clear pictorial statement, I kaleidoscope together individual images into a whole. This means that - rather than the individual image having a definitive reality – each image is a piece of reality and the world, as I see it, is the coming together of all these different realities. The world consists of many different layers and realities which are superimposed upon each other. And in just the same way, each visitor to my installations uses his imagination to create his own reality. My role is not to communicate some definite reality – because each human being develops his own.

And the spatial organization of media draws the human being into a cognitive process, which he himself can further influence through the way in which he moves.

Have you experience of visitors feeling overstretched by these spaces?

No, because I also make a great effort to draw everything together into an aesthetic whole. I don't tell my visitors that they should work hard in order to understand my message. As long as one avoids giving the impression that the visitor owes it to the artist to understand the central message then the visitor can simply enjoy wandering through the spaces and make what he wants of them. And the aesthetic arrangement and manual treatment of the

Hagen environmental office, Kunst am Bau 2000

material should help in this. The visitor does not have to understand everything straight away because there are many aspects of science and primitive peoples which I also don't understand. That's why I produce art. I want my art to help us rediscover the sensual qualities that have been lost to us – and central to this is the ability to accept that there are things that one cannot understand.

Now let me imagine that you create such installations as media façades in public spaces with the aim of establishing links between a concrete location and very distant worlds. Must this always remain an artistic space which speaks for itself or could you imagine your approach being used for projects whose content is also defined by the intentions of a client or of the city?

Here is an example from my work with "Kunst am Bau" – the program for promoting art in public places: the local Environmental Office in Hagen invited me to participate in a competition. At the presentation of the project the leader of the Office invited the participants to visit the laboratory in order to discover what his organization actually did. This demonstrated the difference between my artist colleagues and myself: I was very soon alone with the client because my colleagues had no real interest in the organization itself. I found this very strange. I found the work of the Office fascinating – principally because it was so different from what I had been expecting. It was already clear to me during the guided tour what I wanted to do. I was given the opportunity to use the analytical techniques of the laboratory to create images – although naturally I developed my own very personal approach to using the technical possibilities. I won the competition easily.

This naturally confirms to me why I came especially to see you. This way of working is perfectly aligned with my view of mediatecture. At the end of the day your position is always that your installations are a medium for doing something.

Exactly, my colleagues wanted to create sculptures or mechanical objects and these had absolutely nothing to do with the Environmental Office. Naturally, being interested in one's client is, alone, not enough. One must also produce an aesthetic result.

But that also means that the basic driver is curiosity – which is also seen in your general interest in science and the medicine men. And this is how you create these meaningful connections.

I would like to build here upon my personal experience with projects for media façades. Enquiries about such projects usually – unfortunately, start with technical questions about a precise surface for a precise façade. But I always try to start by finding a content-based approach because I have no idea whether a particular surface or a particular technology is the right one for a particular project. But our society is so fixated with technology that questions of content must often be initially put aside. Only right at the end does someone ask what this medium will actually be used for. And then, as it is not easy to get building consent from the authorities for such media façades it is always a good idea if one can give the content some cultural value – a requirement which the client often believes he can meet by employing a famous media artist to develop content. Or he contacts the students from the school of art. It doesn't really matter, just as long as the screens show art.

No – this is pure decoration.

Part of the explanation for this is certainly the tendency to consider art in isolation. What can I do as an artist in society? Which contracts can I fulfill? Here is an important question regarding society and the meaning of art which has to be answered

And I trust that this approach to art which envisages the integration of art into mediatectonic projects will be more openly debated in future. In practice I often sense bitter resistance. What is your experience in this area?

I often feel alone in my approach. Here there is certainly something which must be discussed.

www.mediatecture.info/09

Hybrids – two light/sound installations, Forum für Photografie, Cologne 24 – 31 Aug. 2008

Dirk Pörschmann *was born in Worms am Rhein (Germany) in 1970 and studied history of art, sociology, history and philosophy at the universities of Heidelberg and Bochum before earning his Master of Arts degree (M.A.). Since 2007, as part of his work on a dissertation in the area of history of art, he has been addressing "Manifestations of Cleanliness in the Art of the Neo-Avantgarde". Between 2005 and 2009 he worked as a member of the research staff of the Kassel School of Art and Design. Alongside several university teaching posts and publications covering issues in the areas of art and media he has been working since 1999 – both theoretically and practically – with state-of-the-art methods in the field of artistic education. In 2010 he became a member of the research staff of the ZERO Foundation which was founded in Düsseldorf in 2008.*

Mediatectonic Approaches and Theories in Art and Media.

Prologue

Mediatecture covers the territory between architecture, technology, art, media theory, sociology, philosophy and the science of images. Mediatectonic theory is a discipline whose contents resemble a fluid material which requires a consolidating "binding agent" if it is to be turned into concrete ideas, concepts and theoretical constructs. In both the theory and practice of mediatecture much is in flux and much is possible. The term "mediatecture" can be traced back to Christoph Kronhagel and ag4. In 1993, as co-founder of the "Working Group for Four-Dimensional Construction (ag4)" Kronhagel used the term - a synthesis of media and architecture - as a means of illustrating the complexity of our environment and the consequent open questions in the area of architecture. The media philosopher Vilém Flusser (1920-1991) regards the growing sense of multi-dimensionality of our environment as conditional upon scientific research into the universe and the expansion of virtual space. He defines this virtual space as "that not-yet space in which not-yet realities spend their not-yet time". Here, the cosmic and the virtual encounter traditional notions of space: "Cosmic space and virtual space are beginning to become part of our real space and these spaces are starting to overlap and partially cover each other." Flusser recognizes that this is bound to have consequences for architecture: "This will force the designer of space to think in spatial rather than in strictly geometrical-chronological categories."[2]

The search for potential mediatectonic approaches in the history of art and media since 1900 has, as a result of the necessarily broad scope of such a search, thrown up a number of interesting "hits" which shed light on many aspects of mediatecture. This chapter is intended as the author's record of his search as he approached this inspiring subject. The objective is to present a collection of positions and examples which can be seen as possible pioneers in the field of mediatectonic practice. As so often, the researches made in preparing this text resembled a journey into areas of subjective (lack of) knowledge and (mis)understanding which were either completely or virtually uncharted – at least by this author. Such a research process can just as easily lead to new discoveries as in the past – although today's discoveries are necessarily of a different nature than those made by someone like Alexander von Humboldt (1769-1859). Whereas Humboldt, the curious traveler, geographer and scientist, was able to unearth – and present to an awed public - previously unknown rocks, animals and plants; our pushing back

»The ›digital glow‹ is the light that enlightens us from within and illuminates the yawning, empty night around us. And we ourselves are the spotlights that create new worlds as an alternative to and way of combating nothingness.«

Vilém Flusser, 1991[1]

of the barriers of knowledge appears both more complex and less dangerous. Progress is more incremental, with each new image, text and personal insight expanding the intrinsic microcosmos of knowledge and understanding of each one of us – and yet this incremental progress contains a danger of which Humboldt had already clearly warned on the level of the macrocosmos:

"In defiance of the tendency to ceaselessly subdivide the information that he identifies and collects, the categorizing thinker should strive to avoid the danger of empirical overabundance. A significant part of the qualitative energy or – in terms of natural philosophy – the qualitative expression of the energy - of the material has certainly yet to be revealed. Complete discovery is, for this reason, impossible. Alongside the sense of satisfaction at any new discovery is, necessarily, the pain brought on by the fact that there are areas of knowledge which remain inaccessible to our ever ambitious and inevitably disappointed sense of yearning."[3]

This chapter addresses precisely this tension between the motivating enthusiasm about newly discovered relationships and the resulting insights and this crippling frustration which results from the fact that – to continue our metaphor of the voyage of exploration – it is impossible for our machete to cut more than just a narrow path through the thicket of content-based relationships and causalities. In addition to reading the synthetic and hypothetical parts of this text, the reader should not restrain from further exploration based on the various sources of analytical quotations or examples. Like all new disciplines, mediatecture is starting life as a multi-discipline which is both inspired and influenced by neighboring subjects but which must also, at the same time, differentiate itself from these subjects in order to be taken seriously in its own right.

Every visitor to a Gothic cathedral knows the effect that the colored light flooding through the windows has upon the cathedral's space and atmosphere. This light "emotionalizes" in a way which reinforces the meaning of the architecture. Together with the architecture and the contents of the cathedral, it creates the context of a built, fixed place whose original function was to offer Christian congregations a safe haven for carrying out their religious devotions. In the Jewish-Christian tradition light is highly bound up with the idea of God and with traditional images of the afterlife. Light is a symbol of the Creation[4], the heavenly and the hoped-for deliverance from earthly travails. The biblical scenes portrayed in church windows may well be both entertaining and informative but it is the act of breaking through massive stone walls that gives these huge glazed openings their media-related meaning. The windows make it possible to contextualize because they create a metaphysical opening in the physical protective space of the church. The uncapturable and apparently immaterial light supports the Christian notion of the overcoming of the perishable body, suffering, fear and pain and of the victory over death.

Gerhard Richter (born 1932) designed the façade window in the south transept of Cologne Cathedral which was dedicated in 2007. [Fig. 01 Gerhard Richter, Stained Glass Window, Cologne Cathedral, South Transept, 2007 © Gerhard Richter 2010; Photo Christopher Clem Franken, Cologne 2007] Richter divided the approximately one hundred square meter window into 11,263 squares with sides measuring 9.6 centimeters. The 72 colors that he used, which are laid out according to the modern pictorial tradition of the grid, trigger associations with the colorful pixels of a digitalized media world. The colored squares were distributed quite randomly in one half of the window and the resulting pattern was mirrored as a "recurrence of randomness" in the other half, in such a way that an impression of harmony arises which defies the aleatory nature of the composition.[5] Richter's window is, like all such elaborately composed church windows, mediatecture in its best form, in that it supports the meaning of the architecture while creating a context in which the physical space is experienced as a real place with all its various spatial, temporal and content-based dimensions. Carefully conceived in line with an abstract language of both form and color, the window works together with the changes in light resulting from the movement of the sun and the passing clouds to create a wide spectrum of different impressions within the church. A latent sense of movement impresses itself upon the viewer and this combines with the religious context to create a sense of internal emotion and emotionalization.

A walk in a light-filled winter landscape makes clear the extent to which one can be sensitized by the ever-changing relationship between light and movement. With each ever-changing twinkling and shimmering of each snowy surface the senses are sharpened. The teasing light and its multiple reflections combine to draw the intense attention of the viewer who looks first this way and then that as he loses himself in the bewitching beauty of the surfaces. This is less a concrete search for an object of attention than a form of abstract observation.

Fig. 01: Gerhard Richter, church window, Cologne Cathedral, south transept, 2007 © Gerhard Richter 2010; Photo Christopher Clem Franken, Cologne 2007

Less a search for a material body than a search for the immaterial in the world around us. A search for appearances and for that which can only be indirectly experienced or which is only revealed – and only moves us – upon being seen. This emotionally charged light intensifies our ability – and desire – to see. It beguiles. An enticing, twinkling, shifting light which, with its tireless teasing of our senses, hints that there is more to come.

Edgar Allan Poe (1809-1849) impressively described the fascination of this phenomenon-based, open way of seeing in his short story "The Man of the Crowd" published in 1840. This is one of the first short stories to portray the idea of the wanderer driven by a sense of inner unease. In the wake of Poe, many writers and artists were inspired by this idea of wandering: Charles Baudelaire (1821-1897), who created the typology of the modern wanderer as night-owl and embodiment of modern urban man; Walter Benjamin (1892-1940), who gave the concept a theoretical basis in his unfinished "Arcades Project"; Arthur Schnitzler (1862-1931) in his "Rhapsody, a Dream Novel" (1925), which Stanley Kubrick (1928-1999) transformed into his impressive film "Eyes Wide Shut" (1999); the wandering driver in "Taxi Driver" (1976) by Martin Scorsese (born 1942) or the work of the Belgian artist Francis Alÿs (born 1959), who used walking as his Modus Operandi.[6]

The wanderer moves through the chasms of the big city, driven by a search for the unknown, the tempting, the dangerous and the forbidden. He is both the successor and the opposite of the romantic rambler, who pursued his longing for the primordial by returning to nature in the assumption that here he would also find his spiritual origins.

Before Poe's nameless narrator sets out to follow a man wandering aimlessly through the streets of London at night, he sits in a café in "moods of the keenest appetency, when the film from the mental vision departs" [7], excitedly viewing the street scene through the large window:

"As the night deepened, so deepened to me the interest of the scene; for not only did the general character of the crowd materially alter [...] but the rays of the gas-lamps, feeble at first in their struggle with the dying day, had now at length gained ascendancy, and threw over every thing a fitful and garish lustre. [...] The wild effects of the light enchained me to an examination of individual faces; and although the rapidity with which the world of light flitted before the window, prevented me from casting more than a glance upon each visage, still it seemed that, in my then peculiar mental state, I could frequently read, even in that brief interval of a glance, the history of long years."[8]

With his forehead pressed against the glass he looks in fascination at the lights, the faces and the goings-on in front of the café, which "filled me, therefore, with a delicious novelty of emotion"[9]. The attractive and seductive power of the scene before him have a deep effect on Poe's protagonist, creating in him the desire to lose himself in the ephemeral impression, the moving image before him. The scene becomes a sort of promise which seeks fulfillment. At this point, the window is still a barrier between the observer and the observed and it is the size of the window which determines the size of the scene. But this glass is not just a transparent barrier but also a gateway for impressions and the desires which these impressions evoke. Poe's figure can change sides. In contrast with the mediatectonic barrier of a screen or projection surface he can pass beyond the window and directly satisfy his need for immersion.

The photographers of modern metropolises understood this sense of longing and satisfied it in their lengthy forays with their cameras. Indeed, they were themselves wanderers engaged in a search for the special in the everyday – as exemplified by such figures as Eugène Atget (1857-1927) in turn of the century Paris. The work of Atget inspired the US-American photographer Berenice Abbott (1898-1991) to capture impressions of New York for posterity. Her "Nightview" photographs [Fig.02 Berenice Abbott, Nightview, 1932 © Berenice Abbott/ Commerce Graphics Ltd., Inc., NYC 2010][10], which she shot in 1932 from the top of New York's Empire State Building have an effect similar to that generated by the viewing of twinkling snow or the nightly comings and goings portrayed in Poe's short stories. The eye leaps enthusiastically from point to point, lost amongst the countless light sources of the skyscrapers: a calm contemplation of the night-time scene is simply not possible. Upon looking at the image it is clear to the viewer that seeing and movement are unavoidably linked. The philosopher Maurice Merleau-Ponty (1908-1961) describes this fact in his phenomenological essay "Eye and Mind":

"My moving body makes a difference in the visible world, being a part of it; that is why I can steer it through the visible. Moreover, it is also true that vision is attached to movement. We see only what we look at. What would vision be without eye movement?"[11]

Fig. 02: Berenice Abbott, Nightview, 1932 © Berenice Abbott / Commerce Graphics Ltd., Inc., NYC 2010

In his 1944 publication “Language of Vision”, the Hungarian artist, designer and theoretician György Kepes (1906-2001), whose media façades still deserve detailed examination today, referred to Abbott’s photographs from twelve years earlier and used these, alongside some of his own photograms and others from László Moholy-Nagy (1895-1946) and Man Ray (1890-1976) to illustrate his chapter “The Influence of Artificial Sources of Light”. Here, Kepes addressed the “psycho-physiological effect of the visual organization of light“[12] proposing the now accepted theory that light-based stimulation affects the entire human organism.

Thirty years earlier the German philosopher and sociologist Georg Simmel (1858-1918) had formulated his psychological evaluation of human perception in his paper “The Big City and Intellectual Life”:

“Man is a differential being, which is a way of saying that his consciousness is stimulated by the difference between the current impression and the most recent one: persistent impressions with minimal differences and predictable regular sequences and contrasts require less attention than swiftly compressed images whose abrupt changes can be understood with a single glance, the unexpectedness of evoked impressions”[13]

Fig. 03: Man Ray, La Ville, photogram from the series “Electricité”, 1931 © VG Bild-Kunst, Bonn 2010

The founder of urban sociology enjoyed his own subjective experience of “the unexpectedness of evoked impressions” in Berlin, one of the booming turn of the century metropolises, which was often metaphorically described as “The Laboratory of the Modern”. Rapidly changing visual impressions were generated by the incessantly growing levels of traffic and human activity, the display windows of the department stores and the ever increasing numbers of illuminated billboards whose commercial messages were ever more obtrusively broadcast into the urban space. For Simmel, Berlin was a sociological laboratory in which he could study and analyze not only the newly developed social spaces and actors but also their forms of communication and expression.

A quarter of a century after Simmel’s essay, Walther Ruttmann’s (1887-1941) film classic “Berlin: Symphony of a Great City” (1927) appeared in the cinemas. The experimental documentary was a pictorial representation of the experience of Simmel and that of other city dwellers in the early years of the twentieth century. Using innovative cinematic techniques in his images of work, traffic and cultural activities, Ruttmann portrayed daily life in Berlin. In the early evening, the illuminated advertising billboards come to life, sending their messages far over the city and becoming a brilliant setting for the night and its promise of pleasure and dissipation. [Fig. 03][14] The photogram “The City” from Man Ray’s ten-part series “Electricity” (1931) embodies both the haunting fascination and effects of modern night life[15] and the German journalist and author Max Osborn (1870-1946), writing in 1929, described the visual attractiveness of the big city at night with the words:

"Earlier, architecture went to sleep as soon as it went dark. [...] But today architecture awakes in the evening in a new and fantastic form. Strips of light draw lines in the sky and, while individual details are lost, together they trace the outlines of new, imaginary forms. These are the fairy-tale structures of the modern city."[16]

László Moholy-Nagy, whose sketch for a film script "Dynamic of the Big City" (1921/22) [Fig. 04][17] anticipated the content of Ruttmann's Film,[18] attempted to find visual techniques appropriate to this acceleration of human life. As a tutor at the Bauhaus he instructed the students about his notion of "The New Vision" which could inherently encapsulate speed, changing perspectives and simultaneous impressions and he also published such ideas in books which still inspire today. He gave his best known writings the title "Vision in Motion" (1947)[19] and addressed the current status of visual media in his statement that, "each period has its own optical approach. Ours is the age of film, illuminated advertising and the simultaneity of sensually perceptible events."[20] Such statements become particularly interesting to the media historian when they are compared with the work of the Italian writer Filippo Tommaso Marinetti (1876-1944) and his fellow futurists who had already announced in their Futurist Film Manifesto of 1916 that "our preference, however, is to express ourselves through the cinema, the huge panel with its liberated words and the illuminated moving display."[21]

Fig. 04: László Moholy-Nagy, spread from the film sketch "Dynamik der Großstadt" (1921/22) © VG Bild-Kunst, Bonn 2010

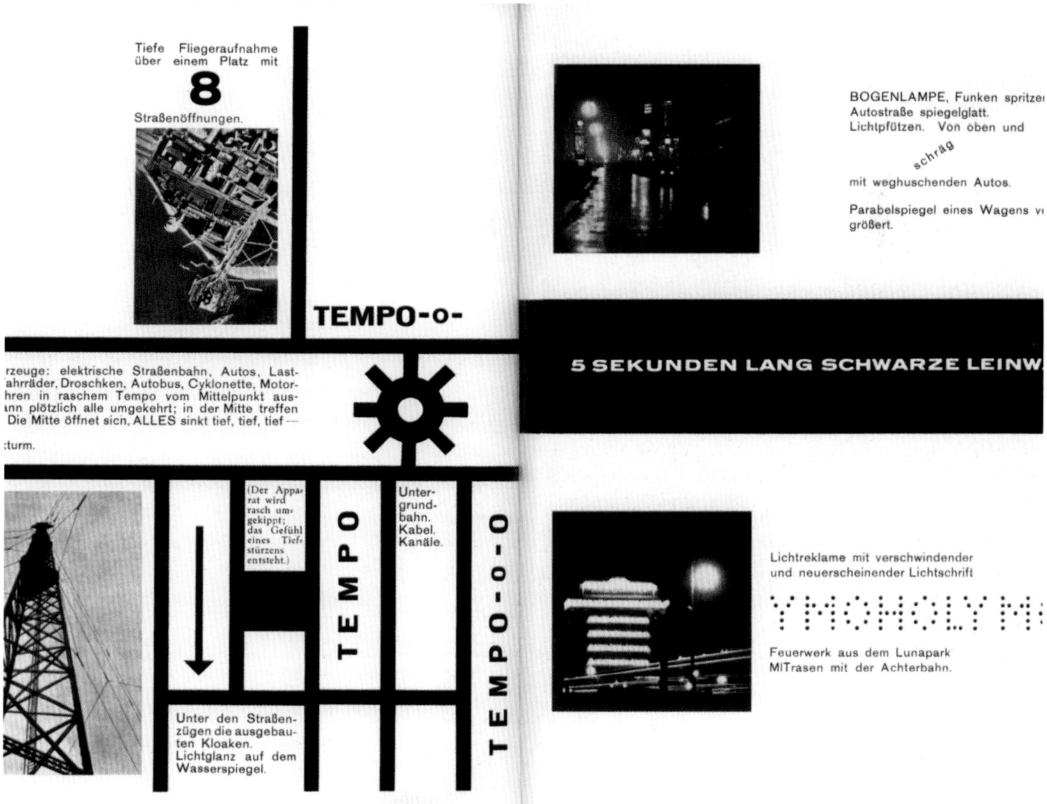

I shall now examine another fascinating example of façade design which takes us back to the mediatectonic potential of the window. As the traditional opening elements in a building volume, doors and windows feature prominently in façade design and, as such, are generally to be regarded as architectural – and only in special cases as mediatectonic - elements. In his Institut du Monde Arabe in Paris, however, Jean Nouvel (born 1945) created a façade which has both an ornamental and a technical role and, in addition, has an effect on the internal spaces of the building which can be fully regarded as mediatectonic. [Fig. 05] The southern façade of the building combines the traditional oriental Mashrabiyya – a geometrically patterned, timber or, more occasionally, metal lattice which covers windows, bays or loggias as a barrier against both climatic extremes and prying eyes – with computer-controlled iris-shaped diaphragms which open or close like blinds in line with external light levels as a means of managing the illumination inside the building. Using this light-mediating technique, Nouvel succeeded in combining technology with tradition, function with content.xxi The Franco-Arab cultural centre has a façade which can be understood as a mediatorial expression of its function. It refers to both the traditional self-image of the Arab world and its basic desire for interaction with France and Europe.

Fig. 05: Window with aperture technique seen from inside, Institut du Monde Arabe, Paris © Photo Boris Doesborg, Brussels 2007

[Fig. 06] The perforated metal façade of the temporary pavilion "switch+", which was created in 2007 as a key infrastructural element in the "sculpture projects münster 07", also carries a regular pattern although, unlike the complex abstract ornaments of the Mashrabiyya, this pattern features a single, simple geometrical form: the circle. With impressive lightness of touch, the pavilion takes the hemispheres with which Otto Piene (born 1929) hid the light sources in his installation "Silver Frequency" (1971) on the façade of the State Museum for Art and Cultural History and reinterprets these - in negative – as holes.[23] As a result, the walls of the pavilion work in a way similar to a Mashrabiyya: atmospheric light effects are created on the inside and a visual threshold is created to the outside. The visitor to the pavilion becomes a voyeuristic observer of the surrounding area and of the adjacent southwest façade of the museum, [Fig. 07] where Otto Piene mounted the 639 aluminum hemispheres of his sequentially repetitive light installation. The artist who, together with Heinz Mack (born 1931), co-founded the ZERO movement which was officially established in Düsseldorf in 1958 and internationally active until 1966, became well-known for his work with light and movement. Starting with experiments with two-dimensional grids whose movements only became apparent upon being regarded by the viewer, his work developed in the direction of three-dimensional space. The special atmosphere and impressions of his illuminated spaces or "Light Ballets" continue to enchant viewers today. In "Silver Frequency", Piene combined the motif of the grid with an orchestration of light in such a way that by switching the lights he created abstract moving images. To the viewer of today, these images provoke associations with aspects of technical media which are both contemporary with the installation (the perforated cards used in early data processing and the fuzzy lines common to cathode ray televisions) and highly relevant today (the pixelated lights of the modern media environment). Leaving the post-war artistic mainstream, ZERO turned to new materials and forms. Lucio Fontana (1899-1968), whose art and artistic theory were inspirational to the ZERO artists, hinted at this new direction in his "White Manifesto" (1946):

Fig. 06: "switch+", Pavilion for "skulptur projekte münster 07", modulorbeat urbanists and planners, Münster
© Photo Thorsten Arendt / artdoc.de

"We are living in the age of physics and technology. Painted cardboard and arrangements of plaster can no longer justify their existence. […] We have to change both content and form. We have to overcome painting, sculpture, poetry and music. We need a comprehensive art that corresponds with the needs of this new spirit."[24]

As early as 1966 Piene succeeded in using light to "dematerialize" a façade. [Fig. 08] In this project, he fitted a kinetic lighting sculpture to the main façade of Cologne's Wormland fashion house and, as in Münster, the aim of the abstract formal language of the structure was create a sense - and indeed a promise – of innovation and modernity. Commissioned in 1966 by Theo Wormland to rebuild and modernize the functional building, the architect Peter Neufert (1925-1999) clad the façade with stainless steel panels which gave the entire streetfront – with the exception of the ground floor - a very hermetic appearance. In addition, this lack of windows – which made it possible to optimize both the air conditioning and the lighting of the goods in the sales areas – together with the narrowness of the street and the height of the buildings made it impossible to create views to the outside which - in any case – would have been highly unusual. The result was that Piene's lighting sculpture has much more than a functional role: while breaking up the solid nature of the upper floors it is also designed to draw the attention of passers-by away from the narrow street and the simple act of consumption towards the "more beautiful and better world" above. This world was dreamed of by Neufert and Wormland, both of whom were collectors of modern art, and the dream was poetically formulated by Piene: "Yes, I dream of a better world. Should I dream of a worse one? Yes, I long for a wider world? Should I long for a narrower one?"[25]

In 1974, Otto Piene was appointed Director of the Center for Advanced Visual Studies (CAVS) of the Massachusetts Institute of Technology (MIT) which had been established seven years earlier by György Kepes. He succeeded Kepes, who had invited him in 1968 to become the first international fellow of the CAVS. The Hungarian Kepes, friend, colleague and compatriot of László Moholy-Nagy, was also a designer of the early media façades which paved the way for the mediatecture of the 21st century – such as the huge programmed façade for the New

Fig. 07: Otto Piene, “Silberne Frequenz” (Silver Frequency), 1971 © VG Bild-Kunst, Bonn 2010; Photo LWLLMKuK / Rudolf Wakonigg 1971; Otto Piene Archive

York office of KLM which he completed in 1960. [Fig. 09]. Kepes and Piene represented the Bauhaus tradition and both had also been influenced by the ideas of László Moholy-Nagy. All three artists were driven by the desire to combine art and life and the conviction that technology should play the role of mediator in this coming together of man and nature. Moholy-Nagy saw designing with light as representing the future of art and architecture: "just as one paints with brush and pigment, we could "paint" direct with light, transforming two-dimensional painted surfaces into light architecture."[26]

To end this selection of examples from art and architecture I have chosen two works which have the ability to create very special mediatectonic spatial effects. The architect Werner Ruhnau (born 1922) designed and built the Musiktheater im Revier in Gelsenkirchen (1956-1959), one of the most interesting new theater buildings of the post war years. [Fig. 10]. Ruhnau's special relationship with Yves Klein (1928-1962), who he invited – along with Norbert Kricke (1922-1984) and Jean Tinguely (1925-1991) – to complement his architecture with works of art inspired the development of ideas for his unconventional "shed" which led to a sustainable enrichment of both the architecture and the art. In the theater foyer, Klein and Ruhnau used huge wall paintings (7 x 21 meters) and sponge reliefs to create a "blue surge"[27] which soared over the space.

Fig. 08: Otto Piene, "Licht und Bewegung" (Light and Motion), 1966, kinetic light sculpture, façade of the gentlemen's outfitter Wormland, Cologne, Architect Peter Neufert, © VG Bild-Kunst, Bonn 2010; Photo Schmölz & Ullrich, Cologne 1966; Otto Piene Archive

Fig. 09: György Kepes, Programmed Light Mural, KLM Office, New York 1960 © György Kepes 2010

The painting of Yves Klein and the “Gelsenkirchen blue” that he developed together with Ruhnau made possible the colored “conditioning of space“ dreamed of by the unconventional founders of a planned “School of Sensibility”.[28] The aim was to arouse not only the visual but also the tactile perception of the theater public.

At this point I wish to refer again to the futurist Filippo Tommaso Marinetti, who in his “Manifesto of Tactilism” (1921) proposed an education of the sense of touch that could prepare people for the development of “The Art of Touch”. His aim was to improve both people’s sense of touch and their abilities to classify and express the resulting impressions. For this reason he proposed, in addition to a series of exercises for educating tactile sensibility, a finely graduated “scale of touch”. Alongside glass paper, a variety of silks, wool, hair and brushes, the scale also incorporated the natural sponge so beloved of Yves Klein, which Marinetti placed in the sixth touch class, labeled with the adjectives “hot, sensual, ingenious, tender.”[29]

Fig. 10: Inaugural evening of the Musiktheater im Revier Gelsenkirchen, 15 Dec. 1959 © Photo Ernst Knorr; Anita and Werner Ruhnau Archive

The joint work of Ruhnau and Klein was rich with such ideas and these resulted in a series of such detailed ideas as an architecture composed of the elements air, fire or water. In 1958/1959 they created their "Manifesto for the General Development of Contemporary Art in the Direction of Immaterialisation (not Dematerialisation)", from which the following passage is quoted:

"The guiding principle here is the use of new materials in order to create dynamic and immaterial architecture. Air, gas, fire, water and air again are such materials and these materials together should, ideally, fulfill the functional roles of protecting against rain and wind and of cooling and warming."[30]

Having begun with an example of the contemporary use of one old medium - Richter's church windows - these thoughts about mediatectonic approaches shall end with another. Since the mid 11th century, historic events have been portrayed on tapestries which have, thus, combined their role as thermal and sound insulation with that of collective memory. With its embroidered portrayal of Norman history the "Bayeaux Tapestry" (approx. 1070) is one of the most important examples of medieval tapestry making. [Fig. 11] Like the Bayeaux Tapestry, "Red Carpet" (2003) by Via Lewandowsky (born 1963) is also a historical representation. Lewandowsky produced the large-scale work for the Federal Defense Ministry. Prominently positioned in the twelve meter high Hall of Columns in the ministry building which was originally built between 1911 and 1914 for the Imperial Admiralty, the carpet transforms this into a place of both memory and warning. From the distance of the upper floors the image is clear: the red carpet, with all its associations with state receptions, carpet bombing and blood, can be seen to portray the ruins of a city. This is Berlin's Tiergarten District, home to the "Bendlerblock". In the Weimar Republic and, subsequently, in the so-called Third Reich, this building was home to a number of key military institutions. At the same time, however, it was also the centre of the political resistance to Adolf Hitler, which led to the attempted putsch of 20th July 1944.[31]

Only from a distance can the visual message be seen: if one is physically too close the image disappears; if one is too close in time, the memory is painful. Victims of aerial warfare do not require media-based messages to understand the destructive power of human fantasy and of its technical implementation. Yet, for the inexperienced visitor to or, indeed, employee of the Defense Ministry, whose only experience of the realities of war come from the media, the carpet gives pause for thought. Lewandowsky has succeeded, in the 21st century, in using a pictorial medium which, unlike the new electronic and digital media, demands distance rather than proximity. It is a healthy anachronism that, in the age of these obtrusive media, a space can be both filled and rendered tangible by a carpet. One almost gets the impression that, by opening up the floor of the Hall of Columns, one could gain an archaeological insight into the layers of history below the Bendlerblock. A simple aerial view, reproduced on a red carpet, enables a space and its turbulent history to become tangible.

If media are promissory – which is to say that they are promises of eternal life without physical pain, of variety and entertainment, of innovation and modernity, of new perspectives, of consumption and satisfaction, of insight and perspective, of diversion and distraction, of truth and knowledge, of a better future or, simply, of more - then Lewandowsky's carpet is a promise of the human ability to learn and to develop in a more humane way. And this is a promise which, like most others, can never be honored.

Fig. 11 Via Lewandowsky, “Roter Teppich” (Red Carpet), 2003, carpet, hand-tufted, 5x10 m, Federal Ministry of Defence, Berlin

1 Vilém Flusser, “Digitaler Schein”, (1991), in Vilém Flusser, Lob der Oberflächlichkeit. Für eine Phänomenologie der Medien, (Bensheim & Düsseldorf, 1993), p. 285.

2 Vilém Flusser, “Räume”, (1991), in Jörg Dünne, Stephan Günzel & Hermann Doetsch, eds., Raumtheorie. Grundlagentexte aus Philosophie und Kulturwissenschaften, (Frankfurt am Main, 2006), pp. 277, 280 & 278.

3 Alexander von Humboldt, Kosmos. Entwurf einer physischen Weltbeschreibung, Vol. 1, (Stuttgart & Tübingen, 1845). Based on a quote from the American Memorial Edition (Philadelphia, 1869), p. 36.

4 “In the beginning God created the heaven and the earth. And the earth was without form, and void; and darkness was upon the face of the deep. And the Spirit of God moved upon the face of the waters. And God said, Let there be light: and there was light” Genesis 1, 1-3. (King James Bible)

5 See Stephan Diederich et al, Gerhard Richter - Zufall. Das Kölner Domfenster und 4900 Farben, (Cologne, 2007)

6 See Stephanie Gomolla, Distanz und Nähe. Der Flaneur in der französischen Literatur zwischen Moderne und Postmoderne, (Würzburg, 2009) and Elisabeth Bronfen, Tiefer als der Tag gedacht. Eine Kulturgeschichte der Nacht, (Munich, 2008), pp. 381-415.

7 Edgar Allan Poe, The Man of the Crowd, (1840)

8 Ibid.

9 Ibid.

10 Picture Credits: Elsbeth Kearful & Astride Waliszek, eds., Berenice Abbott, (Cologne: Könemann, 1997), (Aperture masters of photography), p. 57. © Berenice Abbott/ Commerce Graphics Ltd., Inc., NYC 2010

11 Maurice Merleau-Ponty, Das Auge und der Geist. Philosophische Essays, edited & translated by Hans Werner Arndt, (Reinbek bei Hamburg, 1967), p. 15ff. English translation from Galen Johnson, Michael Bradley Smith, The Merlaeau-Ponty Aesthetics Reader: Philosophy and Painting (Evanston: Northwestern University Press, 1994)

12 György Kepes, Sprache des Sehens, (Mainz & Berlin: Neue Bauhausbücher, 1970), pp. 128-136.

13 Georg Simmel, “Die Großstädte und das Geistesleben”, (1903), in Georg Simmel, Gesamtausgabe, Bd. 7/I, Aufsätze und Abhandlungen 1901-1908, Volume I, edited by Angela Rammstedt, Otthein Rammstedt & Rüdiger Kramme, (Frankfurt am Main, 1995), p. 116ff.

14 Picture Credits: Roland Penrose, Man Ray, (London, 1975), p. 100. © VG Bild-Kunst, Bonn 2010

15 See Renate Heyne & Floris M. Neusüss, “Die Rayographien”, in Emmanuelle de L‘Ecotais & Alain Sayag, eds., Man Ray. Das photographische Werk, (Munich, 1998), p. 119ff.

16 Max Osborn quoted in Berlins Aufstieg zur Weltstadt. Ein Gedenkbuch, edited by the Berlin Commercial and Industrial Association on the occasion of the 50th anniversary of its founding with contributions from Max Osborn, Adolph Donath & Franz Maria Feldhaus, (Berlin, 1929,) p. 206.

17 Picture Credits: László Moholy-Nagy, Malerei, Photografie, Film, 3rd ed. (Berlin, 2000), p. 126ff. © VG Bild-Kunst, Bonn 2010

18 See Krisztina Passuth, “Moholy-Nagy und Berlin. Berlin als Modell der Metropole”, in Gottfried Jäger & Gudrun Wessing, eds., Über Moholy-Nagy. Ergebnisse aus dem internationalen László Moholy-Nagy-Symposium, Bielefeld 1995, zum 100. Geburtstag des Künstlers und Bauhauslehrers, (Bielefeld, 1997), pp. 37-44.

19 “Vision in motion is simultaneous grasp. Simultaneous grasp is creative performance – seeing, feeling and thinking in relationship and not as a series of isolated phenomena. It instantaneously integrates and transmutes single elements into a coherent whole. This is valid for physical vision as well as for the abstract.” László Moholy-Nagy, Vision in Motion, (Chicago, 1947), p. 12.

20 László Moholy-Nagy quoted in Petra Eisele, “László Moholy-Nagy und die »Neue Reklame« der zwanziger Jahre”, in Patrick Rössler, ed., Bauhauskommunikation. Innovative Strategien im Umgang mit Medien interner und externer Öffentlichkeit, (Berlin, 2009), p. 245.

21 Filippo Tommaso Marinetti et al, Der futuristische Film, (Milan, 11. September 1916), quoted in Wolfgang Asholt & Walter Fähnders, Manifeste und Proklamationen der europäischen Avantgarde (1909-1938), (Stuttgart & Weimar, 1995), p. 124.

22 For comparisons of the cultural use of the Mashrabiyya in East and West see Hans Belting, Florenz und Bagdad. Eine westöstliche Geschichte des Blicks, (Munich, 2008), pp. 272-281.

23 The extension to Münster’s State Museum for Art and Cultural History was demolished in 2009. The extension from the 1970s is to be replaced by a new extension. Otto Piene is to recreate his installation according to his own plans elsewhere in the museum.

24 Lucio Fontana, Manifesto blanco (Weißes Manifest), (Buenos Aires, 1946), quoted in Charles Harrison & Paul Wood, eds., Kunsttheorie im 20. Jahrhundert. Künstlerschriften, Kunstkritik, Kunstphilosophie, Manifeste, Statements, Interviews, 2 Volumes (Ostfildern-Ruit, 2003), Volume 2, p. 781.

25 Otto Piene, “Wege zum Paradies, ZERO 3,” (1961), in Heinz Mack & Otto Piene, eds., ZERO 1, 2, 3 (Reprint), Massachusetts Institute of Technology (MIT), (Cambridge/Mass., 1973), p. 146.

26 László Moholy-Nagy, “Light Architecture,” in Industrial Arts, I/1, (1936), quoted in Marion Ackermann, ed., Leuchtende Bauten. Architektur der Nacht, Exhibition Catalogue of the Stuttgart Art Museum, (Ostfildern, 2006), p. 99.

27 Quote from Werner Ruhnau during a conversation with the author in Ruhnau’s atelier in Essen-Kettwig on 3. November 2009

28 See Volker Rattemeyer, ed., Wie das Gelsenkirchener Blau auf Yves Klein kam. Zur Geschichte der Zusammenarbeit zwischen Yves Klein und Werner Ruhnau. Exhibition Catalogue from Wiesbaden Museum, (Spangenberg, 2004), pp. 11-29.

29 Filippo Tommaso Marinetti, Der Taktilismus. Futuristisches Manifest, (Milan, 11. January, 1921), quoted in Wolfgang Asholt & Walter Fähnders, eds., Manifeste und Proklamationen der europäischen Avantgarde (1909-1938), (Stuttgart & Weimar, 1995), pp. 219-223.

30 Heiner Stachelhaus, Yves Klein, Werner Ruhnau. Dokumentation der Zusammenarbeit in den Jahren 1957 – 1960, (Recklinghausen, 1976), p. 41.

31 See Bundesministerium für Verkehr, Bau und Stadtentwicklung, ed., Kunst am Bau. Projekte des Bundes 2000 – 2006, (Bönen, 2007), pp. 55-57.

Aspects of Mediatecture

The artist ***Nicolas Schöffer****, who was born in Hungary in 1912 and died in Paris in 1992, left behind a body of work which one can fully regard as mediatectonic. The Lab(au) group of artists in Brussels, who see a strong relationship between their work and that of Schöffer, arranged for me to visit Nicolas Schöffer's wife in Paris. Eleonore Schöffer engages herself energetically with her husband's artistic legacy, still living in the former studio in Montmartre where she continues to keep many of his installations and sculptures in working order.*

A Visit to Eleonore Schöffer, 6 March 2010

Eleonore Schöffer has always hoped that a museum would one day be devoted to her husband's work. And it is thanks to the astonishing openness of this 86 year-old lady to contemporary media that this museum now exists – in Second Life! She invited me to visit and explained that she also organizes events in Second Life at which up to 300 guests have been known to appear.

As Nicolas Schöffer's key work was engaged with the idea of the "cybernetic city", it is harder to find a more fitting testimony to this engagement than the existence of this virtual museum.

The work of Nicolas Schöffer must be seen in the context of the 1960s. He was without doubt influenced by the optimistic spirit of the age and its belief, that one could change the world. From the standpoint of today, the theoretical buildings which Schöffer created in his texts and works are undoubtedly difficult to access. His idea of substituting democracy with a "chronocracy" - as a means of creating a new society in which people have time for each other - or his concept of a Center for Sexual Recreation are simply difficult for us to assess.

The studio in Montparnasse in its current state

I am personally a little too young to have truly experienced these times, but I am sometimes envious of the sixties generation and of the free and easygoing atmosphere in which they could imagine their new world.

What is quite extraordinary is the level to which Schöffer's work predicted the future. He imagined an architecture which has great formal and technological relevance today. An example: "Let us consider apartments, fully conditioned by perfect systems, whose interiors offer a constant but dynamic balance between all the senses: noise, light, temperature, smell, air pressure, air-flow and humidity. With their remote controlled programming devices, the occupiers can constantly manage these sensations." What would Werner Sobeck and Norman Foster say to that!

But what really interests me is the way in which Nicolas Schöffer speaks about a programmed architecture. And not only this – he also programmed his own light installations at a time when there were no PCs to help him. He developed his own combinations of mechanisms and switches which controlled the small motors and these in turn controlled his lights. But – and this is particularly important to me – he did this in a non-linear way in order to ensure that interaction was possible.

Nicolas Schöffer Museum in Second Life

Eleonore Schöffer showed me his fully functioning installation Chronos 10. The light from a series of spotlights is broken up by mirrors and prisms to create light structures on a projection surface and Eleonore Schöffer underlined that no single moment of this visual experience was ever repeated!

Schöffer planned to develop this same idea further in his "cybernetic light tower". The idea – developed during the presidency of Georges Pompidou – was to build this 303 m high construction in Paris La Défense district and only the president's death prevented its realization.

Installation in car dealership in Genf

Chronos 10

Projections from Chronos 10 on a projection panel

The light tower would have been a truly fascinating example of cybernetics. Based on the principle that "everything is connected to everything else", it first simplified and then symbolically presented cybernetic theory. The theme of interactivity which is so important today was central to the concept which sought to incorporate all possible influences and input from the surrounding area and its inhabitants.

There are many points in this book at which the reader will understand how impressed I was by this early example of the process-based medialization of a public space.

Chronodynamic façade, 1959

Chronos 4, 1961

Cybernetic lighttower, Paris La Défense 1961

Second Order Cybernetics – A Paradigm for Mediatecture?

Cybernetics is one of those incredibly intriguing terms; when it comes to using it, however, one constantly fears to be treading on thin ice because any questions inquiring about its actual meaning can only be answered vaguely. That is why I will start out simply by asking Wikipedia:

"Cybernetics (from ancient Greek κυβερνήτης kybernétes = steersman) is defined by its founder, Norbert Wiener, as the study of control and communication in the animal and the machine. Cybernetics explores the basic concepts for the control and regulation of systems, irrespective of their origin. This makes such different fields like machines, humans or organizations comparable. It was developed in the 1940s by scientists from the most different disciplines, and it has since inspired diverse new fields of application.

One typical feedback system is a heater regulated by a thermostat. The thermostat compares the actual value of a thermometer with the one it is set to as the desired temperature. A discrepancy between the two values causes the switch in the thermostat to regulate the heating in such a way that the actual temperature approximates the desired temperature. Cybernetics models the regulation of the body temperature of an organism or the market situation of a society in much the same way."

It appears that cybernetics is concerned with the description of processes that touch on very different areas at the same time and influence each other. Naturally, a mediatect like myself is interested in this.

The undefined, open character of cybernetics offers some advantages for the conveyance of the idea of mediatecture. I myself only encountered cybernetics through working on this book and soon came to realize that I wasn't the only one in my field. Through my colleagues at LAb[au], I came across Nicolas Schöffer, who even titled one of the main books from his body of work: "Cybernetic City".

Reading it, you feel overcome by a slight dizziness, which you then try to generously explain with the exuberant spirit of the 60s – at least that's what I did.

The search for other protagonists of cybernetics inevitably leads to Heinz von Foerster. The media scientist Prof. Bernhard Pörksen, Tübingen University, held long conversations with Heinz Foerster and published these under the title:

Truth is the Invention of a Liar

Conversation for skeptics, Bernhard Pörksen in conversation with Heinz von Foerster (Wahrheit ist die Erfindung eines Lügners. Gespräche für Skeptiker, Bernhard Pörksen mit Heinz von Foerster im Gespräch) © Carl-Auer-Systeme Verlag: Heidelberg, 1998.

Heinz von Foerster, born 1911, is held to be the "Socrates of Cybernetics". The trained physicist first worked in different research laboratories in Germany and Austria. In 1949, he immigrated to the USA with a quantum physical theory of memory in tow. There he worked with the pioneers of cybernetics – the mathematician Norbert Wiener, the neurologist Warren McCulloch and the inventor of the computer John von Neumann – eventually founding the meanwhile legendary Biological Computer Laboratory (BCL) at the University of Illinois, which became a research center for cognitive science. In the inspiring atmosphere of this institute, physicists and philosophers, biologists and mathematicians dedicated themselves to questions of cognitive theory and worked on the logical and methodological issues raised by the cognition of cognition.

It is Heinz von Foerster's merit to have repeatedly drawn attention to the inevitable blind spots and idiosyncrasies of the observer who approaches the supposedly independent object of observation. He introduced epistemo-critical doubt in cybernetics, thus upsetting what would have to be called the mechanistic ideas of early cyberneticists who, with wholly unbowed enthusiasm, spoke of building an "artificial brain".

Reading the conversation about cybernetics in this book, I was once again overcome by this strange dizziness. I am surprisingly familiar with this feeling. The complexity of mediatectonic trains of thought and the corresponding discussions often make you feel like you are getting lost. A mediatectonic concept can always be examined from different sides, and the result is always something different. Generally, I have learned that mediatecture is only possible if you are prepared to give up a certain control and know how to work with paradoxes.

This article mainly consists of excerpts from the conversation between Bernhard Pörksen and Heinz von Foerster. It is absolutely impossible to distill a quintessence from it; you will just have to open yourself to it. With this I am hoping to be able to convey an idea of what I would boastfully like to call the "mediatectonic flow".

Bernhard Pörksen:
Anyone searching for your name in scientific literature will find entries in connection with cognition and constructivism, theories of the memory and the concept of self-organization. However, you are probably best known as a cyberneticist. First of all: What could a definition of cybernetics, which was incidentally also fundamental for the development of artificial intelligence, look like? What is cybernetics?

Heinz von Foerster:
I'm not particularly happy about the search for a definition because it draws up conceptual boundaries. You might as well ask me: What is a table? And my answer would be: A table has four legs and a flat top that children can jump onto. Now we have to clarify what the difference between a table, a pony and a horse is. And eventually it will become necessary to talk about the difference between living beings and non-living entities. That serves us just right. For me, every definition has one fundamental flaw: it excludes, it confines.

Bernhard Pörksen:
Still, it makes sense to bring some order into the variety of definitions that are currently circulating. To give a few examples: The mathematician Norbert Wiener defined cybernetics as the study of control and communication in the animal and the machine. The business and management consultant Stafford Beer defined cybernetics as the science of organization; the neurophilosopher Warren McCulloch talked about cybernetics as a cognitive theory dealing with the production of knowledge through communication. An explanation from the American Society for Cybernetics reads: "Cybernetics is a way of thinking, not a collection of facts." And Gordon Pask put it even more generally: "It might be an art, or a philosophy, a way of life."

Heinz von Foerster:
First of all, I would like to say how very happy I am that such a small collection reveals such a colorful and broad range of attempts at a definition. This is evidence of a way of thinking that allows a variety of different approaches. The range of the possible and this openness form an excellent stimulus. It becomes clear that it is about a mindset that every one of these people interprets in their very own way. That is what is fascinating about cybernetics: You ask a few people for their definition and learn very little about cybernetics itself but a great deal about the person giving the definition; about their expertise, their relation to the world, their joy in playing with metaphors, their enthusiasm for management, their interest in communication or information theories. This way I get to learn even more about my friends Stafford Beer, Warren McCulloch, Norbert Wiener and Gordon Pask. That is just wonderful.

Bernhard Pörksen:
Is there a connective principle that all these ideas on cybernetics share?

Heinz von Foerster:
The fundamental principle of cybernetic thinking is, I believe, the idea of circularity. That's where everything starts, you have to continue thinking from there, that's the foundation. The principle of circularity yields enormous implications if you fully think it out and combine it with epistemological questions. You suddenly enter forbidden territory and find yourself dealing with self-referentiality, a concept frowned upon by logicians. However, it takes its time to fully fathom the implications of circular causality. I remember I once pointed out that circularity was the essential principle in cybernetics and that it needed to be explored much more fundamentally. I wrote about a new dimensionality of arguing that supersedes linear causality. My paper, which was conceived as an introduction to the reports from a series of cybernetics conferences, was sent back to me, and I was asked to instead write about the water closet, the thermostat and the Maxwellian regulator in steam engines. I replied saying that I wasn't particularly interested in the water closet and, while the thermostat made life more comfortable, I didn't find it particularly important.

Bernhard Pörksen:
The examples that were being talked about back then really didn't encourage any epistemological reflections: thermostats, water closets and the steering of boats were at the center. Could you still take one of these examples to explain the principle of circularity?

Heinz von Foerster:
Best we talk about the steering of boats because the term cybernetics, which Norbert Wiener coined and in 1948 used as the title for his book, goes back to the Greek word for steersman (kybernetes), which in Latin becomes "gubernator" and in English "governor". So if one were to follow the history of the word, an American governor would have to be a cyberneticist. But back to our example: What does a steersman do who wants to maneuver his boat safely into the harbor? He doesn't carry out a once-and-for-all fixed program, he constantly adapts it. If the boat yaws to the left because the wind is so strong, he will assess this course deviation so that he continues to drive toward the harbor. He tries to correct the mistake. And maybe he countersteers a bit too much. The result might be a course deviation to the right – and the necessity to countersteer again. In every moment, this deviation is corrected in relation to the desired goal, the Telos, which could be a harbor, for example. Consequently, the use of the steering wheel, a cause, produces an effect, the course correction. And this effect becomes yet another cause because a new course deviation is detected. And this cause produces yet another effect, another course correction. Such steering processes are a wonderful example of circular causality.

Bernhard Pörksen:
What happens here is basically a process of information analysis that makes you change your behavior. You notice a course deviation and act accordingly, by countersteering.

Heinz von Foerster:
The early cyberneticists – Norbert Wiener, Claude Shannon, Warren Weaver, Ross Ashby – repeatedly emphasized exactly this aspect. They pointed out that the steersman, for example, must "inform" his motoric system how and to what extent it should move the steering wheel. And this information about the manner of the movement in relation to a certain desired goal can be seen as a process of information analysis.

See my attempt in "The Network (R)evolution" to explain the tasks of the mediatect and the corresponding significance of regulatory systems as the subject of conceptual work.

And I believe that the image of circularity is very helpful in order to work with the term "mediatecture". It describes the difference to the popular term "media architect", which by comparison, however, denotes a far more linear understanding because it simply adds up two disciplines. As soon as these disciplines seriously engage with each other, they start to circulate, making it impossible to separate them; the result is something new. And at the end of a mediatectonic project, you cannot tell anymore whether the architectonic part has become media and the media part spatial, or both, or neither – exactly, a mediatectonic flow.

Bernhard Pörksen:
In what respect are such analogies and metaphors ("electronic brain" and the "memory of the machines") legitimate and helpful?

Heinz von Foerster:
It is quite possible to characterize an operation carried out by the brain with the help of an automatic mechanism. That's ok. Naturally, I can also use a metaphor and say that the brain, when I'm feeling cold, turns up the warmth button. And, accordingly, that the thermostat senses a certain outside temperature – and also activates a kind of warmth button. Of course, it is legitimate to use a metaphor to describe a phenomenon or a group of phenomena, but you always have to bear in mind that every description, whether formal, mathematical or poetic in nature, is only ever a comparison. However, if one were to reverse the metaphorical relation and say: the brain works like this machine; that's when it becomes dangerous. You think you understand the brain because you have grasped the automatic mechanism you're taking as a point of departure. You think you understand the memory if you use a metaphor to describe it as a storage mechanism and maybe start looking for the place where certain information is supposedly "stored". The result is a blindness to the miracles of the brain. I recognized this danger quite early on, and, consequently, I have repeatedly criticized such metaphors and analogies. But I wasn't heard – I was talking into a void.

Bernhard Pörksen:
You have to bear in mind that metaphors consist of an image-giving and an image-receiving element. If I say that the brain is a machine, then our ideas of a machine are image-giving. The image-receiving element is the brain. If, however, I claim that a certain machine works like the brain, then the machine becomes the image-receiver and the brain the image-giver. This means: In the case of the metaphorical relation between the brain and the machine, first the brain has the primacy, then the machine, whose characteristics are transferred onto the brain.

Heinz von Foerster:
You can say that, yes. What I feel is crucial is that computers, or any other machine for that matter, are produced synthetically: We built them and, consequently, we also know how they work. And if you infer back from the functional principle of such a machine to the brain or a human being, then the result is the misconception that you now also understand the brain and the person. That's the problem: You infer back from something that is known and understood to something that is unknown and not understood, carelessly believing you now also understand this. What is being overlooked is the fact that there are certain systems that principally cannot be analyzed. Anyone who has come to understand this will be suspicious of all these metaphors.

Bernhard Pörksen:
The disciplines of artificial intelligence and robotics research, which build on cybernetics in many respects, are infected with the claim that it is possible to decipher the human mind, that it will be possible already in the near future to fully understand it. Here, the human being – as part of an idea characterized by the machine metaphor – is conceived of as an "information processing system", the process of thinking as "data processing", and the brain was once described by the great star of artificial intelligence, Marvin Minsky, as a "meat machine".

Heinz von Foerster:
These statements reveal the suggestive power of such metaphors: As you can see, they have lost their descriptive function and taken on a life of their own. The result is a grotesque worship of the machines. And such images and ideas encourage certain steps. At a certain point in time, for example, neurophysiologists received large amounts of money from the most different foundations to search for engrams in the cellular tissue of the brain: These were supposedly experiences a person made at a certain point in time and, following the metaphor, has stored in their memory ever since. The question was: Where is the little word "and"? Where is the memory of my grandmother or the schnitzel I ate today? The search for engrams was and has remained unsuccessful. The memory cannot be located at a specific point in the brain. However, I

still believe that such ideas and metaphors are dangerous. They lead to trivializing observations; they make certain steps appear possible that might be entirely inappropriate to the miracle of human life.

Bernhard Pörksen:
This trivializing use of metaphors we're talking about here can also be read in epistemological terms: It is the linguistic indicator of a certain epistemological position, of a mechanistic and, basically, naive-realistic thought style. If you understand the human brain as a machine or the memory as a storage unit containing information, then you are always acting on the assumption of the full transparency and reproductability of a phenomenon.

Heinz von Foerster:
The fallacy of these illustrious and highly gifted men was to believe that the models to understand the brain would be constantly improving. What was being overlooked was that you need a brain to understand a brain and develop models based on it. You actually have to explain and understand yourself in order to understand the brain. The structure of the theory I'm referring to must fulfill the requirement to describe itself: Symbolically speaking, that's the Ouroboros, the snake that bites its own tail. This is where the phenomenon of circularity comes back into play. When I arrived in America with a suitcase full of European traditions and started to work with the early cyberneticists, I tried to point out that the concept of circularity is also fundamental in epistemological respects and has far-reaching implications.

Bernhard Pörksen:
What are they?

Heinz von Foerster:
The result is an entirely different attitude toward that which you are trying to explain. You get into a loop that connects you with the respective subject and object of the observation. You don't only have to explain the brain of someone else but also your own with which this explanation is being worked out. All of a sudden cyberneticists are talking about themselves, all of a sudden you get a cybernetics of cybernetics or a cybernetics of the second order: The cybernetics of the first order separates the subject from the object; it refers to a supposedly independent world "out there". The cybernetics of the second order or the cybernetics of cybernetics is circular in itself: You learn to see yourself as part of the world you're observing. The entire situation of the description slides into a different area where you suddenly have to assume responsibility for your own observations.

Bernhard Pörksen:
It was your reflections that brought the observer into cybernetics and signaled the end of naive-realistic and mechanistic positions.

Heinz von Foerster:
What is crucial is that the whole language changes in the cybernetics of the second order; the references to an observer-free world are replaced by references to yourself, the observer. The descriptions are always also self-descriptions. However, it is not only the language that changes. The reflections on the whole purpose of the observation – carried out for whatever reason – gain a different dimension, too; you start to realize why you want to know or learn something. From a perspective of the second order, an epistemologist dealing with the issue of cognition always also asks what good the cognition of cognition is, how meaningful the attempts to shed light on the process of cognition are. He justifies the epistemology in an epistemological fashion. The idea we have of an experiment also changes from the perspective of the second order: You understand that the question you're asking already contains the possible answer you might get.

Bernhard Pörksen:
Naturally, it is possible to speak of a cybernetics of cybernetics or of a cybernetics of the second order to linguistically signal the observer-dependency of all cognition. These terms, however, are especially valid within the discipline of cybernetics; that's its point of reference. But the idea of the observer-dependency of all statements is, of course, far more general in nature: There is always this connection between the subject and the object, this close relation.

Heinz von Foerster:
Certainly. Accordingly, it is possible to differentiate between the terms of the first and second order. The terms of the first order are based on an apparently objective observation of the world that becomes externalized. The terms of the second order can be applied onto themselves; they don't allow this strict separation between the subject and the object, between the observer and that which is being observed, anymore. You concede that someone who is talking about issues of consciousness or cognition requires a consciousness and cognitive skill in order to do so. It becomes clear that an observation requires an observer. The perception of the world requires a person who perceives it.

Bernhard Pörksen:
Can you illustrate the concept of the second order with an example?

Heinz von Foerster:
Sure, just think of the experiment with the blind spot. My eye is focused on a star – and the black dot that is still visible for some time disappears at a certain distance from the eye. It becomes invisible. The physiological explanation says that the black dot at a certain distance falls onto an area of the retina that doesn't have any light-detecting photoreceptor cells and where the optic nerve leaves the eye. However, what this physiological explanation cannot account for is the question of why we can't see this blind spot and why we sense nothing of its existence. We always imagine our field of vision as being closed; there are no invisible spots. We don't see that we don't see. We are blind to our own blindness. This is an example of a problem of the second order. The not-seeing is applied onto itself. However: this double negative (the not-seeing of the not-seeing) doesn't result in a positive. The fact that we see that we don't see does not mean that we can now see. And this means that the logic of the terms of the second order is not compatible with orthodox logic. Because, accordingly, two negatives would have to equal a positive.

Bernhard Pörksen:
In the example you have chosen, the concept of the second order is explained ex negativo. It is a problem of self-application from the perspective of the negation. We don't see that we don't see.

Heinz von Foerster:
Exactly. Of course, it is also possible to find an illustration ex positivo. Just think of the term of the goal or purpose in the Aristotelian sense. From the perspective of the second order the question is: What is the goal of the goal? What is the purpose of the purpose? Why is the idea of the purpose introduced at all? When I'm traveling by plane, I often observe the person sitting next to me and how they lace up their shoes. When the plane takes off, people take off their shoes. Once we're approaching our destination, they put them back on. They thread the laces through the eyelets and perform a strange dance with ten fingers: They create a bow. The observation of the persons sitting next to me has shown that they all dance differently, that every person performs different finger movements to tie a bow. A physicist would now write down a differential equation of the finger movements, receive all sorts of different equations and, consequently, label the phenomenon of the finger dance as inexplicable. From a teleological point of view, the finger dance can be easily explained: It serves to tie a bow and to lace up the shoe. From a perspective of the second order, the idea of the purpose has created an enormous simplification and unambiguousness of explanations. That's the purpose of the purpose.

Bernhard Pörksen:
Consequently, the terms of the first order are located on a level where perceiving and acting are not yet epistemologically reflected on. The terms of the second order allow the addition of epistemological dimensions and make the special interests of an observer visible who, for example, uses the term of the purpose. Together, we could now climb further up the ladder of abstraction and, from the perspective of a third order, ask the question: What good are the terms of the second order? What can be observed on this level?

Heinz von Foerster:
You have already hinted at the answer. The terms of the second order reveal insights into the process of observation that are not at all possible on the level of the first order. On this level, you just act; you use certain concepts, presumptions and theories that are not reflected on. It is only on the level of the second order that the possibility for self-reflection arises. Nothing is just there anymore, nothing is self-evident. What is crucial is that the observer is responsible for their observations, their talking and acting. They are intrinsically tied to the subject and object of their description. The epistemological and logical area of one's own statements is taken to a new dimension.

During this passage, I constantly have to think of the interview with Bazon Brock. Could the media façade be an architecture of the second order? After all, Bazon Brock emphasizes that, from a historical point of view, a media façade's potential for interactivity in architecture is something truly new because the façade is becoming mimic; it makes contact with the people on the street by interacting with them. Humans orient themselves in their behavior on their spatial surroundings, and now these surroundings are changing because they react to the people. What is going to happen now? Will this now influence the people again? If so, the result will be a circular connection between the people and the architecture.

Bernhard Pörksen:
This is probably due to the fact that the descriptions of the second order are always also self-referential. And if you transfer self-referential statements onto the area of classic Aristotelian logic, you get paradoxes.

Heinz von Foerster:
Yes, just think of Epimenides, who came from the island of Crete and said: "I am a Cretan. And all Cretans are liars." You can shorten this statement to: "I am a liar!" What do you do with a person who says: "I am a liar!"? Do you believe him? Then he can't be a liar, so he must have told the truth. If he spoke the truth, then he lied because he says: "I am a liar!" What has been driving logicians crazy ever since is that Epimenides's statement does not comply with the Aristotelian postulation that "a meaningful statement is either true or false." This statement becomes false if you think it's true and true if you think it's false.

Bernhard Pörksen:
Epimenides's attempts to confuse us have something absurd about them. You never know if this Cretan is telling the truth or lying.

Heinz von Foerster:
Careful! What we can establish is that self-referential statements were characterized in this manner: They were claimed to be absurd. Because they force us – that's the argument – to constantly go from a yes to a no, from a no to a yes. Just think of the famous paradox of the barber who lives in a small town and shaves all men who do not shave themselves. Does the barber shave himself? If he shaves himself, then he cannot shave himself because he only shaves those who do not shave themselves. And if he doesn't shave himself, then he must shave himself because he shaves those who do not shave themselves. Self-referential statements are effectively forbidden already in the introduction to Bertrand Russell and Albert North Whitehead's "Principia Mathematica". These excellent logicians are fully aware that it must be the self-referentiality that causes this strange paradox. This means that, strictly speaking, even the little word "I", which constantly produces this self-referentiality, could not be used anymore. That is grotesque, of course.

Bernhard Pörksen:
May I sum up? To start off, you talked about circular causality as the basic cybernetic principle, pointed out the circularity of all cognition and outlined the contours of a cybernetics of cybernetics. Finally, the question of a new logic arose that doesn't forbid but allows the self-referential statements of the second order. It was always about considering the idea of circularity with all its implications.

Heinz von Foerster:
Well said! This is a fitting ending to our conversation about cybernetics and circularity. And this ending is not an end, but another beginning that will not lead to a definitive end but rather to another beginning. As Heraclitus once said: "Everything flows." To which he sometimes added: "It is impossible to step into the same river twice." I would like to paraphrase this as: "It is impossible to look into the same face twice." The face once seen will never be seen again, it is – just like everything else – gone forever. But I can look into the face of Uncle Theobald twice because it is language that halts the stream of time. Nothing is static, there is no definitive beginning or end. These somersaults that are being performed here can be learned, indeed, I would even say: You can enjoy it mid-tumble.

Can we now imagine an architecture that expands in such a way that its appearance is shaped by the social processes within and outside of its spaces? An architecture whose spaces, as part of the corresponding social system, become self-referential and whose sensual appearance can never be predicted with certainty? And one that directly interacts with people, and exactly because of this conveys a feeling of security?

Aspects of Mediatecture

FAÇADES

Media Façades

What is a media façade? What role can it play?

This chapter demonstrates the highly varied methods and approaches associated with this new way of designing façades. But are we not missing the point if we simply focus on designing the outer skins of buildings? Is our objective to give architecture the impression of being more dynamic or to allow it to become a medium? A comparison of very different design approaches will allow us to examine the design potential of this new métier.

... the Pixel as an Element of the Façade ...

Since 1997, the brothers ***Jan*** *and* ***Tim Edler*** *have worked together on projects at the interface between space, information and communication, firstly in Berlin between 1997 and 2000 as part of the trans-disciplinary group of artists "Kunst und Technik" which they themselves had founded. In 2000, the duo jointly established* ***"realities:united"*** *as a "Studio for Art and Architecture", with which they have, in the intervening years, realized artistic and hybrid installations in, primarily, urban and architectural contexts at an urban and architectural scale. The focus of the interest of the brothers is on the development of situation-specific methodologies in the area between art, architecture and technological research.*

STRATEGIES FOR MAKING ARCHITECTURE MORE DYNAMIC

Jan Edler & Tim Edler / realities:united, Berlin 2009

The age of media architecture has not yet begun, despite the many new LED screens which are based on such science fiction films as Bladerunner[1].

And the main reason for this is that the role of the media façade as an architectonic and urban element has not yet been sufficiently researched and developed.

In contrast with the huge technical progress of the last ten years, convincing concepts for the use of such screens are still scarce. The majority of current urban media installations are unimaginatively cobbled-together arrangements of television monitors in XL format. And discussions about media façades and media architecture tend to be equally uninspiring. One group would like to ban media façades altogether in the interests of protecting both "the city" and "architecture": a misjudgment which is perhaps as understandable as it is misguided. A second group treats the elements of media architecture as fetish objects, seeking to use the apparently futuristic qualities of such objects as a means of giving their – in truth - conventional projects the appearance of being progressive. This exploitative approach is a real problem because it merely prolongs the life – and debases even further – a series of existing clichés, and results in media façades which recall the plastic and aluminum foil décor of the 1970s – and which age just as fast. The third group of practitioners, rather than wasting its time philosophizing about architecture, simply seeks out opportunities to enhance the cityscape with as many, huge "utilization windows" as possible with the aim of optimizing the presence of advertising screens in public places.

From the architectural perspective, none of these three approaches is convincing because each tends to hinder rather than help architecture in its current process of transformation.

And this process is a considerable one. The ongoing media-based (and, in future, the increasingly geometrical) mutability of buildings makes it not only possible but also essential for architecture to develop further from a static to a dynamic form of expression. Distant comparisons can be drawn with the enormous technical and aesthetic challenges which accompanied the transition from still to moving images: and even a cautious assessment recognizes that this is a project of enormous significance.

What we need is time to stop and think; Time for modest but systematic developments rather than dramatic yet empty gestures and words. Time for consciously addressing all the details: not only aesthetic and technical details but also questions of content, none of which have so far been adequately addressed – let alone fully mastered.

This, in any case, is the position of realities:united. The motto of the studio: establish ambitious targets but then address the issues involved with continuous and radical investigative energy rather than too much haste. Look carefully. Evaluate the results. Shift the frontiers.

There follows a presentation of eight projects in this area chosen from the work carried out by the studio over the past few years. Six of these have been completed; one is currently in the detailed design phase (although realization is still uncertain) and one is currently under construction, with completion planned for early 2011.

BIX

For the studio's first media façade – the BIX[2] light and media installation on the Kunsthaus in Graz – there was, initially, neither a commission to develop such an installation nor an execution budget - and nowhere in the world was there a technically comparable reference project. Nor, indeed, had realities:united an adequate project background, with the natural conclusion that the chance of realization was around zero!

In these circumstances, the team from realities:united developed an "asymmetrical" design, meaning that, from the (usual) catalogue of requirements for a media façade, just a few key requirements were selected and then fully developed - with little regard for all the other (seemingly less important) requirements.

In the case of BIX, the key criteria were identified as size, affordability and aesthetic integrability. This explains the adoption of a system-based solution using individually controllable, off-the-shelf fluorescent lamps with the result that the solution, quite deliberately, failed to meet state-of-the-art technical standards for television-like LED façades and viewing screens. The system allowed no use of colors and – particularly seriously – the size and relatively small number of light sources extremely limited the image resolution and the range of content which could be presented.

BIX had been motivated by modifications to the original competition scheme for the Kunsthaus from the team of architects led by the British Archigram legend Peter Cook. A key element of this original project had been the notion of a buil-

ding being enveloped by a homogenous skin, but this idea of a membrane with communicative translucency had been a victim of the programmatic and budgetary realities of the design process.

And this led to BIX. The idea was that the installation would be a way of reintroducing the notion of translucency and transparency to the design and, at the same time, be a speech laboratory – a development platform with which the building could develop its own "language".

The project was realized after gaining the last-minute support of the owners, the architects and the designated Museum Director.

The debate set in motion by both the fundamental investigation and the detailed design process in Graz had much wider repercussions and played a key part in the development of the further projects presented below.

The confrontation with architectonic scale led to an installation with an extremely low resolution and a "pixel" measuring around one square meter, with the result that it is impossible to avoid seeing - and evaluating - these objects as independent architectonic elements. BIX transforms the idea of the screen from that of a surface – defined by its height and width – into that of a series of independent configurable objects. One can get to like this idea: but is

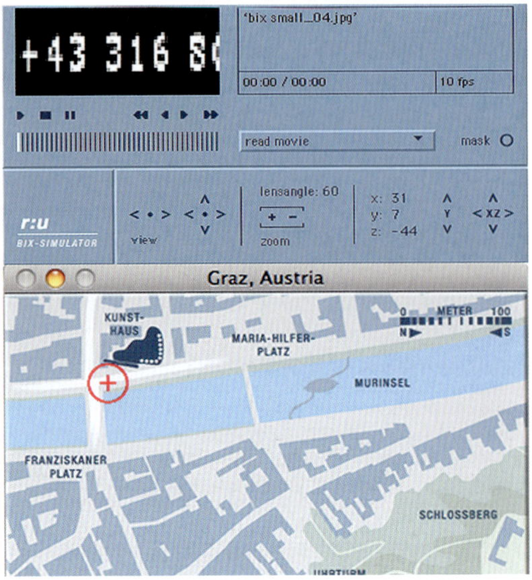

there an aesthetic – as well as an economic – justification for such an abstractly low pictorial resolution? One approach would be to ask why, while architecture has been remorselessly developing in the direction of abstraction ever since the 19th century, the subject of media façades seems randomly excluded from this tendency towards the abstract and, instead, appears driven by a naïve desire to cover buildings with the latest, highest resolution images that technology allows? Is the apparent knee-jerk tendency to adopt each latest technology and product appropriate or would it not be better if – even in the area of media façades – the same sort of reflective process took place that (in architecture) often leads to "older" materials and construction technologies being given precedence over newer ones?

The BIX development process as led by realities:united led to a measure of both sensitization and thematic emancipation: if screens are really to become true architectural elements then their design must be addressed at every level. And there are more levels than one imagines.

Further (media façade) projects of the studio have enabled the standardization of this contemplative process.

Each basic characteristic of a screen is questioned and, where appropriate, altered. This is basic research and it is leading to the subtle, gradual reworking of the rectangular, orthogonal, grid-based screen. The drivers of this design process are, successively, the architectural design and urban context, the overall objective or constructional task of the installation and the proposed content. And the basic goals of the research are the synthesis of the medium and the object and the understanding of the relationship between built substance and moving signals, etc.

www.mediatecture.info/10

SPOTS

In SPOTS[3], a temporary installation on the Potsdamer Platz in Berlin, a basic orthogonal grid was used to create different pixel types which were then, in turn, used to create middle-format arrangements on the screen.

Both the technology and the basic dimensions (approximate diameter 40cm and approximate separation 82cm) of the pixels were taken from the forerunner project BIX. The project also addresses the nature of the edge and of the transition between the screen and the rest of the building – a design question which had been resolved at BIX with a sort of soft edge or chamfer. This project, however, contains clear geometrical figures which have the effect of defining a two-part asymmetrical screen.

The attempt to define the screen as a self-sufficient formal element can also be seen in the layout and the colors of the semi-transparent, silkscreen-printed façade cladding which is designed to better distribute the light at the level of the façade while also ensuring that the image content can be better read. The layout of the pixels, the asymmetrical form and geometry of the screen and the permanent, colored impregnation of the image by this film ensured that the installation – both as a projected image and as an object with its own inherent characteristics - was ever present.

At the same time, the geometrical figures created by the individual lights and the film laid over the glass façade were also designed to meet the requirement that the building should have a specific daytime appearance (in addition to its nighttime appearance as an illuminated object). Such a requirement is known as Building Tuning. The task of the screen-based installation throug-

Sensor4, an interactive media design by Carsten Nicolai for SPOTS

hout the eighteen-months of its design life was to transform this central but relatively discrete office complex into a Berlin "landmark" with a firm place in the public consciousness.

The starting point for this use of media as a means of raising awareness of both the Potsdamer Platz itself and the wider location was the development of a comprehensive program of art. Having established a range of possible themes, three curators or curator teams – Andreas Broeckmann, Ingken Wagner and realities:united themselves – commissioned international artists and designers to produce location-specific works which could be shown on the façades for four to six weeks.

Andreas Broeckmann's curatorial starting point was, for example, the notion of the "fictional city" which used the building as a "looking subject". To investigate the notion further, Broeckmann commissioned works from such artists as the US-American Jim Campbell, the Mexican-Canadian Rafael Lozano-Hemmer and the Berlin-based Carsten Nicolai.

While Campbell and Lozano-Hemmer adapted existing works, Nicolai's "Sensor[4]" was new. A reactive work, Sensor used software to translate dynamic environmental data collected from the location – movement on the square, the density of people, brightness, loudness, etc. – into a continuous, complex light modulation representing an abstract image of life on the Potsdamer Platz. Despite an artistic concept which, at first glance, appeared almost simplistic, Nicolai's work was particularly convincing. The translation of life on the square into abstract images on the façades of the buildings seemed to give the place an almost magical specificity and this, in turn, reinforced its urban quality. The same was not true however of all the artistic material shown on SPOTS. Projects which simply used the installation to conventionally project more complex images on the screen tended to appear more superficial and were unable to establish the same convincing relationship with the architectural and urban context - a lesson which has, incidentally, informed the way in which the BIX installation is currently being used as a platform for a range of artistic works.

Jim Campbell on SPOTS

In the view of realities:united, the Berlin experience merely underlines the importance (particularly for architects) of developing an architecturally adequate language for the contents for such systems as a means of retaining some sort of influence over the installation while preventing, as often happens (and particularly in the case of advertising use), other players from taking complete control of the screen. For one thing is clear: the use of dynamic elements in architectural contexts is not to be prevented. The question is rather who will assume long-term control of the design of such buildings and building surfaces.

www.mediatecture.info/11

Crystal Mesh

In the Crystal Mesh[5] installation for the "Urban Entertainment Centre" in Singapore by WOHA Architects[6] this idea of the ornamental molding of the individual lighting elements was taken further. During the course of the design process the lighting installation which the architects had originally conceived as being located behind the glass façade[7] first broke through and then became integrated into the glazing. These two systems – lighting and glazing - evolved into a single permeable layer of ornamental, interlaced crystalline objects which could be seen by day and by night.

The lighting elements are arranged in a regular honeycomb pattern with the pixels set out at 56cm centers. However the homogeneity of this grid is challenged, both by the regular wall openings as a result of which pixels are regularly missing and by the distorting effect created by the three-dimensional topography of the individual objects.

The "design" of the individual lamps, the static arrangement of the façade elements and the continuously changing screen matrix are engaged in a simple, constantly overlaid dialogue. The associations which the viewer is invited to make are clear – with the ornamental, permeable "screen" façades of the traditional local architecture as much as with the light-bulb aesthetic of the prototype entertainment district of the 20th century city. The Crystal Mesh project is the first significant element of the government-sponsored project to create the "Bugis" nightlife district in Singapore. Further buildings

are to follow, and this explains why this project adopted such a simple, tested technical and aesthetic approach. It is hoped that the combination of the simplicity and the scale of these installations with the raw power of the potential lighting effect will give this project a real chance of holding its own against the competing projects which will appear on neighboring sites using the latest LED technology during the course of the next ten years. The key to this will be the development of adequate content. As in the cases of BIX and SPOTS, it will be important to commission a properly curated program of artworks which have the aim of creating a visual connection between the activities going on in the cinemas, multifunctional theater, night clubs, restaurants and shops of the (programmatically necessarily) windowless building and its urban context[8].

www.mediatecture.info/12

AAmp

At the same time, on a site a few hundred meters away in the same city, another project, AAmp[9], was realized in partnership with the same architects. This project presented another opportunity to address basic design questions but, in this case, the challenge was very different because the key architectural decisions had already been taken before realities:united came on board. Here, a high definition LED screen was fixed to the nose of a building. Seen as being economically independent, the screen was regarded as a technical means of transforming the building's prominent location into advertising time and, hence, revenue.

realities:united was commissioned to address the issue of integration; to develop a link between the advertising screen and the rest of the new building. The proposed solution was AAmp - an additional, deliberately low-resolution (550 pixel) element which, together with the screen (with its 600-fold higher resolution), forms a whole in

which the advertising messages in the center are artistically reworked as they are carried over onto the neighboring façades. The bright rectangle of the central screen is surrounded by a constantly changing aura and, as a result of AAmp´s use of a conventional double-skin glazed office façade, this aura is often interrupted by a sudden glimpse into the "real" workings of the building. LED projectors at the level of the outer façade project the aura onto the inner glass façade and its automated solar protective blinds which, after nightfall (and only after occupants have left their offices), descend to become a projection screen. The central building management system automatically controls both the closing of this screen and the LED projectors. Importantly, in contrast with the projects described above, the production of content in the case of AAmp was clearly controlled and, for the first time, determined exclusively by the use of real-time algorithms. This time there was no platform.

The Berlin office Thismedia[10] was commissioned by realities:united to develop a special software, which, in normal operation (Direct Mode), carries out a real-time analysis of the commercial video images on the main screen and uses a changing combination of algorithms to create the aura – a visual echo of these images which acts as a mediating element between the screen (with its total disregard of the architecture) and the building. Again, the scale of the installation plays an important role but so do the aesthetic cross-references between the two screen elements – whether these are abstract references to color, graphic form or editing speed or the more pictorial references exemplified by the software-supported "sampling" of pictorial content based on motif recognition.

The Dream Mode, which kicks in when advertising time on the commercial LED screen cannot be sold (a situation in which advertisers would normally fill the gap before the start of the next advertising block by projecting "pretty pictures") is very different. Here, the artistic software also takes control of the main high-resolution advertising screen and generates a video collage across the entire façade in the form of a dream sequence abstracted out of the advertising images which have gone before.

www.mediatecture.info/13

C4

The project C4[11] is the current project which has moved furthest away from the concept of uniform pixels and orderly grids.

The project in question is the competition entry for the "Espacio de Creación Artística Contemporánea" – a building for the production and exhibition of contemporary art – by the Spanish Architect Nieto Sobejano[12] which is currently being built in Cordoba and which should be completed by early 2011. On the outside of the building, the architects had originally planned a low-resolution screen formed of cross-shaped "BIX" type fluorescent lamps as a communications tool for the artists working in the building. A first test of this approach made it clear however that it had to be adapted. The modified concept takes as its starting point the subtly varying sequence of interlocking deformed polygons which form the floor plan of the exhibition space. In close cooperation with the architects, this same polygonal structure was superimposed upon the several hundred meters of the longitudinal façade of the building which soars high above the river and can be seen over great distances. The polygons of the façade were then transformed into a mosaic of illuminated prisms cut into the glass-reinforced concrete surface and it is these prisms, which follow no recognizable pattern and no two of which have the same size or form, which are illuminated from behind in a carefully controlled sequence in order to create a pixel effect.

In addition to this, the prisms in the façade panels can be illuminated to three different intensities and this, in an analogy of the way in which the retina works, makes it possible to create areas of varying visual acuteness in the façade. The idea of combining various levels of resolution in the façade resulted from the observation made in both real projects[13] and also numerous simulations that, in reality, it is not the eye but the "brain" that sees. In the case of very low-resolution images it is the previous knowledge of the brain that determines if a motif is legible and a motif which was previously shown in a higher resolution is often automatically recognized even when it is shown with the resolution reduced.

In the further design development process, realities:united looked in particular at the surface quality of the prisms as it was clear that one result of the indirect lighting of the individual prisms was that light radiated out in all directions – including directions in which it was neither needed nor welcome. This led to the development of computer algorithms which define the specific microtopography of the surfaces of each prism, making it possible to steer the light leaving the prisms in the direction in which it is required in order to optimize the "legibility" of the façade installation. In this case, the light is steered towards the viewer on the riverbank opposite the building, minimizing both unnecessary light pollution and energy consumption.

Museum X

In the area of the development of "communicative façades" the "Museum X"[14], developed in 2006, is a special case because it gets by with a single image.

The background to the project was the decision to close Hans Hollein's[15] Abteiberg[16] Museum for a whole year for refurbishment. realities:united was commissioned by the Museum Director[17] to develop a sculptural representation of the museum for the time during which it was closed. The proposed location for this sculpture was a decrepit theater located in the pedestrian zone just a few hundred meters from the museum. This theater – an example of post-war modern architecture which had stood empty for years – was highly significant for the people of Mönchengladbach because the city government had recently, after years of hefty debate and in the teeth of massive protests from citizens against its demolition, approved plans for the replacement of the important monument (and symbolic heart of the city) with a shopping center built by the ECE[18] Group. The idea behind the city's decision, that such measures can reverse retail drift from the city center, is a widely held cultural-political idea in such cities, and one about which opinions are divided.

For this reason, realities:united decided not to create, as commissioned, a sculpture of the museum but rather a "Museum as Sculpture". With the use of a few simple techniques the theater was transformed into a detailed simulation of a museum as it probably would have looked had it been built in the city a few years earlier than Hollein's epoch-making postmodern building of 1982: A curtain wall façade consisting of 60 steel framed textile panels printed with a photograph of the washed concrete from Werner Düttmann's[19] Akademie der Künste in Berlin, a "MUSEUM" sign, a rooftop flag and a museum foyer created in the theater foyer with museum personnel and six days a week opening. But: the

interior contained no art! Hidden behind the façade were just the inaccessible, empty 65,000 cubic meters of the former theater.

In addition to acting as a temporary "representation" of the permanent museum, Museum X thus also became a catalyst for discussion about the future and the design of the city center – in just the same way that Hollein had envisaged the Abteiberg Museum as a catalyst for urban development.

NIX

Another way of assessing the potential of dynamic architecture starts with the very basics of our architectural understanding. This approach assumes that a building is understood as the sum and combination of all its potential elements, systems and processes. These include the growing number of "intelligent" and networked sensors and actuators that are becoming established in buildings as part of the growing trend towards building automation. These are typical of the sorts of building elements whose role in the future will be more than just functional and which will have a key effect on the aesthetic appearance of buildings by doing much to define their "aesthetic behavior". The booming, competitive cities of today require new, dynamic forms of communication with unique aesthetic qualities on an incomparable (urban) scale and lighting is one of the first areas in which state-of-the-art technology is able to turn such forms of communication into reality.

The NIX[20] concept, which realities:united first presented at the Architecture Biennale in Venice in 2006[21], is a concept at the interface between large buildings, transparent glass façades, reliable network technologies and modern lighting systems.

NIX takes the individual ceiling lights in large office buildings and combines these commonplace elements into a dynamic, three-dimensional lighting sculpture covering a whole building – or even several buildings. As soon as the building control system records that the last occupant has left a certain area of offices, control over the lighting in that area is handed over to a second management system which then combines all these ceiling lights into a three-dimensional lighting choreography. The concept is called NIX because this choreography requires no extra hardware.

In the case of large buildings or groups of buildings, the constant dialectic between areas of offices which are still occupied and those which are now unoccupied and under the artistic control of the NIX system is a key aesthetic element of the project. The irony is that it is those who are working late who destroy the illusion of a perfectly coordinated super organism.

The technical paradigm change behind NIX is accompanied by a huge artistic and cultural challenge. The development of a new visual language based on dynamic, spatially moving volumes of light requires a new artistic approach freed from the restraints of the two-dimensional screen.

NIX started life in 2005 as a realities:united internal research project. In 2006-07 the studio was commissioned by the European Central Bank to finalize the concept to the point at which tenders could be invited for its implementation in a tower designed for the bank by Coop Himmelb(l)au[22] in Frankfurt. The realization of the system in Frankfurt was, however, abandoned in 2007 due to a combination of financial uncertainty and lack of courage. Since then, realities:united has carried out discussions with numerous architects and developers regarding the implementation of the system in a range of projects across the globe.

Contemporary Architecture

The "Contemporary Architecture"[23] Installation developed in 2007 for the "Artistsspace"[24] gallery in New York gives a first sense of the potential of such a double-use of basic lighting infrastructure which, until now, has been used only functionally. By being arranged in seven-segment displays forming two digits, "Overstrength" standard lights become, simultaneously, both information carriers and a lighting source. In its "display" role, the installation shows the exact time in minutes while, at the same time, providing the pragmatic lighting for the space. And as the information on display hardly changes from minute to minute, the lighting in the room appears to stay effectively constant.

1 Bladerunner, 1982, directed by Ridley Scott (Warner Bros)
2 BIX, Communicative Display Skin for the Kunsthaus Graz, 2003; Team: Rainer Hartl, Juan Ayala Cortes, Wolfgang Metschan, Timm Ringewaldt, Cornelia Neumair, Carla Eckhard, Margarete Pratschke, Gesine Borcherdt; Collaboration: Peter Cook & Colin Fournier/ spacelab, Architektur Consult ZT GmbH (Architects), John Dekron, Jeremy Rotsztain/ Mantissa, Ulrike Brückner/ Musterfirma, Peter Castine (Software Development); Link: http://realities-united.de/#PROJECT,69,1
3 SPOTS. Temporary light and media installation, Potsdamer Platz, Berlin, 2005; Team: Ulrike Brückner, Carla Eckhard, Erik Levander, Malte Niedringhaus, Stefan Tietke, Christoph Wagner, Jean-Philipp Wittrin; Collaboration: Marc Fiedler/ Café Palermo Pubblicitá (Project initiator), Eckhard Fischer / Ansorg & Horn Engineers (electrical design), Dr. Michael Kiel / HHP (Fire protection expert), Dr. Steffen Müller (Lighting emissions expert); John Dekron & Jeremy Rotsztain/ Mantissa (Software development), Bernd Hiepe (Photography); Ina Froehner/ Pleon GmbH (Project communication); Link: http://realities-united.de/#PROJECT,81,1
4 Technical Production and Software Design: John Dekron, Berlin
5 Crystal Mesh, Ornamental and granulated light and media façade, Singapore, 2009; Team: Mason Juday, Gunnar Krempin, Daniel Mock, Stefan Ludwig Neudecker, Malte Niedringhaus, Ulrich Pohl, Hendrik Poppe, Christoph Wagner, Markus Wiedauer, Christoph Witte; Collaborators: WOHA Architects Singapore (Architecture), John Dekron & Erik Levander/ thismedia (Software development), Timm Ringewaldt (Video); Link: http://realities-united.de/#PROJECT,138,1
6 http://www.wohadesigns.com/
7 As in the case of SPOTS, the commissioning of realities:united to carry out the Crystal Mesh project envisaged the copying of a previously realized façade installation. However, as is often the case with the studio during the course of the design process, the design process was accompanied by a partial redefinition of the original commission.
8 Exact details were not known at the time of writing.
9 See project AAMP, Media art installation mediating between & merging a commercial hi-res billboard and architecture, Singapore, 2009; Team: Johann Christoph Bätz, Wolfgang Metschan, Daniel Mock, Malte Niedringhaus, Stefan Tietke, Christoph Wagner, Markus Wiedauer; Collaboration: WOHA Architects Singapore (Architects), John Dekron & Erik Levander/ thismedia (Software development); Link: http://realities-united.de/#PROJECT,140,1
10 www.thismedia.de

11 See project C4, Centro de Creacion Artistica Contemporánea de Córdoba, 2011; Team: Marie Banâtre, Johann Christoph Bätz, Christoph von Mach, Daniel Mock, Malte Niedringhaus, Ulrich Pohl, Stefan Tietke, Christoph Wagner, Markus Wiedauer; Collaboration: Nieto Sobejano Arquitectos (Architect of building and project partner); Link: http://realities-united.de/#PROJECT,77,1
12 www.nietosobejano.com
13 See SPOTS and AAmp
14 Museum X, Urbane Installation als einjähriges Interim für das Städtische Museum Abteiberg, 2007; Team: Marie Banatre, Nikolina Licina, Milena Monssen, Malte Niedringhaus, Stefan Tietke, Christoph Wagner; Kollaborationen: M:AI Museum für Architektur und Ingenieurkunst Nordrhein-Westfalen, Initiative StadtBauKultur Nordrhein-Westfalen, Museumsverein Mönchengladbach (Projektpartner); Hardy Unterberg (Statik); Natalie Czech (Photografie); Prof. Gerhard Comelli, Norbert Köllschen (Video); Link: http://realities-united.de/#PROJECT,78,1
15 www.hollein.com
16 www.museum-abteiberg.de
17 Susanne Tietz, Direktorin des Museum Abteibergs seit 2004
18 Die ECE Projektmanagement G.m.b.H. & Co. KG ist ein von Werner Otto gegründetes in Hamburg ansässiges Unternehmen, das gewerbliche Großimmobilien entwickelt, realisiert, vermietet und betreibt, und als eines der größten Shopping-Center Entwickler Europas gilt: http://de.wikipedia.org/wiki/ECE_Projektmanagement
19 www.de.wikipedia.org/wiki/Werner_Düttmann
20 NIX, Research on the artistic potential of synchronized lighting systems in high-rise buildings, unrealized; Team: Johann Christoph Bätz, Wolfgang Metschan, Malte Niedringhaus, Christoph Wagner; Links: http://realities-united.de/#PROJECT,142,1 & http://realities-united.de/#PROJECT,178,1
21 "Convertible City", 2006, German contribution to the tenth international Architecture Biennale in Venice, kurated by Gruentuch Ernst Architects, Berlin
22 www.coop-himmelblau.at
23 Contemporary Architecture, Art installation on growing sophistication of functional lighting, 2007; Team: Mason Juday; Link: http://realities-united.de/#PROJECT,76,1
24 www.artistsspace.org

... Algorithms and Light ...

***LAb[au]**, laboratory for architecture and urbanism, is a Brussels-based interdisciplinary group of artists including Manuel Abendroth, Jérôme Decock and Els Vermang. The work of the group addresses the interface between architecture and controlled lighting, which is explained by the fact that its members include specialists in both fields. This expertise enables the group to create highly complex algorithms as a means of setting up interactions between installations and urban space.*

ARCHITECTURE OF LIGHT

Manuel Abendroth

Architecture has always had a close relationship with technology and technological progress. The industrial age has, for example, created not only new building materials and forms but also a wide range of new building types. In addition to this, the emergence of such new means of transport as the railway and the car has led to basic changes in urban structures. When one compares cities before and after the industrial revolution one becomes very aware of the extent to which technical progress has influenced both architecture and urban planning.

The international style and "new ways of building" were expressions of these changes and represented the new self-image of architecture. But as well as merely responding to fundamental changes in society and in its economic, political and social structures, modern architecture also sought to influence these changes: to make a statement. This relationship between technology and architecture thus raises a basic question: what is the extent to which technology changes the way in which we live and what is the extent to which architecture can be a physical expression or part - or a sign or symbol - of this process?

The world of artistic production is constantly witnessing new signs and symbols (semantics) and methods (praxis) which have been triggered by aspects of technical progress. The very term "design" for example, as it emerged from the Bauhaus at the beginning of the last century, is a symbol of the transformation from a pre-industrial to an industrial society. Its meaning lies in the new definition of artistic creation at a time of fundamental technical and social change and seeks to form a bridge between such changes and the notion of art itself. Since then, our understanding of design has – like technology itself – experienced further dramatic developments: from industrial design to visual communication; from process design to cybernetics ... and the result of such developments is that technology has come to determine many of the basics of artistic production - its forms, methods and content.

Framework notation, LAb[au] 2009, 1.4x1.4 m digital print

The emergence of the computer signaled a further wave of far-reaching social change. Just as the invention of the steam engine had heralded the industrial revolution, the computer symbolized the beginning of the information age. Both inventions led – and are still leading – to both fundamental transformations in all aspects of life and to new ways of thinking. Our understanding of material, bodies, space and time has fundamentally changed and is today increasingly related to the notion of information. Information-based structures, processes and systems widen our understanding: our understanding of space, through the use of such parameters as immersion and ubiquity; our understanding of time, through real time and entropy; our understanding of materials, through smart and nanomaterials and our understanding of biology, through genetic technology (to name but a few). In such a way, information influences our daily life and our very social and cultural being. In view of such changes it is only natural that architecture should address the issue of information – with the result that it is common today to speak about information architecture, architecture as interface or medium or, even, architecture as code. This is all an expression of a new architectural understanding and of architecture's new role.

The conclusion from the above is that, in the information society, architecture must address the development of software and hardware and the issue of content management as a means of integrating new forms of communication, information processes and information flows into spatial artifacts. And the influence of such issues on architectural design is, in the future, set to increase. One thinks, for example, only once about the meaning of designing a library in the network age. But the issues involved are much more than the network integration of buildings – just like the design of a library has to do with much more simple questions of storing and retrieving books. The issue is much more one of creating a new form for these new functions and processes in order to establish a signal: just like the way in which a library embodies our knowledge-based society and humanistic way of thinking.

When addressing issues of network technology it is important to pay attention to both the creation of new artistic forms of expression and layers of meaning and the examination of the deeper questions of contemporary social symbols and social change.

Parameter Design

This new understanding of architecture finds expression in LAb[au]'s "Parameter Design" method which is based upon the interdependence of media-specific and architectural parameters and demands that artistic design and thought processes employ methodical and conceptual ways of dealing with media and technology.

This interdependence is established by a control system and program; an algorithm, which is based, on the one hand, on a language or general logic and, on the other hand, on the functionality of the chosen technology. Parameter Design depends upon the integration of a series of semantic systems (those of architecture, software and hardware) in the sense that it both accommodates and gives a particular aesthetic treatment to media-specific structures and processes, as a means of binding these into a cultural symbolic canon. In following these principles this design method creates both a form and a symbolism for the construct itself, which is once again consistent with the word "design" and its derivation from the Latin word "designare" or "to describe". The way in which a work of art is executed is derived from the specific ap-

Binary waves, cybernetic installation, Saint-Denis canal, Paris 2008

pearance of the context – or "setting" - which is transformed into both the content and starting point of the work. Most important here are concepts and ideas which are regarded as of equal value for the artistic work. Accepted ways of seeing things, terminologies and connections are challenged and reduced to parameters as a means of creating new rules, connections and associations. The result is that Parameter Design leads to a coherent language of artistic forms and creates a methodology and an aesthetic for technology-based design.

"One of the roles of art and artists is to be witnesses of their time, which is to say that they should investigate the artistic aspects of the use of new technologies - whether in connection with new materials or the development of new ideas."
E.L.Schöffer Commentary about LAb[au] _ Catalogue of the bi-annual art grandeur nature 2008

In summary, one can say that the projects of LAb[au] are about the creation of signals in urban spaces. And yet a signal is only seen as such when it employs a universally understood language or primary code. In order to be understood as symbols of the age such projects also have to employ the latest technology as a means of establishing a clear relationship with contemporary social and cultural processes. Parameter Design is one concrete way of dealing with technology in the creation of artifacts whose semantics operate on such a meta-level. Its methodology is derived from the worlds of communication and information science, process methodologies (industrial design) and the spatial concepts of architecture.

Light as Medium

The following description of various projects realized by LAb[au] in the past few years is intended to shed some light on this relationship between architecture and technology with a particular focus on the special aspects of light architecture. There are many ways of addressing media façades and their relationship with architecture and urban context. A first approach is to consider the largest common dominator of these constructions – light – as part of the architectural content. This approach of light is illustrated by the "chrono.tower" project which was developed by LAb[au] for the Dexia Tower in Brussels.

Chrono.tower _ midnight, Dexia Tower, Brussels, Belgium 2008

Chrono.tower _ 22h36m2s, Dexia Tower, Brussels, Belgium 2008

Chrono.tower / Chrono.prints

Each of the 4,200 windows of the 145 m high tower can be individually illuminated by a RGB-LED lamp. But instead, as is often the case with such interventions, of treating the resulting light façade as a huge screen with a resolution of 45 x140 pixels, this project attempts to highlight the spatial and temporal aspects of the lighting of the tower.

The lighting concept is based on a simple relationship between the basic units of time and the primary colors of light, namely: hours=red, minutes=green and seconds=blue.

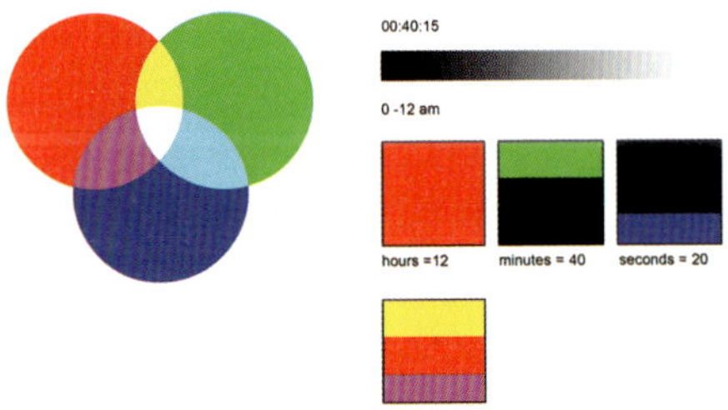

The result is a process in which three colored surfaces progressively illuminate the façade of the building. With each second that passes an extra level of windows is colored blue; with each minute, one in green and with each other hour, one in red. The result is that the three primary colors overlap –which results in the creation of areas of the secondary colors: yellow, cyan and magenta.

As midnight approaches, the overlapping of these colored surfaces increases and the illumination of the building becomes ever more intense. At midnight, when all three RGB colors fill every window, the tower appears to be completely white. Then the process is reversed, with the result that, as areas of color progressively recede, the illuminated area climbs slowly to the top of the tower.

This time-related change in the intensity of the light lends the process a symbolic character. Nightfall is generally associated with a loss of light – or darkening – whereas midnight represents the start of a new day and a resulting increase in lighting levels. This project reverses this relationship in that the lighting intensity of the tower increases as night advances until the white burst of light at midnight heralds the start of the new day. From this moment however, the light recedes until just a thin illuminated strip can be seen at the top of the skyscraper: the artificial light climbs towards heaven where the light of day gradually appears.

The process behind this project is based on a widely understood relationship between the parameters of light and time, a relationship which we experience as our circadian rhythm. The basic unit of time underlying this biological rhythm is 24 hours and its light component is the light of the sun. Modern and, above all, metropolitan man may feel increasingly separated from such natural forces, but this must not mean that he has lost his relationship with light and its rhythms. Quite the opposite in fact: this project uses a color-light cycle rather than an illuminated clock precisely in order to lend the relationship a new, appropriate and urban expression.

In conclusion, the project is an artistic response to the relationship between light and architecture. Light becomes the actual content of architecture – it becomes medium. And as a medium it communicates nothing other than its own construct.

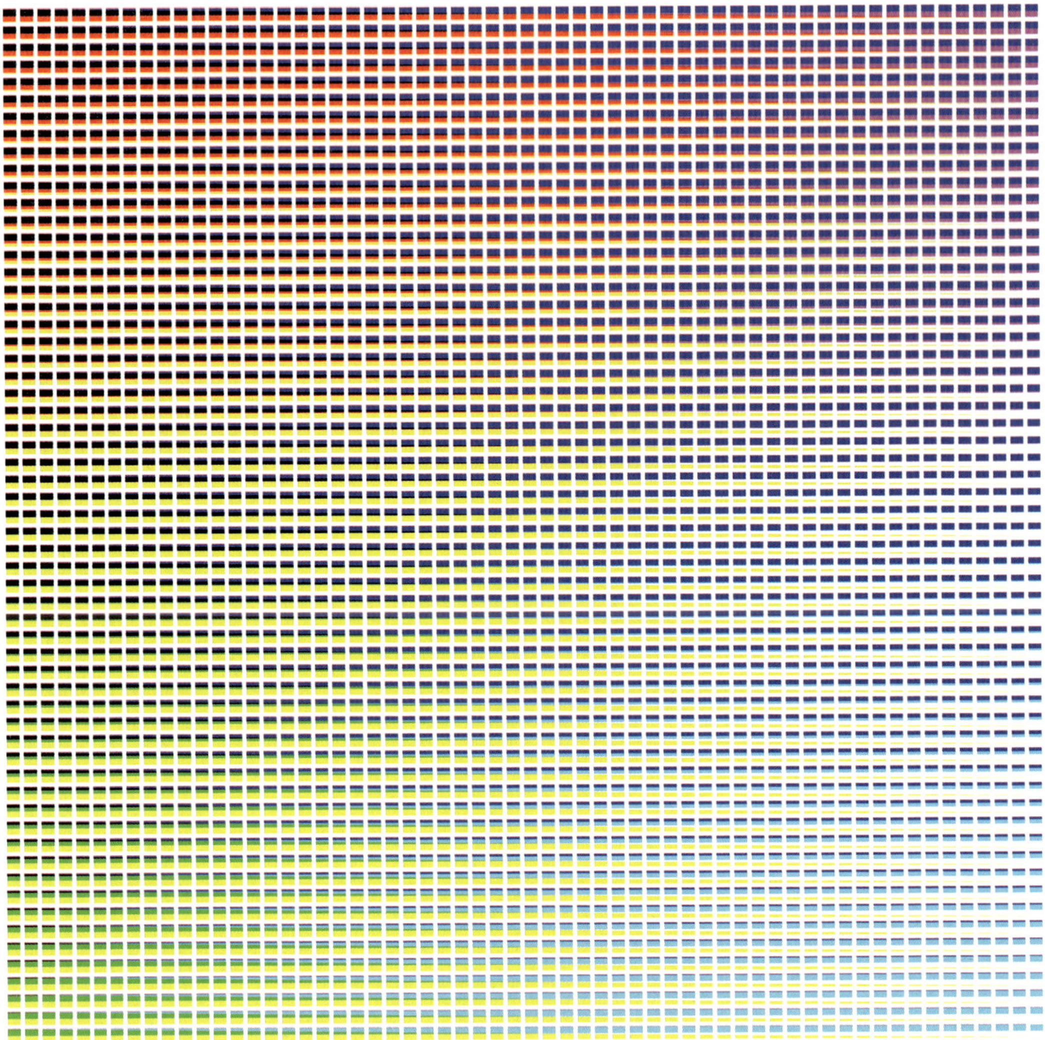

Chrono.print _ 15h-16h, LAb[au] 2009, 1.1x1.1 m digital print

Chrono.prints

The "Chrono.prints" are a variation of the process which lies behind the chrono.tower. Each of the 24 prints represents the development of one hour of the day starting, top left, with its first second and carrying on to the 3,600th image in the lower right hand corner. Each of these 3,600 squares is further divided into clearly denoted colored surfaces which correspond with the time-based additive color system. To the viewer, these prints appear to be chromatic motifs, which one can best describe as chromatic time textures.

These prints are the results of a computerized calculation process based on very simple parameters. The theme of the images is light and its relationship with time. In this sense, the prints represent not just a formal examination of the traditional medium of painting but also of its central themes: color and light, which not only places them in the tradition of the abstract, minimal concept-art of the 1960s but continues this tradition with the help of a computer. The programming of such a system, which is based on a program language – that is to say, on something logical – creates a pictorial language made significant by the creation of rules. These "chrono" works demonstrate that Parameter Design is just such a method: and one which can be employed using various media (in this case the lighting of a tower or computer-generated printing).

Touch _ a Light Symbol in the City

In order to create light architecture one can either – as in the chrono.tower project – use widely understood parameters in order to create a symbol in the city or one can choose another approach which involves the entire public realm in the process.

"Touch": the project which inaugurated the illuminated façade of the Dexia Bank tower in Brussels, serves as an example of such an approach. The objective was to shift perceptions of media façades away from corporate design and towards collective urban light art. The challenge was to use interaction to encourage public participation and identification in order to transformed architecture into a form of communications design.

A pavilion was built at the foot of the tower in which passers-by could use a multi touch screen to interact, both individually and collectively, with the light in the tower windows. Not only simple, static inputs but also dynamic gestures were recorded in order to generate and animate points (single windows), strips (between the building edges) and areas (entire façades) of lighting. These geometrical forms appeared on the otherwise monochromatically illuminated façade in real time in either black or white light, depending on whether the information was input positively or negatively (up/down, left/right ...). A simple touch of a finger on the interface could transform the entire appearance of the 145m high tower. Such forms of interaction establish a new relationship between architecture and the individual. And in the case of this project, the reduced and geometrical formal language made its easier to both recognize and experience the architecture.

One could also use the interface to take, at any one moment, a snapshot of the "tower composition" set against the Brussels skyline – a photo, which could be sent as an e-card from the pavilion. The composition created by the user could be seen from right across the city and, as such, influenced the entire image of the city; this was "his/her" image, which could be viewed at anytime (on the website) and shared with others. This chain of different forms of communication made it possible to hold a real-time conversation – a dialogue between a newly created building and the inhabitants of the city – and then to capture the moment, create an image of it and place this image, via a social network, in the internet. The use of technology made it possible to experience and to identify with the public space - to define architecture as interface.

www.mediatecture.info/14

CROWNE PLA
HOTEL

The Design of the Architectural Interface

This made it vital to pay special attention to the design of the Interface. The basic idea was that the multi touch screen was a two-dimensional representation of the tower on which each coordinate represented a specific window. This virtualization of the tower created an intuitive interface between the user and the pixel façade which gave the user the impression of touching the actual building. To achieve this, the interaction was reduced to a very simple input procedure and the direct feedback from the light tower was optimized in order to focus the attention of the user on the tower rather than on the interface. The reduction of the tower imagery to an abstract game with forms also had the intention of confronting the viewer with an architectural vocabulary and focusing his attention onto the resulting dialogue, rather than allowing him to be distracted by illustrative or – even worse - figurative images.

The first input from a user – in which he touched any point on the screen – resulted in a dramatic alteration of the tower lighting. A point (window) would become a vertical line (window column) and then spread like a curtain across the entire façade with the color being defined by a combination of the coordinate on the screen chosen by the user and the visible light spectrum. The aim of this opening sequence was to emphasize the closeness of the relationship between user and tower. The development of the required technology (a weather-resistant multi touch screen measuring 0.8 x1.7m) demonstrated the extent to which a technological development can result from the development and execution of a project idea (and even influence the project title!)

Touch interactive urban installation, Interactive station, LAb[au] 2006-2007, Dexia Tower, Brussels, Belgium

But the development of the interface was not only limited to questions of software and hardware but, as a result of the urban scale, also required that the design of the pavilion was addressed. This pavilion was a spatial element formed of three folds: the first fold created a "desk" where the interaction could take place, the second created a projection box which made it possible to follow, from street-level, the symbol on a screen with the finger and the third directed the attention of the user towards the tower. This principle of folds was drawn from the image of the tower on the screen and hence became the leitmotiv for the entire project. The implementation of the project extended the notion of interface to a spatial object, in the sense that it took the architectural and media-specific parameters of the adopted technology and used these as the content – the very theme - of the lighting concept. As is the case with architecture, these parameters communicate their meaning on the meta-level, which makes them media in their own right. The conclusion to be drawn from this is that the latest communication and information technologies offer us the opportunity to fully rethink such classic architectural themes as the design of façades, the creation of public space and the creation of representative symbols in this space.

The projects described here required the development of a feature which is vital to the implementation of media façades and other similar forms of lighting infrastructure. This feature is "Content Management". These projects, for example, use a lighting concept which LAb[au] developed from the very start in close cooperation with the client, the Dexia Bank. The focus of this development process was the relationship between art, architecture and light as embodied by interactive, performative and generative lighting projects. And, in addition to this, design criteria for the artwork were developed which addressed this relationship on an abstract and conceptual or, so to say, meta-level. In short, it was necessary to establish both a detailed timetable and a clear artistic position for the tower project. Such a "repertoire" is perhaps something which we associate more with cultural establishments such as theaters or museums - and it is certainly seldom so clearly reflected in the physical architecture of a project. But this is appropriate in this case because the bank understood that the project was much more than an element of a marketing strategy and saw itself much more in the role of an architectural or artistic player on the scale of the city. The result was that the tower was able to become a landmark in its urban context.

In the case of LAb[au]'s projects for the Dexia Tower, the lighting design had no direct influence on the built form of the architecture of the tower. And although the projects were based on the relationship between light and the architecture of a building in its urban context, this cannot hide the fact that a "global" architectural concept is missing – a fact that is particularly obvious during the day when the lighting system is switched off and the banal corporate architecture of the tower is once again apparent.

The conclusion of this experience is that mediatecture must be based on the complementary relationship between media and architecture. In order to achieve this, architecture must be understood as a medium and, vice versa, media must be designed as architecture. Parameter Design is a method which shows how the characteristics (the parameters and processes) of the chosen media can be related to the parameters of building and urban design, as a result of which it should be possible to treat media and architecture as equals.

The following projects not only highlight these parameter-based borderlines between media and spatial artifacts but also provide a snapshot of the moment at which the borderlines completely disappear.

Binary waves _ Cybernetic Installation

The installation "fLUX, binary waves" measures flows of traffic (passing cars, etc) and communications (the electromagnetic fields caused by mobile telephones and radios, etc) and transforms these into light, sound and movement. The installation was designed in 2008 for the forecourt to the railway station in the Paris suburb of Saint-Denis, a transport hub used by around 60,000 commuters every day. The station forecourt opens onto the Seine Canal but the steps from the station down to the canal end facing a blank wall on the opposite side of the canal. The two bridges flanking the steps give the whole space a stage-like theatrical atmosphere.

In this setting, a kinetic canvas was created consisting of 40 illuminated panels measuring 3m high and 60cm wide which rotate about their vertical axes. The individual panels rotate at varying speeds creating a wave-like effect which is reinforced by their alternating black and white surfaces. The waves oscillate from one side of the installation to the other gently slowing down until the moment when a new pulse sets a new wave in motion. In addition to these panels, red and white lighting strips are illuminated at irregular intervals accompanied by short tones. The kinetic principle of the installation is derived from the water and the connection with the context is reinforced by the interplay between the reciprocal optical effects of the panels being reflected in the water at the same time that the water is reflected in the panels.

www.mediatecture.info/15

Binary waves, cybernetic installation, day photo + wave pattern, Saint-Denis canal, Paris 2008

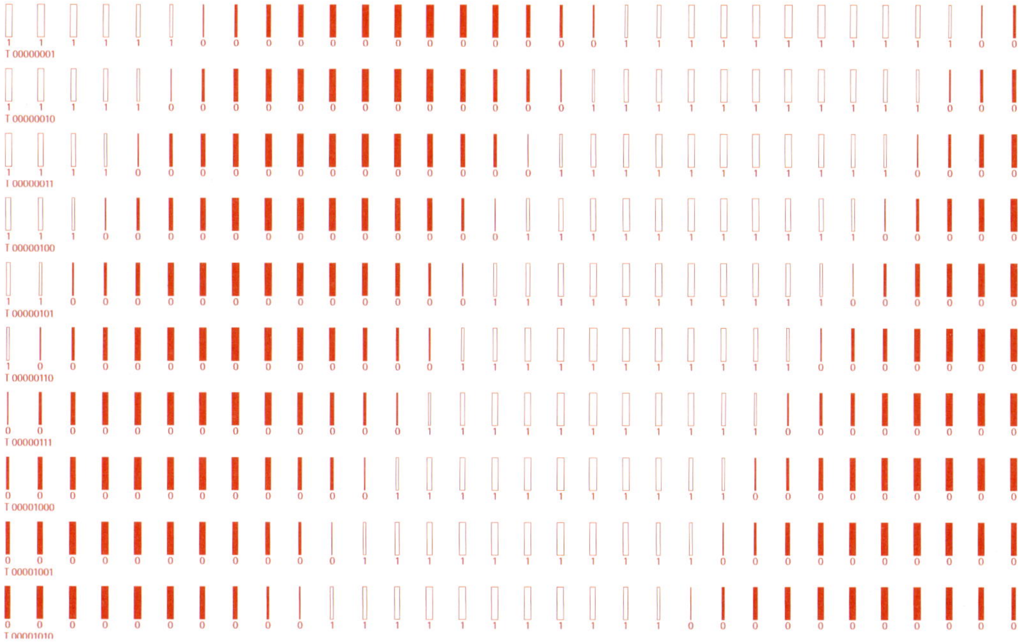
T 00000001
T 00000010
T 00000011
T 00000100
T 00000101
T 00000110
T 00000111
T 00001000
T 00001001
T 00001010

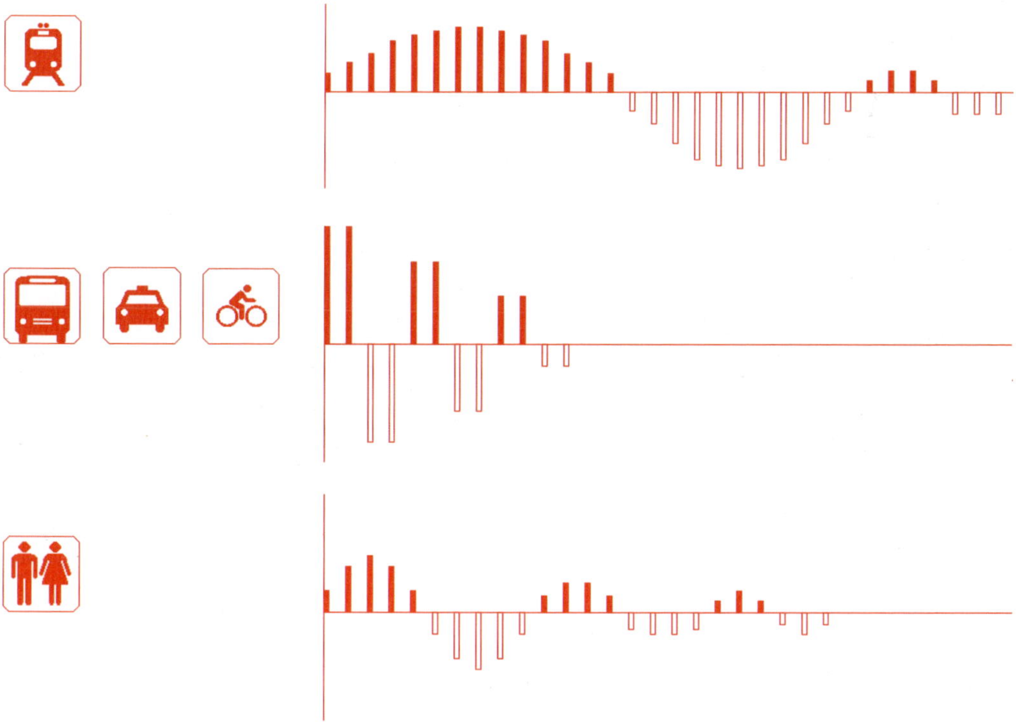

The sequence of movement and the light signals are determined by measurements of the local traffic and communication flows. The speed at which the panels rotate is, for example, determined by the traffic flow whereas the red light, which illuminates one side of the panel in the form of eight horizontal lights, indicates the intensity of the local electromagnetic field and the white light, which illuminates the edge of the panel, reflects the frequency of passers-by on the bridges. The intensity and frequency of the lighting signal varies in line with traffic density and the strength of the local electromagnetic waves.

This analogy between the spreading out of waves, kinetic behavior and the light effect is derived from the concept of combining single urban events into a collective image. The installation reveals all the rhythms of the big city in that it even records both individual movements in the urban space and those invisible electronic waves which are, nevertheless, vital to urban life. This makes the installation both the mirror image and the spitting image of the public space.

Hence the "fLUX" concept (which also features in the name of the work) represents the transformation of flows (flux) into movement, sound and light (lux). The installation employs a control-system which requires the linkage of measured signals, the programming of patterns of behavior and the design of output equipment (hardware). It is this information architecture which determines the actual meaning of the work. The creation of such a "meta" system as that which lies behind this work has its roots in the cybernetic art of the 1960s and, above all, in the work of Nicolas Schöffer. The cycle is thus also an official homage to this pioneer of an art form located somewhere between science, technology and architecture.

Binary waves, cybernetic installation, night photo + impulse/waves, Saint-Denis canal, Paris 2008

framework f5x5x5

Amongst other works in the cycle known as "16n" is the interactive installation framework f5x5x5. This kinetic lighting sculpture consists of 5 modules, each measuring 2x2 meters, which combine to form an architectonic element with a length of 10 meters. Each element is further divided into five horizontal and five vertical square "English" frames creating a total grid of 5x5x5 = 125 frames. The title of the installation "f5x5x5" corresponds with the variable nature of the installation which is derived from the "decomposing grid".

Each frame contains two further motorized frames which can be rotated about both their vertical and horizontal axes. In addition to this, red and white LEDs are fitted to the visible edges of the frames. Each vertical column of the grid is marked in the base of the installation by an infrared sensor which reacts to any object which cuts across the infrared beam by sending a piece of binary information. These sensors note the position and direction of movement of any viewer. If he moves along the installation then he will be followed by a series of illuminated frames. As soon as he stands still all frames will be directed towards him like a vector field.

In this way, constantly changing lighting patterns are created by a combination of measured signals, binary information and complex kinetic behavior. Hence, an important part of the project was the establishment of parameters for determining when and on the basis of which information a particular program or mode of operation would be triggered. These rules for governing rules are the meta-structure or "framework" of the installation.

In informatics, the term "framework" denotes an outline program or a structure which the programmer uses to create an application. It is

Framework f5x5x5, LAb[au] 2007-2009, detail

based on a diagram which shows the flow of data and the interfaces between individual programs and calculating processes. In the case of this installation the term "framework" describes the logic inherent to the interaction as well as such information as which mode of operation will be triggered by which signal in order to create the movement and lighting patterns designed to catch the attention of the viewer.

The starting point of this project is an information architecture based on the interpretation of binary signals. And, by employing the binary principle in its detailed execution, the installation also gives this principle a physical appearance. The installation is, for example, black (1) on one side and white (0) on the other, which makes it possible to contrast a white surface with an opposing black one. It is precisely this contrast which allows the visitor to properly recognize the various movement patterns. The binary principle is also used to determine the positions and lighting conditions of the frames: open/closed, rotating/still, light on/light off, red/white etc.

The spatial division of the structure also rejects such classical principles of architectural design as proportion and is based instead on a technical figure, the resolution of the pixel image. The 5x5 grid and its colored layout are divided up in a way which makes it possible to represent letters and symbols in a similar way to the rotating displays which one knows from airports and railway stations – although these are just black and white (which is to say binary). This determines the design of the information system, while the decision as to whether this information is abstract, graphic or symbolic in nature will be defined, once again, by the "framework".

This demonstrates a further aspect of "Parameter-Design": the demand that the parameters of the chosen technology should be both clearly presented and concretely used. This principle is followed in all the projects described in this article:

For example, primary colors are the parameter defining the use of light in the "chrono.tower" project while the modification of this light and the resulting animation of the tower use the parameter of time. The "touch" project similarly uses geometrical light forms to highlight the architectural characteristics of the building and these characteristics are communicated very clearly to the user by the way in which the multi touch screen responds to his touches and gestures. In doing so the "touch" project appropriates the spatial division – and both projects appropriate the pixel resolution – of the tower. Windows are reduced to colored points of light and the input of users is recognized by means of infrared technology before being transformed into geometrical forms. Both projects involve the incorporation of the light spectrum into a spatial concept.

This concrete design of the parameters of the "setting" also occurs in the "siloscope" project. Here, vectors are represented by the straight elements which form the hyperbolic figure and similar parameters govern the functioning of the sound analysis and the light concept. The same is true of the black and white panels and kinetic and light waves which embody the binary signals in the "binary waves" project.

From which the conclusion can be drawn that mediatecture should be based on the complementary relationship between media and architecture. In order to achieve this, architecture too must be seen as a medium and, vice versa, media must be designed as architecture. Parameter Design is a method which shows how the characteristics (the parameters and processes) of the chosen media can be related to the parameters of building and urban design. As a result of which it is possible to treat media and architecture as equals and combine them into a common formal language.

www.mediatecture.info/16

... the Media Façade as a Dress ...

The New York Architectural Office ***Asymptote*** *built this unique hotel – which is unique in that it is built over a Formula 1 track – in 2009. A flowing construction loosely caresses the contours of the hotel. The structure for this construction is enhanced with a lighting installation which refers back to the constructional grid – the individual pixels of which can be individually controlled as a means of creating variable effects.*

YAS HOTEL IN ABU DHABI

This is the world of the super rich who meet here. The architecture itself is a star of the gala performance and appears in a luxury sequined dress. The form of the construction has just one objective in mind: elegance. And the objective has been faultlessly achieved by Asymptote in cooperation with the engineers – and in particular the lighting engineers – of Arups.

It is naturally interesting to examine the mediatectonic aspects of this project. Rogier von der Heide, who led 2010 Arup Lighting until 2010 emphasized that lighting diodes were chosen for the lighting design of the installation for the simple pragmatic reason of supporting the aesthetics of the construction.

But does the façade – leaving aside this unimpeachably credible justification – also have a medial character?

The architecture itself is without doubt medial because of the way in which it embodies the longing of Arab high society to combine its own culture with Formula 1 – the most emotional symbol of western high-tech culture.

But what is the role of media technology? Normally the hotel glows at night in clear colors whose medial dynamics are represented by gentle changes of color. During races, however, the lighting installation demonstrates what it is capable of, accompanying the dynamics of the race with the appropriate animation. It is hard to be clearer about the medial role of a lighting installation. The architecture changes like a chameleon as a result of the things going on around it. The irrepressible potency symbolized by the Formula 1 cars arouses the smooth, flowing body of the host. What a joy for a culture in which any other public erotic display is strictly forbidden. And how successful this is as a theater of power[1].

1 See "Showplaces of Power" by Bruno Schindler

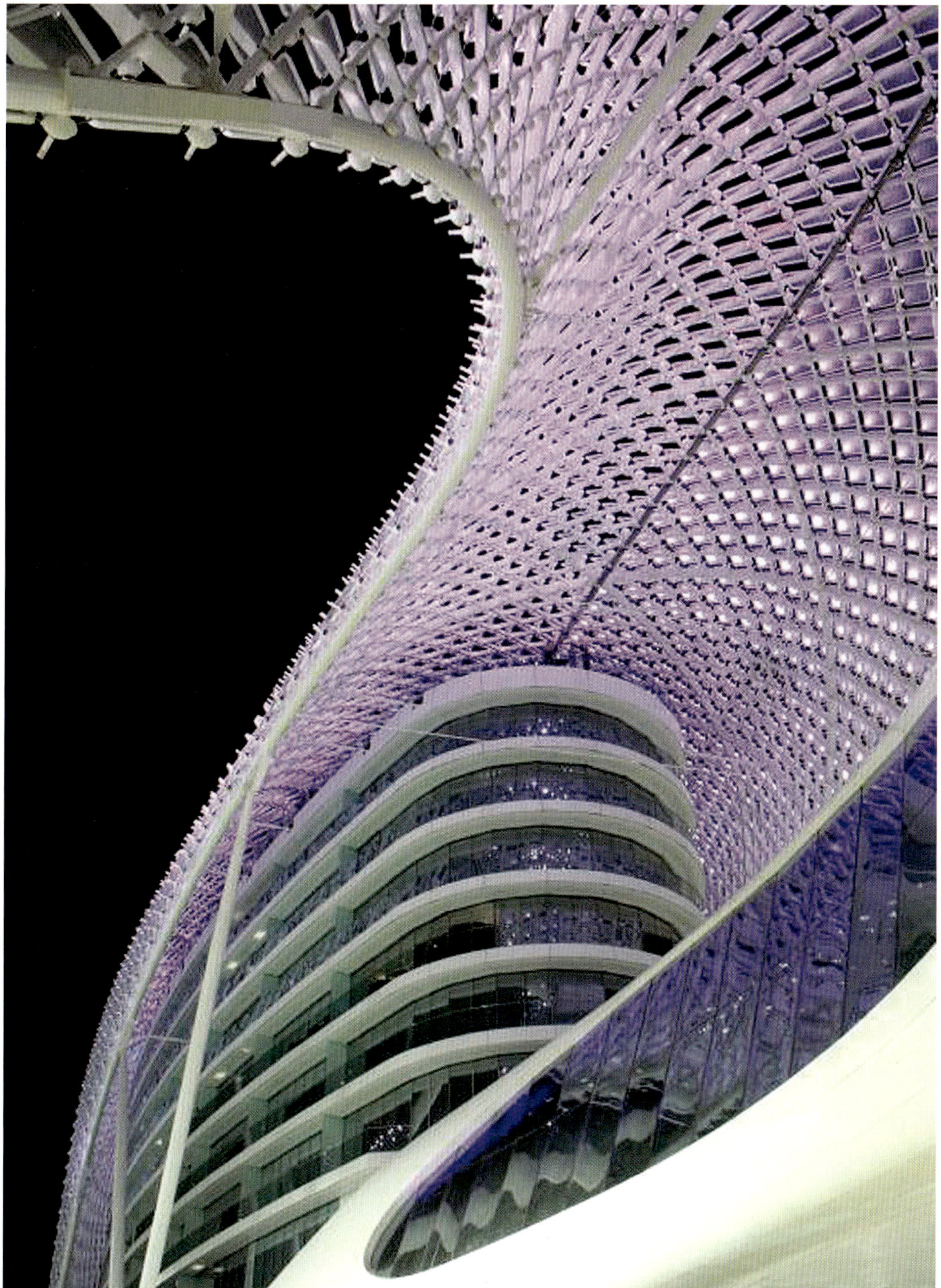

... the art of projection ...

Till Botterwerk, Manuel Engels *and* ***Thorsten Bauer*** *form the core of the interdisciplinary group* ***UrbanScreen*** *in Bremen. They lead a network of architects, musicians, stage designers and artists who give façades new life through the use of video projectors.*

A CONVERSATION WITH URBANSCREEN

But rather than treating smooth areas of façade as traditional screens, UrbanScreen projects its images onto architecturally more complex façades – and has developed its own technology as a means of doing so. It has even been granted a patent on its "process for creating tailor-made projections for specific objects", which allows the focus of a video image to be specifically adapted in line with the various surface characteristics of a façade. In this process, the diverse material and constructional qualities of a façade are programmed into a mask on the computer screen and the specific images are then adapted to the various areas of this mask. The result is that when it is projected onto the actual façade, the image is perfectly adapted to the façade structure which, in turn, becomes the architectural stage set for a theatrical performance.

In comparison with the other Mediatects whom I introduce in this book, the members of UrbanScreen are those whose productions have most do with the emotions. Their projections tend to start life as concrete events aimed directly at the public in an open space and making full use of sound. Their more recent work however reveals something of a trend away from this straight-forward "event" character.

The group's first works were all about reaching the public and bringing fun and a sense of delight to the public space. The project "Jump" was based on a surreal desire to present architecture as a climbing frame. The extent to which this project was theatrical in nature can be seen in the very concrete way in which the video was crafted. The group recreated the actual façade as a stage set in a studio and invited dancers to use the set as a framework on which to express themselves. Their movements were recorded and then, thanks to the special UrbanScreen process mentioned above, projected so precisely onto the façade that, for example, the figures appeared to really climb in and out of the windows.

This anarchistic desire to strip historical architecture of its mystique was then taken to such extremes that UrbanScreen even transformed one façade into a pin-ball machine.

In the last few years, the group's work has become ever more intensively involved with architectural content. The usual approach here is to use media to draw out the character of a piece of architecture and then use theatrical gestures to interpret this further – an example being the Kubik project in Hamburg in which the stonework

Projections on the Hamburg Gallery for Contemporary Art. The project "555 Kubik" won the silver prize in both the Cannes Lion International Advertising Festival 2010 and the ADC Awards.

grid of Ungers' museum façade was transformed into a machine. Such interpretations are naturally dependent upon a certain amount of artistic freedom and, as they are essentially temporary interventions, they cannot really be considered as part of the architecture and, as such, are rarely discussed with the architects of the buildings in question. As to the question of whether such installations could become permanent parts of buildings; there are, on the one hand, a number of technical aspects of media production which must first be resolved and, on the other hand (as UrbanScreen is very aware), the unavoidable fact that the power of an intervention diminishes the more often that it is seen. The aim of such works is that they elicit intensive public attention and, the greater the emotional impact, the more they are likely to be regarded as temporary in nature. One unquestionably interesting area for future exploration however is interaction – although this will necessitate an escape from the tyranny of the linear storyline and a move towards process-based projections[1].

The commission received by UrbanScreen to develop a concept to mark the opening of Zaha Hadid's museum in Rome is of course particularly fascinating, not least because the architecture of Zaha Hadid itself so strongly addresses the issues of architecture and movement. One can of course critically question whether – in the case of such a devotedly movement-based building – the possibility of mediatectonic intervention should not have already been addressed in the initial design process – but such criticism is easily answered with counter questions regarding the unresolved division of responsibilities and process of coordination between architects and mediatects. Architects are of course used to addressing immaterial – and, in the case of certain building types, highly emotional - content in their designs. But such content tends to be dealt with on the associative level rather than in the form of concrete images. Architects certainly invest much creative energy in attempting to ensure that viewers of their buildings also understand the deeper reality that they are trying to commu-

Bauhaus Dessau 2009 "BAUHAUS celebrating 90th anniversary"

Rethespeicher industrial complex in Hamburg, Dockville Festival 2009

nicate but the truth is that, while architecture is indeed an art form capable of magical gestures which awake the fantasy of the viewer, this fantasy is often enhanced by a sense of secrecy and the fact that the architectural language is actually carrying no clear message. (I realize of course that this is a great generalization which is quite untrue in the work of, say, Antonio Gaudi)

Gabriele Detterer reflected very much on these questions when she wrote about Hadid's new work in the Neue Zürcher Zeitung (20 November 2009): "One enters the museum in Rome with great expectations: How much sense of movement will the building actually communicate? Will one really get the sense of being able to lose oneself in this bold architecture just as if it is were a field of multiple associations? Hadid's design has a strong physical impact and challenges normal spatial perceptions."

If a result of the emergence of mediatecture is to be that the capacity for playing with the emotions is to become integrated into architecture in the form of media technology, then this will inevitably take a very different form from the interventions of UrbanScreen, because the group's very concrete, theatrical approach is surely be too much of a potential threat to the level of free association at which architects tend to operate (see "Spatial Concepts and Augmented Space"). This recalls the dialectic between closed space and open space, in which the instinctive reaction is to place architecture in the open and mediatecture in the closed categories – although this is only true of image-based mediatectonic interventions which are closed in the sense of relying on constant repetition.

And, in addition to this, it is important to remember that the very art of moving pictures is that these can elicit quite individual reactions on the part of the viewer and that the emotional impact of an image can be much more significant than its simple objective contents. Moving images can create emotional reactions quite different from those created by architecture, which is to say that the role of the mediatect can be an extremely powerful one.

All of which leads me to conclude that there is much to be gained from seeking to extend the emotional and experiential horizons of architecture and that this will best be achieved if architects and mediatects work very closely together. If architects, for example, wish that their

buildings truly create the "impulse" for a city to change, then one approach would be for part of this architecture to be given an extra, theatrical, mediatectonic dimension or, in other words, to be mentally programmed to create this "impulse" in a way that is more than merely traditionally architectural. This mental programming could, for instance, be tailored to the social context of the urban district and directly refer to the history, hopes and visions of the local people.

Such an expanded role for architecture could only be sustainable if this mental element – and the mediatectonic interventions – were constantly renewed by, for instance, being actively integrated with other media. Such integration naturally raises a series of operational challenges, but it seems to me that the close relationship which people could develop with their urban environment as a result of the high potential for identification offered by such theatrical interventions makes the taking on of such challenges more than worthwhile.

Just how close this relationship can become is shown by the success of UrbanScreen, which is now overwhelmed with potential commissions from all corners of the globe. The longing for such a sense of identification appears clear.

Many of these potential commissions still arise from situations in which a city, an institution or an advertising agency are looking to maximize the impact of an event. UrbanScreen is however determined to move away from such event-based work in order to concentrate on creating more truly theatrical interventions.

1 See "Self-generating Libaries of Images"

... an analog media façade ...

***Ned Kahn** is an artist from California who plays with natural elements: Fire, water, sand and, above all, wind. He constructs wind-facing façades with small freely moving mirrors.*

In 2004, in cooperation with the Santa Monica architects Konig Eizenberg, he created the work Articulated Cloud for Pittsburgh Children's Museum.

ARTICULATED CLOUD

This work by Ned Kahn offers us a perfect opportunity to ask just what a medium is. A medium transforms something in the real world in such a way that our human senses perceive it differently. This transformation can, on the one hand, allow us to perceive something which has previously been hidden in immaterial structures and processes. One of the most curious media is music, because music can articulate feelings. And yet it is a mystery how feelings can be communicated by the mere vibration of the air.

On the other hand, there are media which transform physically existing processes into experiences for our sensory organs. A thermometer measures the temperature of an object. This temperature can also be sensed by the body – but in the case of temperature it is not advisable to rely on bodily senses alone. The medium of the thermometer transforms this experience into a visually observable number-based message. This demonstrates our dependence upon media in our daily lives. And the more complex our lives become, the more we need media to both orient us and manage complex processes. Without media we would be fully incapable of dealing with life. And this may be a reason why the psychological effect of media is seen as being very calming and people can be very emotionally connected to media. I think that this is one way in which the phenomenon of art can be understood

And this appears to explain why it is possible to see the visualization of the wind in the work of Ned Kahn as "poetic". In factual terms, such a construction makes no sense because it should be no surprise to us that small hinged mirrors move in the wind in such a way. And yet the movements of the mirrors create a wavelike pattern and it is surprising to us that the movement of the wind can be transformed by this medium into such a recognizable and legible image. I suspect that the resulting positive feeling comes from the fact that one has understood something. Humans must constantly process an unbelievable quantity of data which is produced by the brain as a result of the work of our senses. As soon as this data-processing identifies a pattern, messengers are released in the brain and these arouse positive feelings. This sequence is very important because the recognition of patterns has a signaling effect of which we become conscious - and this allows us to orient ourselves and make decisions[1].

Ned Kahn's articulated cloud uses very simple means to create a moment of comprehension. In this sense, the installation becomes a medial façade, which has addressed – and seamlessly resolved - a problem of communication. One sees that electronic media are by no means always required to achieve this.

1 With thanks to Professor Lukas, who wrote the article "How media create the reality in our heads" for this book. Here I am once again sticking my head out in areas which are certainly scientifically not fully explained. This possibility of working with opinions is a luxury of my free way of writing.

www.mediatecture.info/17

... designing for Commerce in Public Spaces ...

*In December 2009 I visited **Phil Lenger**, President of **Show+Tell**, in New York. "Mediatecture" – a term invented by me – is known in the United States as "Dynamic Environments" and **Show+Tell** is a typical protagonist in this sector. Why do I say typical? Because the media installations in which **Show+Tell** specialize have a much longer history in the USA than anywhere else in the world and this history has been incredibly influential. These installations are the very image of the "American way of life", a culture which has a generally positive view of commerce and advertising. And even though my – European – view of commerce and advertising is perhaps somewhat more ambivalent, I cannot fail to appreciate Las Vegas and Times Square as icons of a carefree zest for life.*

MEETING WITH PHIL LENGER

Whenever and wherever I talk about media façades, the first reaction is, invariably: "Of course – Times Square". Phil Lenger has set up more installations in Times Square than anyone else and this could lead you to assume that he is a hard-boiled representative of the "louder, faster and bigger is better" school of marketing. And yet the truth is somewhat different: precisely because he knows everything that is to be known about this business, his view is much subtler. Right from the start, for example, we agreed that Times Square has nothing – absolutely nothing – to do with media façades. "How often have I tried to explain that Times Square is a very special place – special because it is more a theater than an urban space? Times Square is an interior! And for this reason a large part of it is now closed to traffic – in the middle of Manhattan! – and a grandstand was built above the central ticket shop: The square is a top media event – a stopper in the city which never stops – and at the same time a hub for getting information about shows and buying tickets. What a combination!"

But it is worth looking more closely at the installations in Times Square. A first impression is very clear: This is a collection of urban screens. The fascination of this place is based on the experience of simultaneity. Everything happens at the same moment - and the experience is similar to sitting on top of a mountain, peacefully watching what is going on below. Several events happening simultaneously usually result in our delicate souls being put under stress. But the enclosed space of Times Square gives you a sense of security, leaving you free to become part of the spectacle. Which, when you think about it, is extraordinary because this place is, in reality, a

GMC
WE ARE PROFESSIONAL GRADE.
THE NEW 2007 YUKON DENALI
IF IT DEFINES GREATNESS
abc
Come Clean.
WRIGLEY'S
LIFESAVERS
JVC
SEPHORA
STARBUCKS COFFEE

cacophony where nothing is attuned to anything else. The question of why this works so well here would certainly be an interesting subject for research!

Regarding the spectacle more critically, I have always asked myself how each single brand in this cacophony of images expects to gain if, together, they merely generate a white blur. Phil Lenger obviously asks himself the same question. He points to his M&M advertisement (which also attracted my attention); saying that it was very important for him to keep the design as simple as possible: Easy, emotional pictures which directly catch the eye rather than complicated contents. I agree with him that this work is one of the few media events in the Square which both rises above the cacophony and has a lasting effect. And, above all, one also remembers the brand, which, unfortunately, cannot be said of all such advertisements. (The Coca Cola advertising in Piccadilly Circus is similarly powerful.)

www.mediatecture.info/18

Of course Show+Tell has many projects – which Phil calls his "bread and butter" work - in which they organize content presented to them by agencies. But projects with a more innovative approach are obviously more important to him and he showed me one which he has just finished. A number of star architects, including Daniel Libeskind, have recently built a huge MGM hotel project in Las Vegas. Libeskind was responsible for the shopping center, creating an external entrance to the Louis Vuitton flagship store. Show+Tell integrated strips of LEDs into this façade and Phil explains how the low resolution imagery generated by these LEDs does much to harmonize the union of architecture and media. The aim is to generate an inviting atmosphere with a special character and this is achieved with the use of exclusively white, gently moving forms rather than realistic images.

Phil explains that he often tries to persuade his clients of the effectiveness of low-resolution installations, arguing that one must not always provide high-definition screen quality to be successful in a public space. But he rarely wins the argument because the idea of advertising effectiveness is dominated by known formats and it is incredibly difficult to convince clients to move in a minimalist direction, even if this could save them a lot of money.

The comparison between the projects for M&M and Louis Vuitton makes it clear that the potential range of commercially generated media architecture is very wide. Amongst some of the architects, artists and designers who work with media there is already a serious debate as to whether it will ever be possible to show real pictures in the public space or whether these pictures will end up contaminating the space due to the fact that the mere presence of pictorial representation both implies an abuse of the space and invites associations with the vulgar world of advertising. It is impossible to completely ignore this fear, but I would rather ask a question – of whether developers simply adopt a certain "position" on this issue, or whether regulations can be used to urge a position upon them (see chapter "Urban Identity").

This issue can be seen very clearly in the Victory Park project in Dallas. Here SHOW+TELL was responsible for both the software programming and the choice of media art. Victory Park is a real estate complex comprising shopping, offices and entertainment facilities. At the heart of the project are two office complexes located on either side of a public urban space. The office façades have been fitted with a steel-framed curtain track system upon which large-scale LED screens can move about. The key to this arrangement is that the display software and screen movements are synchronized which means not only that individual images can jump between screens but also that the individual screens can be mechanically combined into a mega screen. The concept behind the video display envisaged both the generation of revenues from advertising and the supply of entertainment through media art.

You can imagine the technical challenge to SHOW+TELL caused by this combination of kinetic and media structures. Phil explained how he attempted to ensure that the media art remained graphically simple both because this would enable him to keep control of the synchronization of the individual screens and because the simplicity of the images would reinforce the sense of the installation as a single entity. There is no aesthetic sense in merely moving TV screens back and forth. A kinetic installation with simple movement patterns must have clear images and intelligent programming if it is to merge into precisely the magic unity that underlies the special charm and entertainment value of the complex Dynamic Environment. Some of the video art however did have narrative structure which meant that the movement of the screens had to be choreographed to support this narrative.

Fortunately, the American company "Target" signed an exclusive contract with the operators of Victory Park. Combining a wide consumer offer with very good corporate graphic design, Target has developed an advertising approach which, rather than focusing on individual products, creates a basic positive feeling and, in turn, establishes a sense of community. The advertising for Victory Park was tailor-made, ingeniously making use of the potential for both mechanical movement and synchronization with the media design. And, once again, SHOW+TELL was invited to support this process of harmoniously integrating media design and technology.

And yet the project had downsides: Not only were the aesthetic and contextual aspects of the media installation largely ignored but the architectural concept also failed to address their potential physical consequences. The first concrete problem arose when the LED displays passed in front of the office windows. It is quite understandable that employees were annoyed

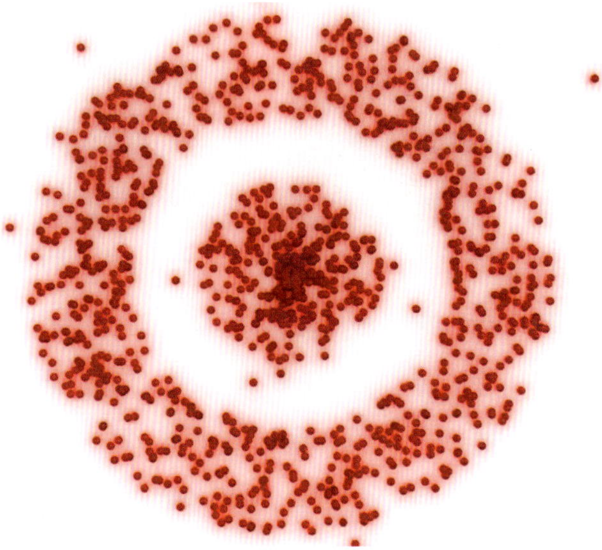

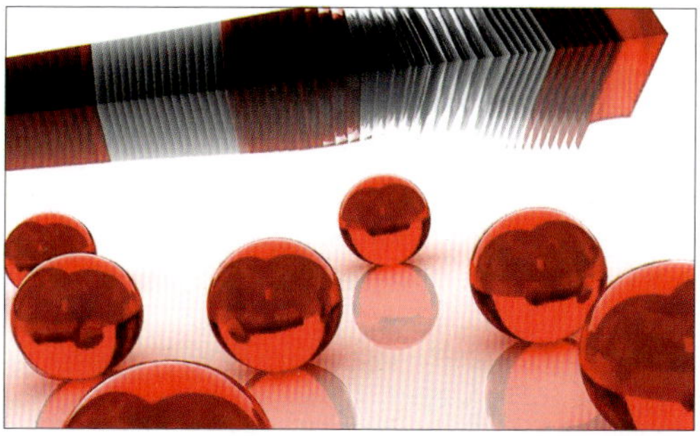

Media design for the brand "Target", custom produced for the installation by Victory Park

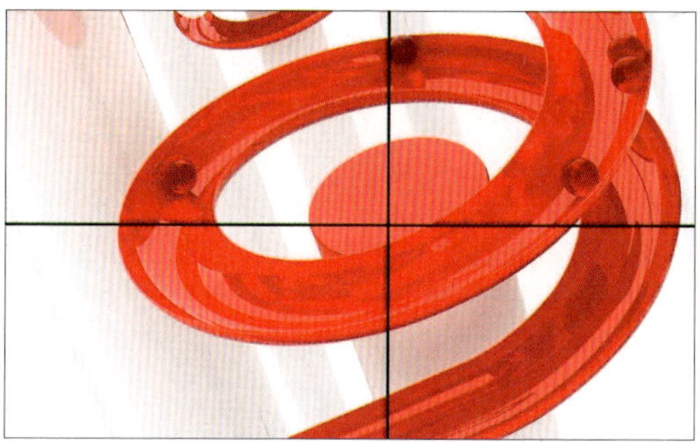

A spherical orbit plays with the "Target" motif

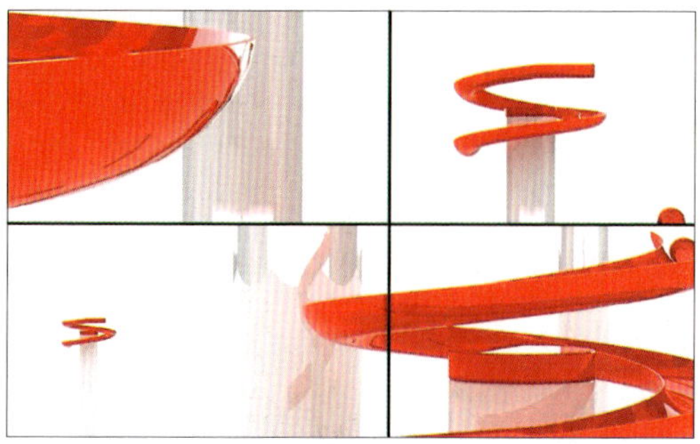

Image programming for 4-Monitor-Split

"Flag Metamorphosis" art program by Myriam Thys

when lighting conditions constantly changed due to displays completely shutting off the daylight in their workplace. The result was that the offending displays were given a fixed position and no longer permitted to move: an outcome which was surely foreseeable?

Such outcomes are frustrating because they illustrate the failure of the real estate industry to integrate media concepts into the design process. One solution here could have been to use a lighter, more transparent media technology. But the result would have been lower resolution images – and this in turn illustrates the problem posed by the American method of project management. Project management consultants are appointed by clients to guarantee that the project gets built. This makes them inherently cautious – which is exemplified here by the way in which they were guided by the standards of the advertising industry, which insists on display screen images of television quality. And this brings us back to the beginning of this article. The painful thing in this case is the extent to which, through its own efforts, SHOW+TELL has creatively shown that media concepts benefit from the use of simple forms and directions. The Victory Park solution could have worked perfectly with such lower resolution imagery. And it would have been even more cost-efficient!

But now the installation is at a standstill.

It gets even worse. The property was sold to another operator, who had no commitment at all to the original plan. Now, normal advertising clips – e.g. for automobile brands – run on the displays.

Architects generally enjoy some form of copyright protection over their works which guarantees that significant changes may only be carried out with their agreement. Until media become an inherent – similarly protected – part of architecture, it will be difficult to for media architecture to become an accepted, sustainable part of the architectural mainstream. The danger that their media concepts could be misused will continue to discourage both architects and the building authorities from becoming involved with such concepts – with good reason! And the critics who claim that media productions do not belong in the public space will be strengthened. If we, the mediatects, do not get involved in the remodeling of statutory regulations in order to improve the status of media, then, in the long run, we risk having the ground taken away from under our feet.

L'UNION FAIT LA FORCE
PACIUGO GRAN CA
3030
103
CAPPUCCINO

... Indoor City ...

ART+COM *translates and dramatizes content in built environments by means of interactive media, making complex themes accessible in museums, exhibitions, visitor and science centers, trade fairs, showrooms and in public space. Originally, Berlin-based designers, scientists, artists and IT specialists founded* ***ART+COM*** *as an association in 1988. From the beginning,* ***ART+COM****'s vision has been to explore the medium of the computer and its possible fields of application.*

ATELIER BRÜCKNER *is a renowned producer of customized solutions for the transformation of content and messages into a three-dimensional language. Following their maxim "form follows content",* ***Atelier Brückner*** *provides the total package, from conception and planning to the implementation of content generated architecture and exhibitions. Their work focuses on scenographic projects for museums, fairs and expos.*

BMW MUSEUM 2008

A contemporary and dynamic architectural language has been employed in the new BMW Museum in Munich: the language of the automobile world. Unveiled on 21 June 2008, it has set new standards in the field of brand museums. Alongside the BMW World, which opened in October 2007, and the BMW Plant Tour, the museum forms the third building block of the BMW triad, which welcomes around two million visitors every year.

In analogy to the BMW brand, which stands for innovative technology and pioneering design, the BMW Museum explores new paths in the interplay between architecture, exhibition design and communicative media. The BMW Group could rely on experienced creative partners in the realization process: Atelier Brückner in Stuttgart was commissioned with the general planning and responsible for the architectural and exhibition design. The media production and interactive installations were the brainchild of ART+COM, Gestaltungsbüro für Neue Medien (design office for new media) from Berlin. The designers from Integral Ruedi Baur in Zurich developed the graphic design.

The BMW board of directors' conscious decision for Munich as the location in 2003 marked the beginnings of the new BMW Museum. In contrast to other building projects for automobile manufacturers, the task was not to construct a new museum building out in the green, rather to integrate the new museum into the heritage-protected building

fabric of the corporate headquarters on Munich's Petuelring. Already in 1973, trend-setting architecture by Viennese architect Karl Schwanzer found its place here. This ensemble – consisting of a highrise building, the so-called "Vierzylinder" (modeled after a four-cylinder engine), with neighboring flat-roof buildings and the "Museum Bowl" baring the BMW logo on its roof – soon came to be the landmark of the company.

With the "bowl", Karl Schwanzer realized the first museum in Germany dedicated specifically to automobiles. It housed outstanding exhibits from the BMW collection, which originated in 1922 and previously had only been provisionally displayed in a room in the Munich BMW plant and in the plant museum from 1966 on.

Schwanzer's guiding principle was the "continuation of the street into an enclosed space". The silver, futuristic construction, with its very enclosed outward appearance, is light and generous on the inside. On a spiral-shaped ascending ramp, the visitor encounters five seemingly free-floating platforms that serve as exhibition spaces. The almost circular floor plan of the museum building widens from an approximate twenty-meter to a forty-meter diameter. This expansive volume can be experienced in its full intensity on platform four and five.

Atelier Brückner was given the task of interpreting this celebrated architectural milestone anew and transporting it into the 21st century. The experience of the bowl's original quality should come alive; it should be freed from later additions and connected with a brand new permanent exhibition space. The adjacent flat-roof building – the west wing of the corporate headquarters which formerly served as a canteen and parking garage among other functions – would be redeveloped for this new charge. The building was totally gutted by the architects from Stuttgart and furnished with a new thematic interior architecture. In this manner, the exhibition area could be extended from a former 1000 to 5000 square meters.

BMW Square – Mediatecture

The highlight of the media installations is without a doubt the BMW Square developed by the media designers from ART+COM together with Atelier Brückner. It is the vibrant heart at the center of the permanent exhibition around which other exhibition houses are situated. Visitors encounter this thirteen-meter-high airspace from different perspectives along their way, crossing bridges in the room and finally coming to admire the vehicles parked here. They seem to be in motion; light

reflects along their contours; architectures and landscapes are fleetingly mirrored in the polish of the roadsters.

The installation sets the space and recipient in motion. The outer walls of the exhibition houses, a total of approximately 706 square meters, are animated with abstract or thematic motifs by means of an LED technique employing 1.7 million light-emitting diodes. With more than 30 possible scenes, ART+COM has created a digital world of images that evokes an atmosphere with a technical and, at the same time, poetic-emotional dimension. Architecture is dematerialized and dynamized.

In an elaborate series of tests, parameters were defined for an impressive, coherent and brand-adequate spatial experience at the intersection of architecture and media technology. The concept was realized with monochrome white LEDs mounted behind double satinized white glass panels. This combination is new and has been put into application for the first time. The glass panels contribute to the impression of a unified and consistent surface – even when viewed from up close – concealing the technology from the visitors. The panels are not perceived as elements protruding into the space; rather the houses seem to glow from within. In combination with a video tracking system, this so-called "mediatecture" can even be set to operate in a reactive mode. The media display responds to the mere presence of visitors, actively involving them in the scenario.

The architecture is dynamized, the square dramatized by a light scenography, and the original vehicles exhibited in the space seem to be in motion.

The more than thirty abstract and thematic displays are composed as fluid sequences, each three to seven minutes long. Depending on the occasion, these choreographies can be adjusted to create a range of different ambiences.

www.mediatecture.info/19

ce the experience of mobility in the mus
ned more than 600 square mete
ic material for façad
experience of mo
ness, Kraft, Kraftent
namik, Charakter, Unverwe
titität, Charakter, Charisma, Ausstrahlun
technology with a classic material for façades t
than 600 square meters of reac
h a classic material f
experience

For this
matic techn
facade
Teamwor
position,
or this project,
ematic tec
ades
Einsatz

Nachhaltigk
pflegen, Den Weg bereiten
Venture onto new ground,
echselbarkeit scha
Create uniq
Bekanntes erneuern, Den Fortschritt
Sicherheit integrieren, Kraft maximieren

... complex Layering as a Medial Principle ...

THE WAX WALL

MerckSerono 2007 Genf by ag4 mediatecture company

In 2005 Murphy/Jahn created a new headquarters building for the genetic technology company Serono on a site in central Geneva. The design of the building, which also incorporates key research departments, had to take account of a series of existing buildings which could be partly hollowed out and newly defined but which, at the same time, established a series of natural limitations to the new building. One of these existing buildings, for example, required a huge fire-wall in the entrance hall and Helmut Jahn's response was create bridges in front of the wall which would optimize the space while helping to connect the various parts of the building. As, despite the creation of these bridges, the view of this massive concrete wall continued to dominate the entrance to the building, Helmut Jahn then developed the concept of hiding it behind a huge waterfall and yet, as the detailed design of the waterfall progressed, he became increasingly unhappy. As we had already enjoyed extensive contact through the various projects which ag4 had developed for Murphy/Jahn, he then asked me if we could not envisage a media installation for the entrance hall which could take account of the fact that, as a result of the bridges, it was no longer an uninterrupted space. Anyone who knows Helmut Jahn's preference for stainless steel will understand his surprise when I proposed to him an installation based on beeswax!

The Concept

At virtually the same moment that the building was completed, widespread surprise was caused by the news that Serono had been sold to the German Merck corporation and was now known as MerckSerono.

However, as the conceptual work for the installation had not been exclusively based on the identity of Serono, the result was, fortunately, equally appropriate for the newly constituted company.

Serono develops and produces medicines aimed at "improving the quality of human life". The treatment of multiple sclerosis and support of childless couples are examples of important areas in which the company operates. Genetic technology plays a key role in this work and Serono is, in all respects, a very science-driven company. A result of this is that it is not always easy for Serono to express its own identity and a key role of the installation in the entrance hall was to help in this task. Such projects are aimed as much at the employees of a company as they are at visitors, knowing that companies which promote their own authenticity are generally able to expect more innovative energy from their employees.

Photo © Arne Hofmann

Photo © Viertelböck

And yet how can one embody such a company in a mediatectonic installation? How can one thematize the improvement of the quality of human life? In such a case one has to ask questions which touch upon the clichés which lie at the heart of the subject:

What are the basic building blocks of the human body?

Answer: Fat, water and information.

We articulated the idea of fat by using beeswax.

Water was presented in the form of computer controlled rain.

Information was represented through the development of the entire installation in the form of genetic code. And, in addition to all this, LED technology was integrated into the wax in such a way that the technology is invisible when the diodes are not illuminated but can be seen through the translucent wax when videos are being projected.

The Installation

In such an installation each detail is unique and very careful technical coordination is required. Almost no standard detail or existing product could be used directly.

ag4 acted as designer and general contractor for the entire installation – including the programming and the content – and the companies Metron and Leurocom were responsible for the construction and the media technology respectively. The model cooperation between these three companies was one of the reasons that such an unusual structure could be realized so successfully.

The first step was the assembly of the individual composite steel and aluminum elements of the vertical load-bearing structure. These elements were given a triangular cross-section because this guaranteed adequate accessible space for the passage of media, water and lighting services and cabling while keeping the general appearance as lean as possible.

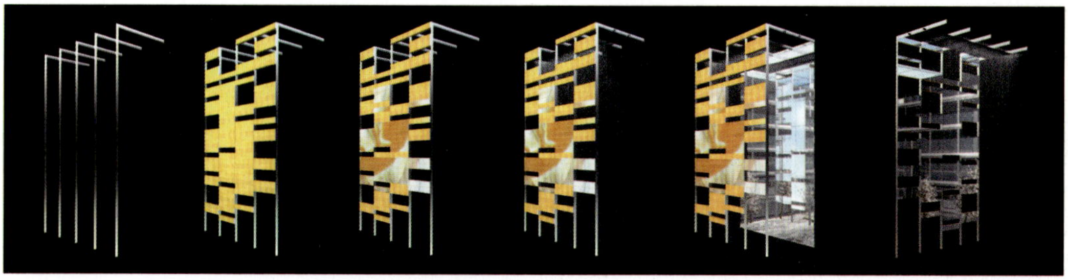

Framework | Wax modules | Wax modules with integrated LEDs | Rain behind the wax modules | Mirrored firewall | Mirrored back sides of the module

The use of beeswax was made easier by my friendship with the Cologne artist Mic Enneper. His work at the interface between art and architecture had brought us together in earlier projects – and he has a very special, craftsmanlike way of handling beeswax. We had experimented on earlier occasions with the combination of wax and lighting diodes and, without this experience; I would never have risked making such a suggestion to Helmut Jahn.

The task of Mic Enneper was to coat the module containing the LEDs (which had been built by Metron) with wax. In order to do this, he required both an entire Metron production hall and enormous discipline, without which the work could never have been completed: he worked 12 to 14 hours a day for 21 days without a break.

In the space between the waxed modules and the firewall we created a basin of water to collect the rain as it fell at differing intensities. Both the firewall and the back of the wax modules were completely covered in mirrors, creating a room of infinite depth.

We made huge efforts to illuminate the falling rain but, due to the optical reality that rain can only be seen against a dark background, this proved impossible. More important to us was the success with which we achieved this mirrored effect – because only thus could the extreme complexity of the installation be made explicit.

And, in any case, the effect of the rain could still be felt, both because it could be heard and also because it could be felt climatically by people crossing the bridge.

MERCK
SERONO

Content

The project for Serono was our first opportunity to create a mix of still and moving images organically developed in line with a clear set of rules. Ideally, our work on such a project would also cover the development of original content but, as the budget in this case did not allow us to produce such material, Serono provided photos and videos related to particular areas of the company's scientific work and we gave these images a specific graphic design.

Yet, despite such restraints, we were able to create a flowing narrative structure with intelligent rules which enabled us to develop a sensually logical combination of these given images with short written statements by Serono – which we merged into the program in a number of ways.

And, in addition to this, we created graphic rules which determined how specific pictorial content could be targeted on specific wax modules in a way which best exploited the form of the modules.

This focusing of the images was a particular challenge because the LED hardware did not permit the creation of an overall image – not only because the sculpture had no clearly defined surfaces but also because of budgetary limitations on our choice of LED technology. The result was that we had to very precisely calculate which image could be seen well from which perspective – paying particular attention to the location of the bridges.

In addition to the pictorial and textual content we also worked with Marius Watz to develop special mediatectonic artistic content. The resulting program-based structure uses colored bars to interactively coordinate individual objects, groups of objects, colors and movements with occurrences on other levels of the installation.

Complex Multi-Layering

All the elements of the installation are layered together into a complex superimposed structure:

Mirror wall
Rain (hearing and feeling)
Light
Wax (smelling)
Scientific structures
Fragments of text
Product logos
People crossing the bridge

This superimposition of such different sensual impulses creates an effect of simultaneity. This reflects the reality of genetic technology, which relies heavily on the interdisciplinary exchange of scientific and technological developments. Indeed, knowledge in the area of genetic technology is expanding so fast that no single person can have a complete overview of this knowledge at any given moment.

Photo © Arne Hofmann

Photo © Arne Hofmann

Our objective in this complex multi-layering was to reflect this reality. The bridges - which would normally have posed a problem for designers in our situation - were central in meeting this objective. And the whole installation is wonderfully rounded off by the fact that people going about their daily business are raised from the level of simple abstraction to become a key part of the performance.

All of which is to say that the truly mediatectonic quality of this piece of mediatecture comes not from the way in which it implements aspects of electronic media but rather from the way in which it combines a series of simultaneous occurrences with a very specific place.

www.mediatecture.info/20

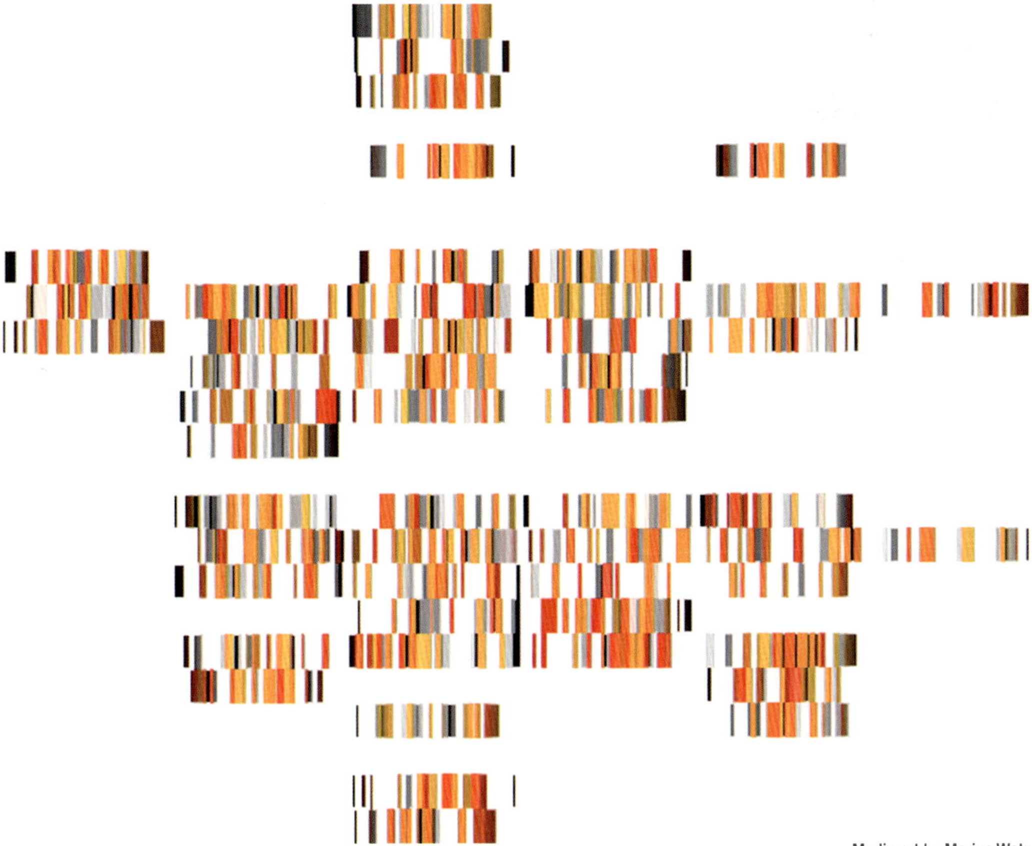

Media art by Marius Watz
Horizontally moved codes, interactively variable
in terms of color, contents and speed
Development of image in real time (programmed by means of processing)
Overlaying with the other layers of content

... electronic Images in Harmony With Architecture ...

THE TRANSPARENT MEDIA FAÇADE

Addressing the idea of the "media façade" is not easy – not least because some mediatects are unhappy about such a term being applied to their work due to the way in which it appears to describe something applied to the external surface of a work of architecture. And this description appears to restrict their scope. The objective of the mediatect is quite the opposite. He wants to identify a way of influencing architecture rather than just cloaking it in a relatively interchangeable dress. This objective is driven by the belief that the medial aspect of such a work emerges from a personal technical-artistic installation and, as such, should be considered as part of the architecture.

I like this approach - which also informed the project which ag4 and I developed for Serono.

Admittedly, I have also been driven on occasions by the notion of superimposing a simple medial component onto a work of architecture as a means of creating a second, medial skin – but in doing so I clearly differentiate between the simple use of media technology in a way which permits the multi-layering of a screen on a façade and the notion of the true media façade in which media technology and architecture converge. As soon as a self-contained screen is simply hung onto a building then this is perceived quite separately from the architecture, and this can lead to the serious danger that the dynamic of the electronic media goes on to dominate the architecture.

The visionary approach, on the other hand, envisages the convergence of the medial image with the architecture. From the very beginning of our work it has been clear that pictures must be dissolved into pixels and that these pixels have to be self-illuminating. Now, however, these pixels can be placed so far apart from each other that the façade remains transparent. A decisive phenomenon here is that everything which is black in the image is invisible because these areas of the image are simply not lit, as a result of which the architecture which is hidden behind comes visually to the fore. Similarly, the illuminated part of the image can also appear transparent if the construction or space behind has the same level of illumination as the LEDs. The image becomes a layer which – like in the photoshop program – can now be superimposed with other layers in order to create an aesthetic effect. The result is a complex multi-layering, which I see as an allegory for social relationships. The effect of this is naturally even more special when the transparent media façade is placed in front of a glazed façade, because now even the activities going on inside the building become part of this complex process of multi-layering. In an ideal case, the lighting of the space is also integrated into the concept of the media façade by means of some form of synchronized control system, making the overall medial effect truly three-dimensional. Ideally, such a projection will make use of objects which, in media design terms, are quite free – because in such a case the image can float. In technical terms, it generally makes sense to organize the media technology into some sort of rectangular frame. In such a case, however, the media design approach described above can then simply make this rectangular framework dissolve.

In order to achieve the desired quality, we at ag4 have spent many years of developing ways of integrating a transparent media façade into architecture. This work has been supported by the know-how and cooperative contributions from many companies and has resulted in the founding of a number of strategic partnerships and in the joint registration of patents.

The roots of such cooperation were already established in 1996 after we were commissioned by Hoechst AG to develop their Corporate Forum in an existing high rise building. A central element of the project was the creation of a raised glass floor with individually controllable blue LEDs (which were then new to the market) which were arranged in a 10cm grid below the glass. This layout permitted centralized and concentric waves of light to fill the space as an expression of interdisciplinary communication. The medial technology for this solution was developed together with Lumino of Krefeld.

Slat Façades

The project for Hoechst launched our work of transparent media façades. In 1999 we developed the idea of integrating a huge grid of LEDs into the horizontal slats of a lightweight construction. A central objective was that, in order to maximize transparency, the slats should be kept as thin as possible. The idea was based on the use of a 20mm wide circuit board, into the edge of which LEDs could be inserted. The result was that the dimensions of the slats could be reduced to 10mm (high) by 30 mm (deep).

We had to exercise great patience in developing this idea with a decade going by between the initial concept and its first implementation. The idea had first emerged from the 1992 competition for the Breslauer Platz in Cologne. (See the article by Harald Singer):

Our first transparent media façade was built in 2004 in Bonn for T-Mobile. A structure of aluminum and steel containing all the electronics and the data and power cabling was placed in front of the glazed conference rooms on the Forum of the new campus. The design of the metal construction as a series of folding frames, each of which carried a section of the horizontal slats with their LEDs, meant that the glass façade behind it could be cleaned.

This slat construction brings great advantages for the transparency of the ensemble. The view between inside and outside is largely free – which is an extremely important criteria when placing a medial installation in front of glass.

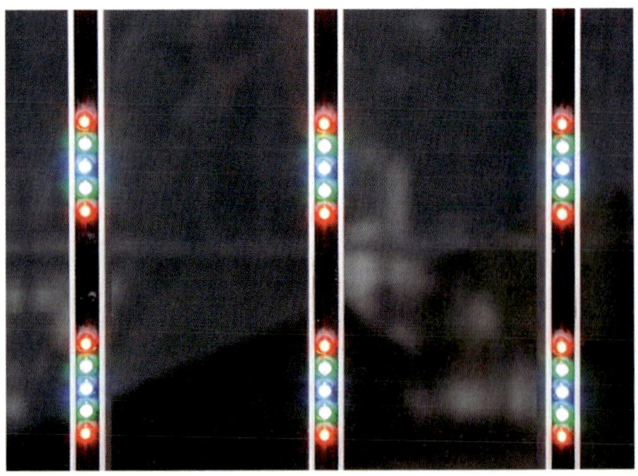

Photo © Arne Hofmann

PSD Bank Münster 2007, ag4 mediatecture company (metal work by metron, LED technic by Leurocom)

www.mediatecture.info/21

Mediamesh® and Illumesh®

For ag4, this project represented the start of the age of the development of professional systems with a modular character. We were besieged by interested potential clients and found that we could only answer these requests by developing such systems. The decisive factors behind these new façades were, on the one hand, that they should be as simple as possible to assemble and, on the other hand, that they should have an architectural quality which has a positive aesthetic effect, even when the façade is switched off.

To achieve these aims we sat down with GKD (Gebrüder Kufferath Düren). GKD is well known for its stainless steel meshes which are used both as solar protection and also for much more purist architectural reasons in the design of façades. GKD has a great capacity to innovate and, correspondingly, is very well known amongst architects. Together, we developed a way of integrating round metal sections carrying LEDs into the mesh that allowed the technology to be updated at anytime – and we named the product Mediamesh®.

At the same time, the development of electronic components for LED technology was developing so fast that a circuit board of just 10mm was now available. This allowed the LED to be fixed to the circuit board and then integrated into a circular aluminum section with a diameter of just 12mm.

A particular requirement which we placed on Mediamesh® regarded the issue of cabling. A "transformer" was developed with which 12 LED sections in a row could be switched – massively reducing the amount of cabling. This design requirement is very important because Mediamesh® is often seen from very close up through a glass façade and such is the success of the solution that Mediamesh® is now used for exhibition halls.

Photo © ag4

Photo © ag4

Mailand Domplatz ag4/GKD 2007

Miami Heat ag4/GKD 2009

Photo © ag4

The pictorial resolution of Mediamesh® is highly flexible but it can be increased to such a high level that videos can be projected onto the surface even in broad daylight.

The use of metal mesh also led to a discovery which opened the way to very large scale projections with the minimum use of media technology. One characteristic of the metal mesh from GKD was to reflect the light from the diodes over enormous areas but, unexpectedly, in a very asymmetrical fashion. Light is distributed over great distances in the vertical direction – but not in the horizontal. We used this characteristic to develop Illumesh®. Horizontal profiles with LEDs at a variable spacing of around 10cm are mounted in front of the mesh at (also variable) vertical separations of 50cm. The image starts to emerge to the viewer at distances of 100m – but only if there are enough vertical pixels – for which one requires a projection surface with a minimum height of around 15m. In the case of really large surfaces, the resolution can be of video quality. From close up, however, the reflective properties offer another surprise: the reflective beam radiates upwards holographically in the form of a three-dimensional arch with a degree of curvature which changes in line with the angle from which it is observed. This close-up image is highly aesthetic - in contrast with normal LED displays where one would not wish to look directly into the beam of light. Illumesh® allows a client, with minimal input, to create a façade which, on the one hand, is lit up in color on one side but, on the other hand, can become a real media performance. Naturally, Illumesh® is also transparent and very suitable for being superimposed upon architecture.

The transparent media façade (slats), Mediamesh® and Illumesh® are, today, key elements for realizing media façades which can be integrated with transparency and sensitivity into an architectural concept. And in doing so, they meet the widespread requirement that media technology should be invisible.

Photo © ag4

Photo © ag4

Torce ag4/GKD 2006

Indemann, Düren ag4/GKD 2009

Photo © ag4

... the Medial Projection of a Corporate Identity ...

ag4 Media façade GmbH developed and realized the Bayer Media Sculpture in Leverkusen, Germany.

*Without **Wolfram Lusche**, this project would never have come into being. For this reason, I asked him to write an account on the project from Bayer's perspective.*

***Wolfram Lusche** is a registered architect, was head of design in Bayer AG's central trade fair department from 1991 to 2002 and responsible for 150 trade fairs annually. Since 2002, he is the head of Corporate Events & Fairs for Bayer AG. Since 2005, he has been responsible for the development of design and content for the Media Sculpture.*

BAYER MEDIA SCULPTURE

There are not many buildings in the history of modern architecture that have such a dynamic past as the former administration highrise of Bayer AG.

Built between 1960 and 1963 according to the plans of the architectural office Hentrich, Petschnigg & Partner as the highest and most modern office building in Germany at that time, it functioned until 2002 as the corporate headquarters, a confident and prominent symbol of the company in the region.

The site was well-considered in terms of urban planning and brought the consequence that a caringly planned Japanese garden with rare trees, sculptures and a tea pavilion had to be relocated by a few hundred meters. In March 1999, the building made headlines around the world: On the occasion of the 100th anniversary of Aspirin, it was completely wrapped as the largest package of tablets in the world.

In the same year, because of the excessive costs estimated for restoration, the decision fell to demolish the highrise and build a new corporate headquarter. Between 2000 and 2002, the new building was constructed on an adjacent site following the plans of the architectural office Helmut Jahn with all intentions to demolish the old (insufficient clearance between buildings). After corporate management's move into the new building, the highrise was still used as an office building until 2007; parallel, preparations were being made for the demolition.

www.mediatecture.info/22

BAYER

The idea

In the center of the Bayer factory, a unique urban ensemble had emerged consisting of:

» *the old main administration building, built from 1903 to 1912 following the plans of the architects Willy Günther and Hubert Amrhein in Historicism style;*

» *the pharma administration building, constructed between 1937 and 1939 following the plans of the architect Emil Fahrenkamp with its strict sand stone façade (both buildings are under monument protection);*

» *the new corporate headquarters by Helmut Jahn with its convex glass façade;*

» *and the administration highrise as a prominent landmark, which spatially concludes the ensemble given its transverse positioning.*

On my lunch breaks, I often walked around this ensemble of buildings, observing them from different perspectives, and played with the thought of how it could be possible to preserve the highrise.

In August 2005, I stopped in front of the highrise and suddenly had a vision: I saw the skeleton of the highrise before me, encased in a transparent media skin.

Photo © Bayer AG

In this way, this important landmark would not only be preserved, but also transformed into the new symbol of Bayer AG with an innovative function.

The task

Departing from this idea, a concrete set of tasks had to be established and a feasibility study conducted. And partners needed to be found and won over for this idea.

I had followed with great interest the global trend and controversial discussions about the medialization of architecture in public space, and I was well-informed about the first planned and realized projects. They reveal that in our society, in diverse countries, there is an increasing need to emotionalize public space. Media façades offer the fantastic opportunity to add dynamic elements to the static character of architecture. They create a communication interface for the observer and the chance for interaction. I wanted to use these chances in the project.

The architectural requirements

There was now the concrete opportunity to create something entirely new – a building reduced to its supporting structure, with no further functions that it must accommodate, enveloped on all sides by a transparent media shell. The available steel skeleton and highly-suited dimensions and proportions of the building formed the basis of this one-of-a-kind starting point. Through an added light installation, the supporting structure could also play a role in the staging of the façades. Hence, transparency became an important issue, not only for the façade design but, above all, to exploit the possibilities of a three-dimensional light display. This dramatic interplay between an illuminated building structure, light design and a multimedia shell around a colossal building volume (21 m x 65 m x 122 m) could no longer be simply classified under the category of media façades. In order to better express the difference, I introduced a new term: Media Sculpture.

During inactive phases, for example during the day, the building should also meet high architectural and aesthetic standards, and through its transparency reflect the various effects of the sunlight at different times of day. But there was another architectural problem to be solved: the creative incorporation into the ensemble and, in particular, its relationship

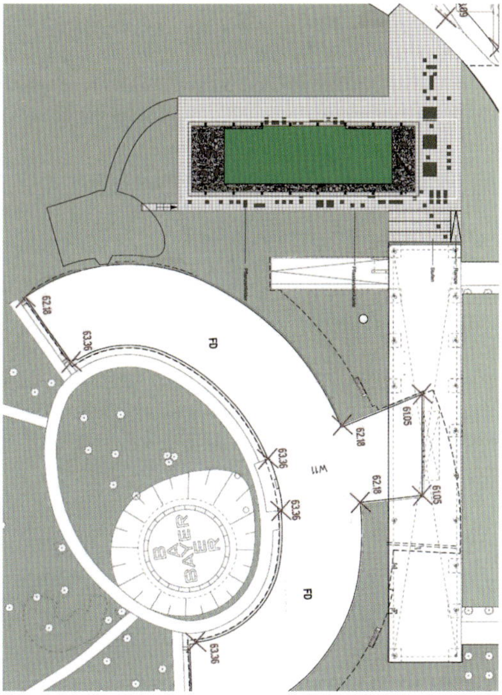

Ground connection between the highrise and the new headquarters by Murphy/Jahn

with the nearby, convex-shaped corporate headquarters.

The content requirements

All media façades realized to date around the world serve either economic (profitable advertising space) or, in the broadest sense, artistic interests. I was interested in something else: Can such an innovative medium also function as a new form of communication for a company like Bayer and, furthermore, also fulfill an artistic claim? Can this new medium and an intelligent and emotionally stimulating display concept modernize the image of a company? Can it become a lighthouse of a new communication culture? Will this positive impulse toward the outside also have an inward effect? When the Media Sculpture is to be understood as a communication medium, then it should also be possible to integrate it into a communication strategy.

Another important point for me – and surely also for the feasibility of the project – was the question: Is there a content management system that can independently generate content according to set criteria and over a longer period of time (24/7, 365 days a year), thereby making permanent supervision unnecessary?

The screening process for the appropriate media façade systems and partners with the corresponding experience led to an answer sooner than expected. The central requirements were:

» a transparent, integrated media façade system
» several years of field trials of installations on buildings
» all necessary building approvals in façade construction

Moreover, a warranty over several years was necessary given the building's exceptional stress loads from weather, in particular, that all sides of the building are open to strong winds.

The product we found was manufactured in the immediate vicinity: MediaMesh®, a development made in a collaboration between the ag4 mediatecture company in Cologne and the stainless steel weaving mill GKD in Düren.

After my first contact with ag4, I was certain I had found a motivated team under the direction of Ralf Müller and a congenial partner in the architect Christoph Kronhagel for the further development of the project concept. This team also made the first visualizations and feasibility studies. I also learned that ag4 already had a content management system available, which was expandable and could satisfy our concept requirements for a self-generating, permanent multimedia display. The need to make one or two employees available for continuous editing and supervision work could have jeopardized the realization chances.

From my experience with multimedia productions for trade fair stands or big screens at events, I knew that images and fixed image sequences can quickly become subject to mental wear-out, and repetition can rapidly lead to oversaturation and repulsion for the viewer. Thus, in addition to the intelligent content management system also a sufficiently large and well-structured image database would need to be developed.

The technical requirements

With respect to dismantling and conserving building components, we could rely on the professionals in our own internal building department at Bayer.

Feasibility studies began with technical analyses, studies of the old building plans, a preliminary static test, a building safety verification, and selective examinations of the building itself (e.g. the condition of the exposed steel beams). The results suggested possible realization. The next step was to estimate the costs and time required for the two major tasks: dismantling the building and refurbishing the remaining parts. In the first costing for the media façade system, it became quickly clear that a high-resolution version with an 18,000 square meter surface area would not be economically viable. But a four-sided, full surface installation was one of the departure points. Thus, we investigated equipping the building with an alternative product, IlluMesh®, where the LEDs indirectly illuminate the steel mesh and images are generated through reflection. In this manner, the large image surfaces could be created with considerably less pixels.

But it was clear to me that additional functions were necessary for the realization of the project, ones that clearly distinguish it from the large light installation of the Bayer cross on the factory site. Built in 1933, this enormous, glowing Bayer cross (51 m diameter) was the largest self-supporting advertising installation in the world; consisting of 1680 light bulbs (today: 1702), it has since then been the symbol of the company and the region, visible from great distances at night.

The decisive additional function for me was thus the possibility for a daytime display. However, this would require a high resolution with direct illumination from LEDs, which was not an option with IlluMesh®. After numerous studies and de-

Ansicht Carl Duisberg Park

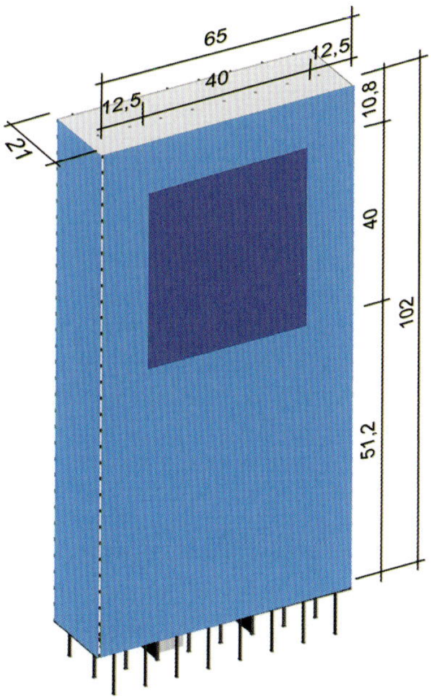

Illumesh V75 H10

Mediamesh V6,75 H6,75 mit Illumesh V75 H10

veloping conceptual approaches together with ag4, a solution was found: Two hybrid surfaces were needed on the long sides of the building. This means that MediaMesh® had to be integrated into IlluMesh® in such a manner that it enabled a homogenous image in the IlluMesh® display without the hybrid surfaces being discernable. After determining the minimum and optimum viewing distance, the pixel pitch was optimized in numerous lighting tests. For the daytime display with MediaMesh®, a pixel pitch of 6x6 cm was chosen with a set maximum surface area of 40x40 m (with a building width of 61 m). The square form was informed by the requirement that our round logo should be able to be displayed on the full surface area of the building and also during the day. After synchronizing the task list and feasibility studies, the concept – which partially was still in progress – was given a concrete form.

The concept

It was clear that the chances for the realization of this visionary project depended heavily on the quality of the concept. Only with a concise and coherent concept would there be perspectives on how to impress and convince internal and external decision-making and approval authorities. For an investment of this scale, also the benefits had to be well articulated. This was not an easy task as there is no reference project of this scope worldwide. At this point, no company has ever employed medialized architecture as a direct communication medium.

In this case, only a holistic view of architecture, society, business culture, brand value, history, and place – both on abstract and concrete levels – could bring the idea further.

The Media Sculpture as corporate architecture

Architecture involves society, time and place and is always viewed in this context with regards to its effect. Corporate architecture is an additional expression of the corporate identity of the enterprise – this means the business culture, the specific value of the enterprise, the brand, the corporate design, as well as the communication of the business – and, as such, is highly influential for the corporate image of the enterprise. Corporate architecture should make these requirements visible and, in the best case, tangible. Moreover, it should create an added value, not only for the place in that time, but also for the people who perceive it (employees, clients, business partners, etc.). Corporate architecture thus has a clear communication task.

In the history of Bayer AG, corporate architecture has played an important role – as illustrated in the aforementioned examples. Several corporate personalities – such as Carl Duisberg, the general director from 1912 to 1925 – even took it on as their personal concern to initiate bold projects and supervise all construction activities.

In this respect, the Media Sculpture as the largest media architecture project in the world carries on the tradition of pioneering examples of Bayer corporate architecture.

Moreover, the building itself is an outstanding example of an architecture that deals with history. By gutting the building to the load-bearing core, the essence, and the subsequent metamorphosis into the Media Sculpture, the building itself becomes a strong, identity-forming symbol of a creative approach to the enterprise's own history. This is succinctly expressed with the motto: "We're taking the best from our history and transforming it into something innovative and future-oriented." This powerful visionary image was one of the important key messages of the concept. With such an unprecedented project, Bayer AG as an inventor company can set a confident and unmistakable sign for innovation and creativity. Comparable with the then spectacular construction of the biggest illuminated advertisement 75 years ago, the Media Sculpture should become a new, appropriate symbol for the company. It should become an attractive visual anchor in press reports worldwide and thereby attain a high advertising equivalent value.

Photos © Bayer AG

The location at the heart of the business premises is ideal, both in terms of its long-distance effect on the important traffic arteries all around (highways, main railways, approach paths to the Cologne/Bonn Airport, boat traffic on the Rhein) and its short-distance effect on the employees, visitors and business partners of Bayer AG as well as those of the CHEMPARK with their approximate 30,000 employees. There is an enormous advertising potential, which splendidly complements the existing lit Bayer cross.

In contrast to the cross, the Media Sculpture, however, can accomplish much more for the company: It not only visualizes the trademark of the company, it visualizes the Bayer brand itself. The medialization of architecture represents – beyond just advertising spots – an enormous potential for the company as it is ideally suited to emphasize the fascination of a brand and load it with emotions. There, where natural limits are set for architecture, the medium can virtually suspend, dynamicize and alienate the laws of gravity, the statics of construction; it can make the inside visible, disintegrate everything and put it back together again. Within lies enormous magic and even poetry. I wanted to use this magic for the Bayer brand.

The architectural concept

For the architectural integration of the Media Sculpture into the complete ensemble, a number of approaches were initially researched. As the was no reference made to the highrise in the design of the corporate headquarters because of the intention to demolish it at the time, now a relationship had to be established, conversely, to the architecture of the headquarters. The building height pergola, a design gesture that Helmut Jahn used to mediate the Historicist façade of the old main administration building across the street, could serve this purpose for the Media Sculpture. A new architectural element was designed: the light frame. It incorporates the height, section and materiality of the pergola frame and, floating freely, encompasses the Media Sculpture. It is equipped with light sources mounted on the inside, which are connected to a media control system, and delineates the surface used for the display from the inactive surface of the lower floors. There is no display at the bottom to avoid any visual disturbance in the nearby offices of the headquarters in the darker times of year.

Additionally, we arrived at an entirely new landscape planning to achieve a distinct connection with the headquarters also on ground level. Paving materials, sight lines and the rhythm of the planted plane trees were taken up in the design. An important design principle was to create a sufficiently large, bright ground surface as a plateau, a kind of passe-partout, and thereby give the light shining down from the light frame a corresponding reflective surface.

Photo © ag 4

The content concept

The stated goal is to trigger positive associations and emotions in the observer of the Media Sculpture, to radiate a high aesthetic quality and thereby serve the image of the company. In the first step, appropriate themes had to be determined. The umbrella theme, under which all other themes had to be subsumed, is our mission claim: "Bayer – Science For A Better Life". In accordance with the communication strategy, the following themes were chosen: Science, Research, Innovation, Future, Human & Society and Nature & Environment. An important matter for me was to place the focus on image and video material that captures the beauty of

The new Bayer logo in full resolution with the integrated Mediamesh® (display also in daylight). Transformation of the colors of the logo by the green heart and blue ceilings

nature, the environment and research in the macro as well as the microcosmos, thereby stimulating new ways of seeing – enhanced by the overwhelming scale of the Media Sculpture and the aesthetics of the display and lighting.

Our logo contains the CI colors blue and green. The original meaning of the colors – competence and responsibility – was only known by a few long-time employees. We proposed to recharge these logo colors with the theme Nature & Environment. It was close at hand to make a connection with the topics of climate protection and sustainability, which in the past years have been increasingly in the foreground of our enterprise's communications. We were convinced that exactly the Media Sculpture as a new communication medium was qualified for this task. In the concept development phase, an interest arose to visualize this not only in the multimedia display but also in the architecture. After many brainstorm meetings with Christoph Kronhagel, the solution emerged. The building stands on the edge of the generously planned Carl-Duisberg Park. A so-called "green heart" within the building should symbolize a green energy stream from the park, redirecting into the vertical, rising and densifying. The green in our logo forms the main color value and is complemented by two lighter and two darker tones of the same color that are artistically distributed. The play of light on a leafy tree in the sun served as an association. This green heart should be built behind the façade out of semi-transparent Makrolon® multi-skin sheets, a Bayer product, and have the dimensions of the set-back ground floor. It will be backlit and integrated into the lighting dramaturgy. In this manner, the rise of this energy can be represented dynamically and with direction. The existing 8-meter-high glass panels of the set-back ground floor façade are to be furnished with an artistic interpretation of magnified organic forms and structures, which harmoniously connects with the overall design of the green heart in its color and rhythm.

The second energy, the blue energy of the sky, symbolically streams from above into the building, visualized through the blue color coating on the ceiling surfaces. From the top floor to the bottom, the intensity of our logo color blue abates progressively to illustrate the direction. Both forces, coming from opposite directions, unite in the building and then – with the help of the lighting and the multimedia display – form the Bayer logo. Suitable image and text statements support the sequence. This unification of two elementary powers has a highly symbolic meaning, which the Media Sculpture illustrates in a fascinating manner, thus shaping the brand values of Bayer AG into a novel emotional experience.

The display concept

It was incredibly important to visualize the early ideas for the display as there were no comparable reference projects. Additionally, not only what was going to be displayed had to be formulated, but also – and more importantly – what should not be displayed. The first question that arises when one speaks with non-specialists about a media façade is: "It's just like a big TV, isn't it?" This association results from the fact that most existing media façades are tagged with terms such as "public monitor", "urban monitor" or "electronic billboard", and the medium's inherent potentials have not even begun to be exploited. This negative image needed be avoided in the presentation of the idea, and the difference had to be clearly worked out.

First of all, it needed to be clear that this Media Sculpture would not be available for all kinds of advertising clips or film and television broadcasts, or as an electronic announcement board, and, above all, it is not for rent, neither by Bayer subsidiaries nor external companies. It was decided that a set of rules had to be worked out and an editorial board set up to supervise compliance with the rules and, when necessary, propose adaptations based on first experiences.

The advantages of this new form of communication are as obvious as they are overwhelming. As it is with any new medium: The freedoms are seductive, but it is crucial to identify the dangers and define rules in advance in order to prevent abuse and nuisances.

The next step was to conceptually determine the what, when, and how of the visualization. Two basic forms of display were defined: a.) according to content and b.) according to time. With regard to the content, it was important for me that there was a distinction between the so-called permanent and the temporary display. The permanent display is the continuous, uninterrupted display throughout the whole year within a set timeframe. However, from the very beginning it was a part of the idea that on certain occasions, perhaps four or five times per year, there would be special events; thus, special forms of display should be possible and a corresponding catalog developed for it. The most important factor was to make the Media Sculpture available to select artists for a certain period of time, or perhaps even to host a media art festival as a fixed regular event. Our company has a tradition of supporting diverse forms of art. Likewise, special seasonal displays are planned as well as interactive ones – for example, a "Games Night" with hundreds of live players and others connected via Internet on the occasion of the gamescom convention in Cologne.

Another differentiation was needed between the daytime and nighttime display. I had already formulated the necessity for a daytime display in the initial requirements. But it was also important to establish a distinction in the content that corresponded with the differences between the two chosen systems: the daylight suited MediaMesh® (3,006 m^2) with a very high resolution and the IlluMesh® (15,066 m^2) with a comparatively lower resolution. Hence, different goals and respective contents were set. The goal of the daylight display: As the new symbol of the company, approximately 75 percent of the display time should be dedicated to the animated Bayer logo and corporate design elements of Bayer subgroups. This can be expanded with visualizations of superordinate themes of the corporation (Bayer Climate Program, Corporate Social Responsibility, etc.). The expressiveness of the above-mentioned themes is supported with corresponding image and video material and additional elements such as a clock that appears on the hour.

Ascending illumination of the green heart
Descending illumination of the ceilings

Presentation of latest information on Illumesh®
running horizontally at varying speeds

The entire range of possibilities was available for the nighttime display. We differentiated this range of elements into:

1. *Image modules*
2. *Video modules*
3. *Graphic modules*
4. *Text modules*
5. *Illumination of the green heart*
6. *Illumination of the blue ceilings*
7. *Illumination of the light frame*

Moreover, an interface for additional effect lighting and an interface for sound were part of the plan from the very beginning. A highly complex masterplan regulates the application of all these elements into a display with a processual character. Following a very precisely defined set of rules, a central computer independently controls the order of the above-mentioned elements, accessing a finely structured content database (in which the first four of the above-mentioned elements are contained). The intention was to create sensible connections in the content and to avoid repetition. The real challenge in the design and the programming was to combine the four elements in the content database with the three lighting elements in such a way that a three-dimensionality develops through the extraction of the motif and the overlap of multiple layers. This is one of the decisive factors for the dramaturgy of the entire production in order to evoke the desired fascination and magic. We want to address the observer directly and communicate within seconds something that they have never seen before and no monitor can offer.

Connection of latest information with the Bayer website
Text and image combination

3D fragmentation of the installation through synchronized control of the green heart, the blue ceilings and Illumesh®

Renderings © Kronhagel Mediatecture

Another idea emerged during the concept development work in the team: The Media Sculpture must be connected to the Internet. We didn't want the effect of this new symbol to be limited to the immediate, visible area. It should radiate out into the world, be perceived and enable global communication. This meant that the Media Sculpture needed to have its own website with a webcam in permanent operation and interfaces for interaction. Herewith, this new communication tool provides the incredible chance for our company to reach totally new target groups and to develop new types of interactive communication with them.

The content database

The construction and structuring of the content database took on great importance. Here it would be defined how and where each data should be systematically stored, classified and indexed. In the beginning, we estimated the needed capacity at 18,000 items consisting of images, video, text and graphic data. For the construction of the database, I relied on Wolfgang Blum, from the company Blickpunkte, and his competence in structuring and administering such image databases. For this reason, we began very early on with the design of the database folder structure, which initially appeared very complicated. But in retrospect, it was left as it was with very few changes until the very end, even though thousands of image files constantly moved around and were manipulated by more than ten designers over the course of many months. The content database consists of three to four levels: In the first level, there are 17 main folders (e.g. logos, texts, image themes, specials, events, pool, etc.). In the second level, there are subfolders which are contained within the main folders and include various corporate and thematic sections. The third level, found in the second level folders, consists of diverse subfolders with theme worlds and images. The theme worlds are divided once again in the fourth level into folders with the selected key topics – such as nanotechnology, optical data storage, biotechnology, etc. – that researchers at Bayer are preoccupied with and which should be visualized. Only on the fourth level is the point where related image and video data are located. The sole exception is the main folder "Pool" where ambient pictures that cannot be thematically ordered elsewhere are located; it is subdivided into just two folders: stills and videos.

All data are given comprehensible paths through this complex and branched structure. This secures the respective reference to origin and theme. However, all data retain the original name for the identification of the source. This elaborate differentiation was necessary to provide sufficient specific selection and linking possibilities for the control software. Additionally, the database should be expandable and function for the years to come.

This work required a considerable amount of time but also formed the basis for the programming because we could only then establish most of the rules. After the structure was completed and agreed upon by all those involved, Wolfgang Blum could then begin with the extensive image research, from both the existing Bayer and commercial databases, and organize the content into the folders. Another advantage is that this database can also be used for other displays: for example, for events, conferences, receptions and the like.

Photo © Arne Hofmann

Back-lit graphic in the foyer, 9 meters high, design: Christoph Kronhagel

Extract from the database theme "Research", database design and organization: Blickpunkte / Wolfgang Blum

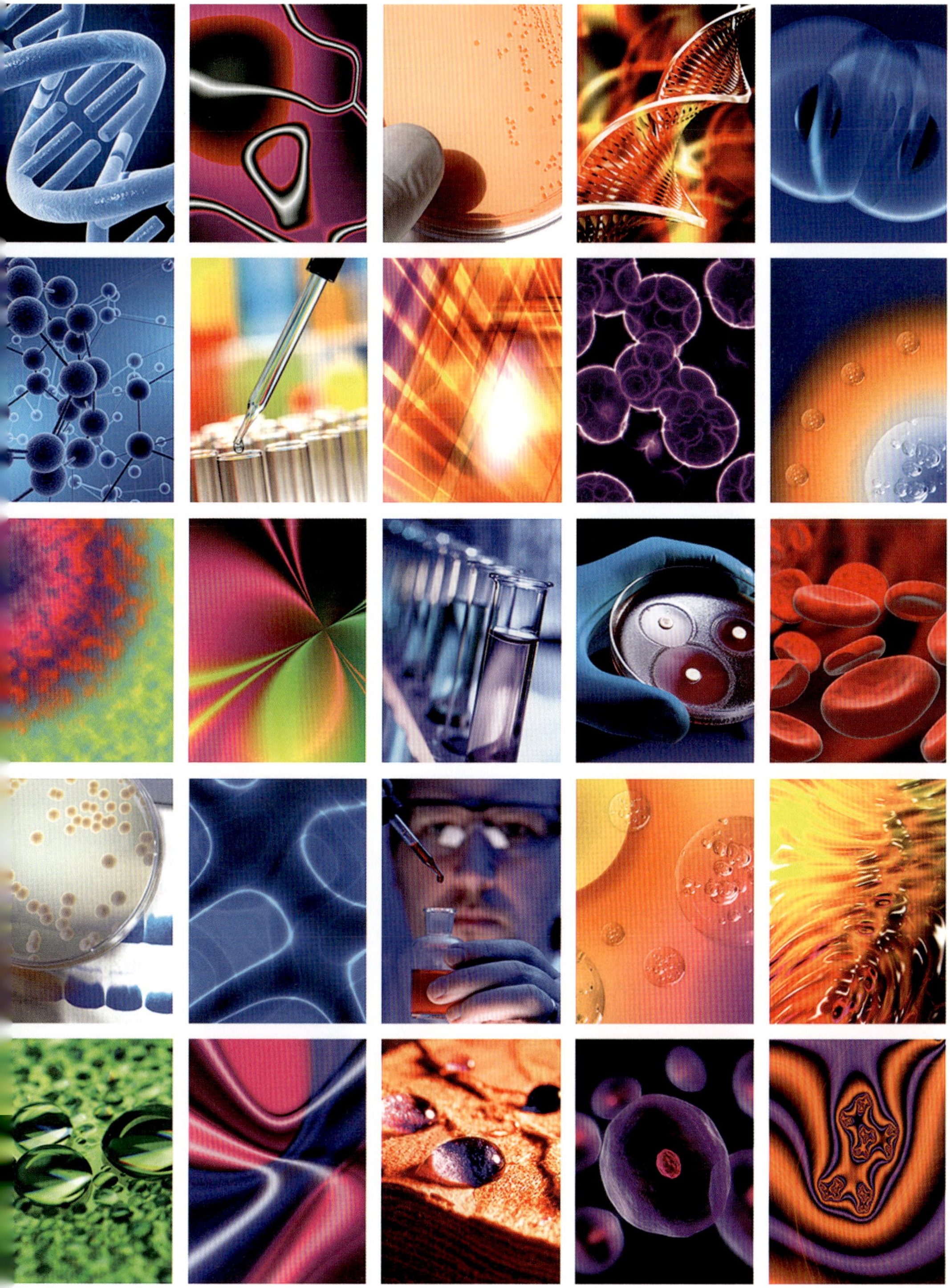

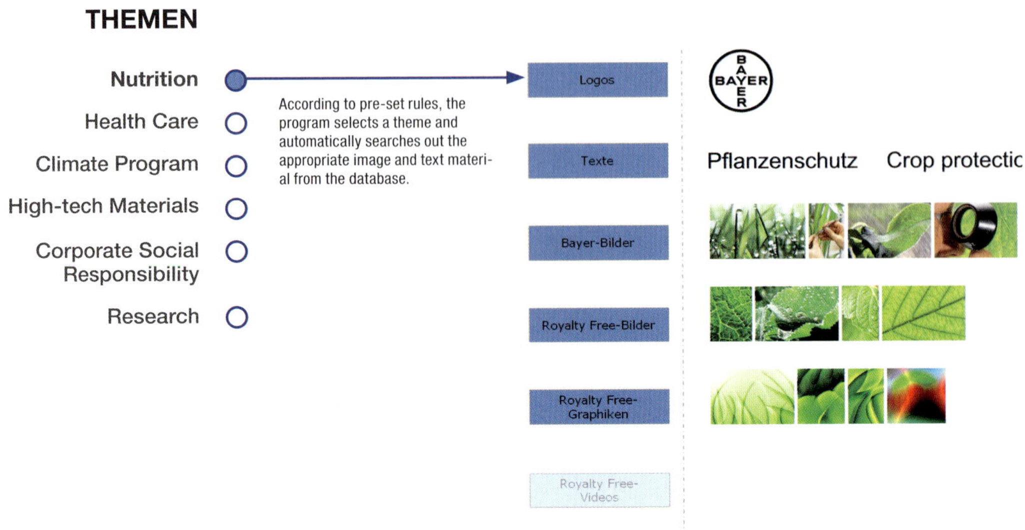

The content management system

After the concept was approved, we formed a content team – consisting of a number of ag4 employees, Christoph Kronhagel, Wolfgang Blum and myself – in order to begin work on the content design. First of all, the display formats were defined as they influence the image formats and thereby also the image selection. Next to the square MediaMesh® surfaces, five other image formats were set for the IlluMesh® surfaces. In combination, this led to 15 types of displays.

The perception of the display was given great importance. I believe that perception is the decisive criterion for the functioning of mediatecture. This has an effect on the selection and preparation of the content. It was clear that a building in public space with this level of visibility should not present narrative content. This means that the content has to be made comprehensible within a matter of seconds and, nevertheless, evoke the fascination of the observer. Additionally, recognizable repetitions had to be avoided in all cases, because a large segment of the target group would have daily contact with the Media Sculpture.

In the past years, I developed the following design principles for Bayer event displays together with Wolfgang Blum, which touch on quick comprehensibility and associative perception:

The basis is formed by photos and videos selected from the company's own image database, which have concrete image content, are thematically sorted and specially cropped for the image format.

They are complemented with appropriate, more abstract images and graphics from image agencies, so-called "ambient pictures", which are used to make connections and transitions between thematic image sequences.

All images are slightly moved, turned or zoomed in different directions, in other words animated.

A selection of animated images and videos are combined with terminology and headlines from the chosen themes, which are also animated and serve as transition elements.

Start rule:
1. Position image from the database on entire display.
2. Layer with logo.

Process rules:
1. Zoom into image and rotate.
2. Create free zone.
3. Position text in free zone.
4. Activate light frame.
5. Activate blue ceilings in free zone.

Process rules:
1. Image in zone 1 – above.
2. Image in zones 2+3+4, exactly beneath the image in zone 1.
3. Activate green heart ascending from bottom to top.
4. Activate the blue ceilings below image in zone 1 descending.

Process rules:
1. Image in zone 1 fades into the next image
2. Move image in zone 1 from top to bottom
3. Rotate text above image in zone 1
4. Activate blue ceilings and green heart

Processual combinatorics for "optimizing plastic"

Renderings © Kronhagel Mediatecture

Processual combinatorics for "biotechnology"

All animated elements comply with a fluid rhythm and continuous dramaturgy, which define the sequences, the type of image composition and decomposition, and various forms of transitions. Only through the combination do the recipients have the desired associations because images are rarely self-explanatory.

It was advantageous that this special visual language was already well-known in the company from numerous events. This linear and two-dimensional form of dramaturgy to date used in event displays could now serve as the basis for the content design. The decisive difference was the omission of the linear narrative structure. For the processual, self-generating control of the image sequences, all content had to be separated into individual sequences, the connection between them precisely defined and all image and video data individually profiled.

Profiling the data

The properties and criteria of the data intended for use had to be defined. They needed to be individually profiled in such a manner that reasonable relationships could be generated by a central computer, which takes over the "editing work" in a sense. We quickly noticed that this is dependent on whether the display concept works. That meant the greater the number and the more precise the single criteria were, the more combinatory possibilities there would be and the lesser number of errors occurred in the sequence. First of all, we set three categories of criteria:

» *thematic criteria, linked to keywords*
» *functional criteria*
» *graphic/aesthetic criteria*

The functional criteria differentiated the possible image positions and all forms of motion on the building, such as rotation, directional movement (horizontal, vertical, diagonal), zoom, image fragmentation and IlluMesh®/MediaMesh® combinations. For the graphic criteria, the image character, among other factors, had to be selected via opposites such as static-dynamic, abstract-real, color dominant-polychrome, and rated with a scale value between 1 and 20. Then the data were arranged into eight color value groups. The goal of the profiling was to sort sequences of data and to identify and evaluate potential successive data. In this way, the largest possible number of transitions should be attained while avoiding unfitting image sequences. When the criteria catalog was completed, it became clear that this process of profiling was not suited to automation. The aesthetic principles – in particular, the combinatorics of the canon of images that I developed together with Wolfgang Blum for corporate events – were very much customized to a mature image culture for Bayer. For this reason, the evaluation and profiling of these desired qualities could only be done by people who could subjectively comprehend this special combinatorics. Thus, there was a need for collaborators who were experienced in image processing and had reliable design skills. These "profilers" were selected and trained by ag4. A special profiling program was developed with masks, tables, and preview functions to make this task easier. A 3D simulation model, including the highrise and adjacent buildings, the street and the park, was created to enable the evaluation of the three-dimensional impact of each image, video, graphic or text on the building. Upon mouse click, each file can be visualized on the building, rotated in every desired position, and the viewpoint and height can be changed. This was one of the most important tools for selecting and evaluating the data. Even the most experienced image editors are pushed

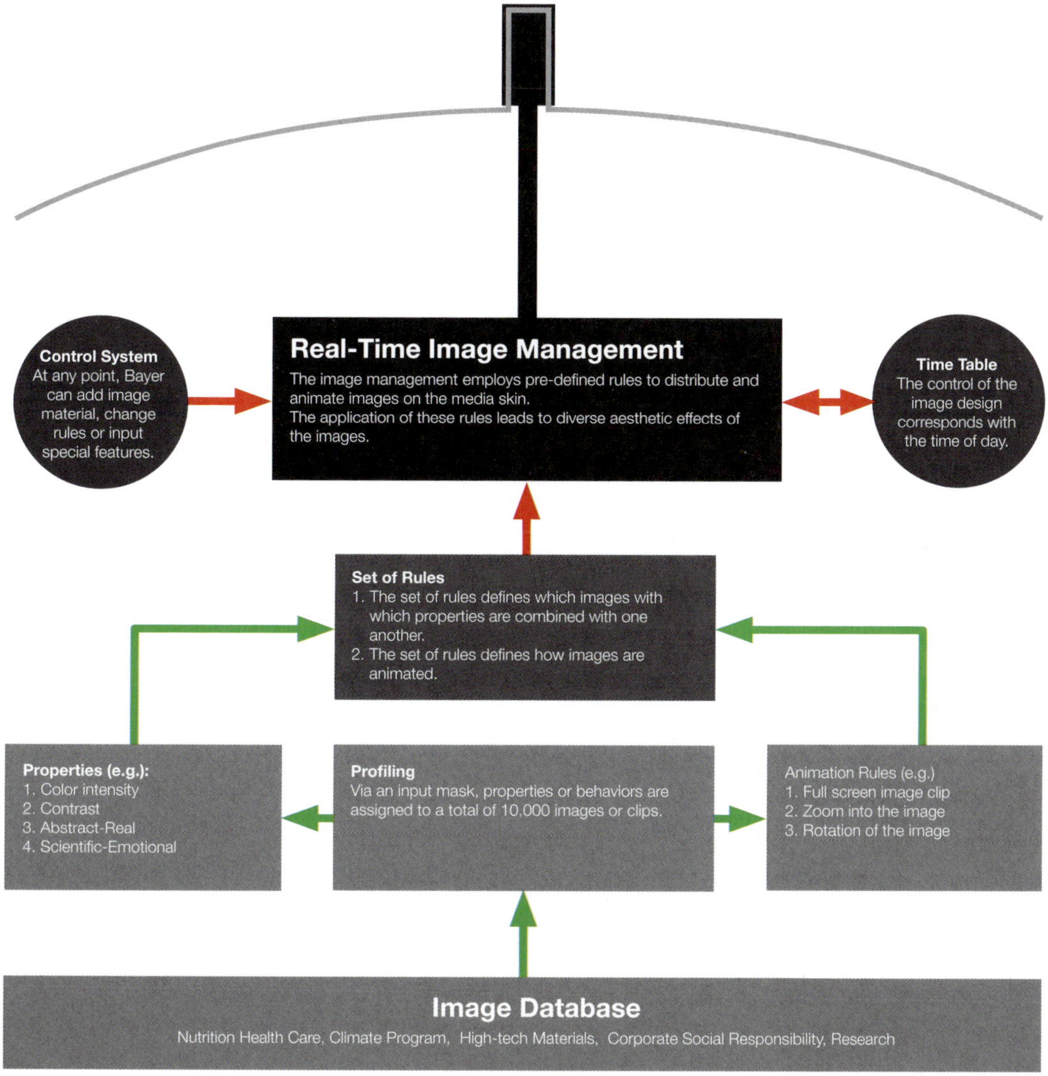

to their limits when they have to imagine and evaluate a motif wrapped around all sides of a building. At the end of the procedure, I had to assess and approve, or send back into the process with remarks for post-processing, each of the meanwhile 10,000 files. This would not have been possible without the 3D simulation model.

The control software

In the next step, a more concrete requirement profile for the control software had to be elaborated on the basis of the above-mentioned concept. During the process, it was constantly updated and synchronized with the current status. In a lot of cases, it was only through the examples on the 3D simulation model that we could find out which rules still had to be changed.

For example, we decided per case which animations are rendered in real-time and which were better to pre-produce in order to at all times ensure the operational safety with such a complex control system.

The control software IMPP (Interactive Media Pool Platform), which ag4 had developed, was upgraded considerably for these requirements. No longer would it only control the image sequences on one medium (e.g. MediaMesh®) but now also those on IlluMesh®, the combination of two as well as the entire lighting system inside the building – all synchronized in time and coordinated with one another.

In close collaboration with the programmers at ag4, the criteria in the requirement catalog were repeatedly controlled to determine which could be taken up in the program and which could better be realized with scripts. Additionally, the software needed to be expandable and adaptable at any moment to new requirements for scheduled special events. A special challenge was to emphasize the three-dimensionality of the Media Sculpture with a fascinating dramaturgy. Specific rules were set and programmed for this purpose. Also in the choice of images and videos, special attention was paid to this aspect. Extracting, masking and fragmenting techniques should help create free spaces in the display for the light dramaturgy. The use of text blocks in such free spaces also had to be taken into consideration. This overlap of multiple layers with transparencies and penetrations into the depth – sometimes even realized with different display modules (e.g. script, image, video and light modules) – results in a very complex visual language that enables diverse forms of expression. This constitutes the enhanced fascination of this three-dimensional Media Sculpture in comparison to the effect of a two-dimensional media façade, not to mention an urban screen. It gains a uniqueness that no other medium can achieve. It can speak to the emotions and aesthetic feelings of the observers while conveying messages in a subtle and intelligent manner.

In this way, the Media Sculpture not only becomes an innovative and attractive landmark but also an extraordinary, novel means of communication – and is, as such, of great importance for Bayer AG.

Renderings © Kronhagel Mediatecture

Example: High-contrast image, abstract, full screen – zoomed and rotated (the parameters for the animation can be adjusted at any time or substituted on special occasions, e.g. holidays or events)

The approval phase

With this concept, I began to search within the company for sympathizers with the project. At this point, my colleagues in the building department were already busy with preparations for the demolition work. In cooperation with Wolfgang Vogel, who I managed to win as an ally, we could change the planning and preparations in such a way that a turn toward rebuilding would be possible without any bigger problems at a later point in time. This bought us time. An official rezoning application (the demolition had already been approved) from the building authority would have been too lengthy a procedure. Therefore, we decided to introduce the concept to the responsible people in the municipality: the Department for Planning and Building Affairs as well as the Lower Monument Preservation Authority. It was essential to make clear the advantages for the city and surroundings and to eliminate worries, especially those of the monument preservation officials. The coherent concept, the urban planning considerations and the visualizations – in particular, the first ideas for the display program – were extremely helpful during the talks. The concern that the old highrise building would become advertising space in the style of Times Square was quickly off the table, and after underlining the artistic aspects, it made way for overt enthusiasm. Support in all approval phases was guaranteed.

Another obstacle emerged on the level of copyright. The designers from the architecture office Hentrich, Petschnigg & Partner were not alive anymore, but the rights had been granted to the successors of the office, which operated under the same name. In the case of demolition, copyright would be subordinate, but this is not the case with adaptation. We decided for a personal talk with the managing director of HPP, Joachim H. Faust, and presented him the concept. He was at once fond of the project and directly promised us his support. Afterwards, to be on the safe side, I wanted an agreement with Helmut Jahn as the designer of the neighboring building, since his concept had been based on the demolition of the highrise. Here, too, I thought that a solution among colleagues would be the best strategy. Hence, we highlighted the ideas concerning the architectural adjustments to his design and built upon his generally positive attitude toward media façades. With Christoph Kronhagel's mediation, we not only managed to obtain his approval but also his best wishes for the success of the project. Thereby, we had dealt with the biggest obstacles in the preparation phase. Clearance declarations from air traffic control and the federal road administration concluded the preliminary approval process. Now the path was clear for the presentation to the board. The board had in fact decided for the demolition of the highrise. However, I kept trying to win supporters for the concept in the company with a presentation. It opened with an animation of an explosion of the highrise, demonstrating the resulting spatial gap within the building ensemble. This small effect was a step toward the goal as was an animation that looked back on the 1999 wrapping action of the highrise.

In the end of 2006, our CEO Werner Wenning heard of the idea and gave me the chance to present it to the board in January 2007. The idea was warmly applauded. However, there were still concerns whether such a spectacular project could be understood as gigantomania and therefore not suit the brand. Therefore, we decided to draft a representative survey, which should clear up how this project would be received by the employees and neighbors, as well as how it would be perceived in connection with the Bayer brand.

The survey

An empirical investigation was quickly set up in the form of personal interviews with 1000 test people. They were divided into the following groups: one third employees, one third residents from the neighborhood and one third people from all over Germany who have no direct relation to Bayer AG. The following aspects were evaluated:

» spontaneous reaction
» general acceptance
» impression (long-distance versus close proximity)
» opinions, evaluation of the idea
» views on the demolition of the building as an alternative
» anticipated effect for Bayer, the location, the region

Along with the interviews, charts were shown, which briefly introduced the idea and visualized it in renderings. The results were outstandingly positive and without any significant differences between the test groups. In the rating of the idea on a scale between 1 and 5, 69 percent chose "good" to "very good". Between 74 and 81 percent of the interviewed persons graded the impression of the project as very innovative, up-to-date and trendsetting, and many of them had not expected this from the Bayer company. 70 to 74 percent agreed that the effect for Bayer would be appropriate. This revealed how much of a positive effect the Media Sculpture had for the brand, even just as an idea.

The results of this survey were presented to the board in April 2007, together with reports of the progress made in the project development, including a time and cost plan. There was then a unanimous decision for the realization of the Media Sculpture project.

Photo © Bayer AG

The three construction phases

At this point, I will only briefly summarize the constructional implications. Our engineers in the building department, with Wolfgang Vogel leading the way, immediately reorganized all plans and commissioned the firms already on-hand with the dismantling of the highrise once it became empty in the summer of 2007. Construction work started on 1 October 2007 with the removal of building elements which were not needed. The initial idea to gut everything down to the steel skeleton could not be realized as the ceilings were made of cast-in-place concrete. The expenditure of time, money and work as well as the noise and dust pollution were not justifiable. Nevertheless, 10,000 tons of building rubble accumulated plus several tons of sorted materials. All non-load-bearing building parts were dismantled, including the eight passenger elevators – only the freight elevator remained. I wanted to preserve the interior furnishings of the ground floor (marble wall panels with fantastic vein patterns, natural stone floors, room-height glazing, custom-made lamps) as historic documentation and for potential usage later on. Several concepts for this already existed.

In the second construction phase, the building refurbishment, all load-bearing elements had to be made weather-proof once the approximate 25,000 square meters of steel had been cleaned with sand blasting. The 90,000 square meters of concrete surfaces were coated with an epoxy resin. Also in this case, preferably Bayer materials were used. The glass and aluminum curtain façade was kept in place as long as possible to protect the construction work from immediate weathering effects. After its dismantling, the process was delayed for several months due to storms and frost on the building components. As the last part of this phase, the "green heart" – consisting of approximately 15,300 square meters of 25 mm thick Makrolon® multi-skin sheets specially colored in five different shades of green – was assembled according to the design plan. The guideline was that no vertical construction components should be visible here

Photo © Ralf Krieger

because of the back lighting. This posed a static problem. The experts from the office Werner Sobeck proposed to insert the sheets between ceiling and floor by bending them. The amplitude was calculated in the wind channel. In additional lighting tests, it then had to be investigated how the disadvantage of the bending could be compensated with the type and arrangement of the back lighting.

The third construction phase, the assembly of the 18,072 square meters of mesh mats, partially with preinstalled LEDs, started on 1 June 2009. The mats were typically 4.03 meters wide and two floors high – meaning 7.20 meters – and were vertically spanned on a steel construction cantilevered in front of the building. The 684 pieces were delivered directly from the weaving mill in Düren in specially designed transport racks and mounted by specialized South Tyrolean assemblers with experience in mountain climbing. The installation was completed in September 2009. The electrical work with roughly 30,000 meters of cables, the installation of the hardware, the wiring of the control system (circa 20,000 meters) and the lightning protection measures took place parallel.

The prefabricated LED tubes, each with a length of 4.03 meters, conceptually have a total length of 79,500 meters. The total number of built-in LEDs amounts to 5,632,905 pieces. Five LEDs, ordered in the color sequence red-green-blue-green-red, make up one pixel.

Photo © Bayer AG

www.mediatecture.info/23

The concept for bird protection

In an early phase of the project, we began to think about bird protection. During research on this topic, we noticed that this problem is very emotionally discussed in Internet forums and publications. In the few cases where there were surveys and concrete figures about dead birds on media façades, the connection between cause and effect was not conclusively evidenced or the cause was not thoroughly investigated and differentiated. Scientific research was missing. Therefore, we wanted to find out something about the behavior and perception of birds with regard to light and color in general, and about their reaction to lit objects in an urban environment in particular. After a longer search, we came across a renowned scientist specialized in the cognitive psychology of birds and very interested in this topic. After examining the situation and the conditions with regard to the Media Sculpture, he differentiated in a first assessment the various reasons that can lead to dangers for migratory birds during migration seasons. He highlighted that migratory birds usually don't orient themselves with light sources rather they employ magnetic compass orientation. Only in weather conditions which disturb this orientation do nonlocal birds look for other options and could eventually be irritated by lights. The danger especially increases when the lights reflect directly, thereby making obstacles unrecognizable for birds. This correlates with the experiences of our company in the case of the mentioned big Bayer cross installation, which is turned off during the main migration times. The birds were not injured by the light bulbs, rather by the fixtures and cables located in the shadow. The scientist proposed a test set-up to attain more precise data. We accepted this proposal, got in contact with the local NABU – Naturschutzbund Deutschland e.V. (Nature and Biodiversity Conservation Union, Germany) – and asked for expert advice. The test will statistically document and conserve injured birds around the building in the autumn and spring migration seasons on a daily basis in the morning hours – while the Media Sculpture is turned off. The results will be used as referential data and later compared with the results of the following flights with the Media Sculpture turned on. Parallel, precise weather data will be recorded as well as information about the operation of the installation (light intensity, predominant colors, times of operation, etc.). When a sufficient amount and quality of data is obtained, the scientists will analyze it, and the results will then be evaluated together. From this procedure, we expect completely new and well-founded knowledge on the operation of media architecture, which could enable a differentiated discourse on the subject of bird protection. It is our goal to determine different threshold values, which we can react to with a change in the display program so that birds are not endangered. If it became evident, for example, that at a visibility of less than 100 meters and a cloud cover below 300 meters, red lighting is particularly dangerous, we could feed these data into the control software. With these weather conditions, it could then switch to a display of images without red elements (thanks to our detailed profiling) – and without human intervention. This method could set new standards of reference for many media installations.

Epilog

At the point when this article was written (April 2010), the project is not yet completed. All constructional measures as well as the content editing were finished on schedule. After the first successful test runs (which can be viewed on YouTube http://www.youtube.com/watch?v=7sS6FbTDDuo), stress tests were conducted with the result that more and more LEDs fell out every day. When it became clear that the failure rate was constantly increasing, the tests were interrupted and a comprehensive analysis phase commenced. As operational safety and sustainability play an essential role in our company, the analysis was carried out by the professionals in every necessary detail and without pressure. After several months with the involvement of experts and research institutes, a number of defects could be localized and further measures were discussed together. It became clear that in a new development of this kind, at a scale and complexity that has never been built before, and without a corresponding level of experience, not all factors can be controlled. The most significant factors were caused by environmental influences, i.e. storms, cold and changes in temperature or humidity. This is why these defects had not been detected during previous laboratory and quality assurance tests, but only after the stress tests on the building. In April, our contractors proposed an action plan: For the correction of all defects, a complete exchange of all LEDs would be inevitable. Taking into account the involved fabrication and test phases – in the laboratory and on the building – it is now the goal of the involved firms to finalize the project by the end of 2010.

Photo © Bayer AG

Identity

Identity and Public Space

This chapter addresses the question of how the identity of social systems can be communicated in public spaces. One consequence of modern building technology is that architecture now tends to be judged by global standards. The result is that public space is becoming the showplace of a global identity. But how can this same space also encourage a sense of regional and individual identity?

Professor Dr. Alexander Nützenadel *(born 1965 in Karlsruhe) studied history, economics, computer science and Romance studies in Göttingen, Berlin and Venice between 1984 and 1990. After receiving a grant to support research work at the Historical Institute in Rome he received his doctorate (1995) and teaching qualification (2004) from the Faculty of Philosophy of Cologne University, following his completion of a year studying at Columbia University in New York sponsored by the Alexander von Humboldt Stiftung. In 2004 he was commissioned by the Volkswagen Stiftung to lead a group analyzing "Globalization as a Historical Process". Between 2006 and 2009 he was Professor for Comparative European Economic and Social History at the European University in Frankfurt (Oder) and in 2009 he became Professor for Social and Economic History at Humboldt University in Berlin.*

How Does Identity Occur?

The creation of identity using architecture is certainly not part of your scientific work. But the creation of identity is, in general, a very important area for historians. As I am dealing with the creation of identity in my work I am very interested in understanding the process whereby identity is created in social communities. How does a historian actually define identity?

The first thing is to question the classical notion of identity. For a long time it was assumed that a community developed common cultural and social notions which made it possible for its members to associate with a collective being. Social communities have a self image which has evolved throughout history - often in step with advances in the community. History presents us with many examples of processes of the creation of identity which, through their close link with ideology, have had very negative consequences. The result is that we have become skeptical of whether the creation of identity really works in such a way. Now we speak of "fragmented" or "plural" identities. We are also not exactly sure how such processes for the creation of identity are triggered. Is this process facilitated by a maximum of homogeneity or is it rather that precisely those societies which lack this homogeneity have most need for such an identity?

Or, in reference to today's situation: do societies in a globalized world require identity to a special degree? Must we always be creating new identities and is this, perhaps, a very individual process, which differs from person to person?

You have already mentioned that identities developed in the past differently to the way in which they develop in today's globalized world. How did earlier cultural and political communities articulate their identity and how is the situation different today?

One crucial difference is determined by the question of whether the context is one of a stable collective structure – a family, village community, regional structure or tribe – or one of greater spatial mobility and social change. Earlier, cultural self assurance came from story-telling, a language and a cultural environment, and this self-assurance is naturally threatened if such traditions are no longer there due to the fact that one has changed one's location four or five times. Where does identity come from? Identity is much more fragmented and fragile today and one could even question whether such a process of the creation of identity can still take place today. At the very least we can say that identity works very differently today to the way in which it did earlier.

Can one therefore say that, earlier, identity was linked to the concept of home?

Of course: home is certainly a key element of the process of the creation of identity. Even nations use the concept of home in the sense of a politically constructed space which is then equipped with particular symbols and narratives. But this is completely different from the more organic sense of home.

Does western culture also not provide a meta-level framework for the creation of identity?

Of course, but the question is whether this notion of the West has developed organically, from within, or whether it was much later created by intellectuals. At the very least this shows us that, earlier, there were clearly different levels of identity and that it is very difficult to determine if an identity is organic or invented.

Which political and cultural parameters have enabled the development of particularly stable identities?

The factor of time is certainly very important. Something which exists for a long time is certain to affect this identity. And experiences – even everyday ones – which are shared at special moments also bring people together. Going to church on a Sunday morning or a National Holiday – these are such special moments. They create the sense of security which comes from being part of a community. Indeed, such holidays are a very important basic element in the process of creating identity.

And yet today's public holidays are, for us, largely empty of meaning. Few of us have any relationship with the idea of how a particular holiday came about.

Yes – and studying this change could just be a way of finding out how this process of creating identity is changing today. If holidays no longer have any role other than offering us some extra free time then it is perhaps other things which have become important instead.

But, at the end of the day, it is very important for everyone to feel that they belong to some sort of community.

Definitely, and the question is whether, for instance, it is consumption which has replaced older identity structures. Maybe it is the act of consuming which holds society together.

This is probably a development which is very unconscious because so many people distance themselves from consumption.

Perhaps – but perhaps not: the rejection of consumption is often accompanied by a very definite sense of enjoying the act!

On a Saturday shortly before my meeting with Alexander Nützenadel I visited the market on the Prenzlauer Berg in Berlin: a very prosperous neighborhood developed out of the old city structure. The local population is largely well educated, ecologically aware - and successful. The market is correspondingly attractive: a world of luxury in an alternative cloak coupled with extraordinary opportunities to browse and purchase in a very healthy atmosphere. There are also, incidentally, a few vegetables. The pedestrian streets around the market are home to highly individualized shops and boutiques. And unbelievable numbers of children – and, naturally, of fathers too, who finally have an opportunity to play their proper role. And, while they may still radiate a sense of the hectic of their work, their outfits are completely cool. The activities for children are without end – not just playgrounds but also highly individual shops – such as utterly cute children-only hairdressers.

One can afford to have children and one does this very consciously. It is impossible to avoid the impression that these children are status symbols. At the oyster stand (I myself am also completely susceptible to the aura of the market and the neighborhood), an appropriately cool and almost shambolic thirty-something male stood with friends and complained with an impressive sense of suffering that he had been chased from his home because a photo-shoot was taking place with his children. It doesn't get any cooler than this.

I think that this is a background against which it is fully appropriate to discuss the notion of the terror of commercialism…

Can globalization really create a universal global identity?

It has long been assumed that globalization is accompanied by a pronounced trend towards homogenization. One spoke of "McDonaldisation" or "Starbuckisation", meaning that a global standardization of consumer products and, particularly, of their presentation was underway. Eating is very important for identity. The suspicion is, that one result of globalization will be the disappearance of regional differences in consumer culture. However, it is becoming clear, that global products are actually adopted in different regions in very different ways. For many years, for example, there was no McDonald's in Rome and, when a branch finally opened in the historic centre in the 1980s, it was fitted out with Roman mosaics. Consumer behavior is also different. One certainly buys standard products but one also complements these with many local products from, for example, weekly markets. And global brands also have different images in different regions.

In my work for corporations I experience how incredibly important it is to be perceived as part of the globalized world.

Naturally: but belonging to the globalized world is only one issue. The question, however, is whether we are really talking here of a truly trans-global society or of some much more specific form of participative communication. For example, within these communications processes, regionally differentiated consumption practices can develop. Hollywood leads to Bollywood. The basic technical situation is the same due to the global context but the question is much more one of cultural differentiation. And this leads further to cultural feedback – as exemplified by the strong influence of Bollywood on film production in the west.

In this context it can be pointed out that globalization has yet to create values of the sort which were always central to earlier processes of the creation of identity. People long for communities and, in this respect, globalization is a disappointment. One accepts the economic realities and is happy to take advantage of the resulting mobility and flexibility but the soul remains unmoved. But is the creation of identity not essential for societies?

Well, one can observe the renewed importance of religion. And, in addition to this, perceptions of globalization in Asia, Europe or America are very different, which is why one speaks of “multiple globalizations”. Perhaps the result is that, once again, different identities will again develop.

Can you attempt to look into the future and address the question of whether future societies will still need superordinate senses of identity or whether these identities will simply be broken up into a series of spontaneous developments? I also ask this bearing in mind the fact that religious communities come together – particularly in big cities – but have little interest in mixing.

The fact that religious communities do not want to mix is certainly not a new one – this was also the case earlier. And whether fragmented identities can lead to a superordinate identity is also a question which I shall not hazard to answer. It is clear that such a desire exists in societies with multiple identities but who knows what this means and whether it will ever be more than just a wish. Perhaps such phenomena as Facebook already represent the basis of a superordinate identity. The result could also be completely new forms of identity.

What is the meaning of the differentiation of cities into individual districts for this question of identity?

A very large one. This is actually a modern phenomenon. One believed for a long time that the sense of belonging to a district was rooted in the past. But this notion is contradicted by the fact that, as cities evolve, new districts are constantly being created and that these districts can develop a collective consciousness which was simply not there before.

Many thanks for this conversation. I hope that this spontaneous collection of thoughts will make a contribution to the development of design in an urban context. Urban design will only be able to support - or even lead to the emergence of – a sense of identity if we have a contemporary understanding of what this process for the creation of identity really is.

Urban Identity

All over the world there is a growing demand for the electronic animation of architecture. There must be a reason for this. Let us first examine what electronic medialization is planning for our built environment:

A fully static structure such as a façade suddenly starts to move. This movement has nothing to do with the functionality of the structure, it is fully free and uncontrolled – it is magical. A spell is cast upon the skin of the building separating inside from outside. And any moment something new could appear.

The media façade offers the chance to emotionalize our living space. Perhaps this is connected with some undefined longing: a longing that our built environment is not only functionally-rationally oriented but will also become a place that is able to touch our complex souls somewhere much deeper.

More recent developments in architecture reflect a growing need for more complexity. Perhaps we need living spaces which are more attuned to our inner tensions and sensibilities. Perhaps we should look for authenticity between our inner and our outer worlds?

We are all involved in a search for meaning. In the context of this search, precise constructions have an immensely calming effect, bestowing a coherent logic on at least part of our world. It is for precisely this reason that architecture and urban design have such an effect upon us. A clearly organized urban space can offer us support which we probably cannot find from within. This explains the widespread nostalgia for the sort of familiar patterns which can often be found in urban layouts from the past. At the same time, however, an urban order which is too closed can also lead to inner unease – in the case, for example, of an urban area which is so nostalgically built that its very existence is clearly in conflict with the reality of our individual lives.

The search for meaning is always related to the search for authenticity. We have the feeling of being in a proper context when we are in harmony with our environment. There is a huge contradiction in our human character: on the one hand, we require clear fixed points which we can hold onto and at the same time we defend ourselves against such order when it appears to no longer have any authentic relationship with our inner selves.

And this explains why everything which remains open, incomprehensible or imprecisely defined is so extraordinarily attractive. This is why we love everything which has a certain magic. Every secret experience hides the hope that our life is not yet fully defined but rather still full of possibilities.

But the contemporary reality of our built environment is that opportunities to experience such magic are rare. Economic pressures are too strict. The cities which have been touched by modern architecture tend to be a somewhat random series of volumes whose general effect seems highly unlikely to be the stuff of experiences. And it is correspondingly difficult for architecture to develop a regional authenticity.

Architecture is naturally full of examples of spectacular buildings whose aim is to give a sense of identity to especially important places or events. An example of such is the stadium in Durban designed by Volkwin Marg for the 2010 World Cup. The process for creating the identity of this project is very well explained in an article by Bartholomäus Grill entitled "Yes, Afri-can!" in "Zeit-Online" (2/2010). He writes, "The fathers of the city who renamed Durban eThekwini wanted an unmistakable icon of African identity, a monument capable of registering in the global consciousness like the Sydney Opera House or the Guggenheim Museum in Bilbao. At the same time, however, the stadium was to express the social cohesion engendered by the growing together of the Indian, black African and Boer and British white South Africans following the end of apartheid. On the other hand the prime minister of the state of KwaZulu-Natal envisaged a rural, mythical Africa: Zulu dancing, lions, wide Savannas – that is to say, a series of symbols which continue the colonial notion of the wild and paradisiacal continent."

The architects from gmp and their South African partners did not even try to combine these diffuse and contradictory concepts of africanness in a single concrete form – this would have been too much like squaring the circle. In the end, they created a sculptural building with many layers of meaning. One can see it as an over-dimensioned Kral. Or as a whale which has ploughed its way out of the sea. Or as a Zulu shield. The parabolic arch over the stadium is formed from two limbs which melt together, quoting the Y from the national flag as a result of which it could be taken as a symbol of the multi-ethnic rainbow nation. Volkwin Marg leaves the interpretation to the viewer. For him it is important "that the clarity of a metaphor does not overwhelm the structural tectonics of the architecture". This stadium is "purely optimized construction which combines engineering and architecture". I ask myself how this building should reflect the identity of South Africa. The implied metaphors can only be seen with a lot of goodwill and their explanation is more of a fig leaf for the claim that the stadium design has addressed aspects of local culture. This is a building which could stand anywhere in the world. Naturally, this is architecture which imbues a sense of magic and makes possible a whole host of associations. But only because it has no desire to exude a sense of individual identity, preferring instead to principally articulate the idea of the modern (see chapter "Spatial Concepts and Augmented Space"). But what does the idea of the modern have to do with Africa?

In Europe, we obviously have the substance of our historic buildings which create an atmosphere with which the locals are happy to identify. This explains why so many cities are happy to concentrate on looking after their cultural heritage. And this is a good thing. But this is no solution to the problem of urban development. Cities are often managed as if they are nothing more than historic substance but this is no more than a symbol of the dilemmas that they face.

A social group can define itself better when it is able to develop some sort of shared understanding based on internal coherence. This codifies modes of behavior and generalizes internal communication structures. Furthermore, this sense of closeness is also apparent to participants in the communication processes who are situated outside the group and it lends momentum to many of the group's economic and cultural processes. One result of globalization is that this phenomenon is becoming increasingly more important. Cities must distinguish themselves if they wish to attract investors. The advantage of a location is often defined by the success with which it attracts well qualified workers and they are obviously flattered if highly self confident cities are attempting to appeal to them. Soft values are unquestionably important!

A medialized urban structure permits the presentation of a special profile which communicates the economic and cultural self image of a city. Investment in the dynamics of electronic media is certainly a very efficient part of this process because such media can be almost spontaneously adjusted to the latest developments.

Many cities have a face which can no longer be understood simply by looking at its urban spaces. Medialization is a perfect method of revealing and then thematizing such aspects of a city's history.

However: communication is a tough business. If façades simply start blabbering, with each façade fighting to project the most attractive content then this will quickly result in a kind of white noise. The messages from a multiplicity of media façades which stand cheek by jowl in a restricted urban space will, if there is no attempt to coordinate the content, generally neutralize each other.

Such communicative and visual overkill can of course be a stimulus in itself, as is shown by the example of Times Square. Here bigger, faster, further is a meaning in itself and competition is the identity. But this recipe cannot be repeated everywhere.

Or perhaps it can? It seems as if Asian culture has absolutely no problem with this idea. I have to admit that the complete overload of signs in Tokyo or Shanghai fascinates me, but that I cannot start to understand them. I cannot imagine that individual advertising elements register any sort of economic return when each can hardly be distinguished from its neighbor. The only possible explanation appears to be a cultural one. It is simply good form to illuminate free areas of façade on a building in such an intensively commercial part of a city. The understanding of this behavioral phenomenon would be a very interesting subject for a cultural research project.

In my deliberations, however, I have to leave these questions aside simply because I can only really address aspects of the European culture from which I come.

Admittedly, we representatives of European (and, equally, American) culture only recognize this (to put it very strongly) Times Square principle of the integration of media in the public realm when we consider the combination of media in urban spaces. Those who place advertising in public spaces appear driven by no more than the ambition to occupy as large a part of human attention as possible. And when a location already contains a lot of advertising, then the next panel must simply be even bigger. This is how the idea of blow-ups developed. This is simple cut-throat competition. Nobody thinks in terms of communicative relationships targeted at the perceptive abilities of passers by.

The difference with Asian culture lies in the power of the city authorities to approve – or reject – the placing of advertising in public space. This is all that happens here. Planning officials are torn between political pressure on the one hand and potential advertising income on the other. Politicians are always well-advised to speak out for the purity of the appearance of the city – whatever that may be. And the paradox is that advertising is as frowned upon as it is longed for. Frowned upon because it is an expression of filthy commercialism and longed for because it is an expression of the economic importance of a location.

The fact is that the design of medial symbols in the public realm is never subject to an overall content-based and cultural concept. The simple question is, rather, whether one is allowed to build them in a particular location or not.

Medial Design Rules

The problematic of the aesthetic coordination of individual objects as a means of achieving a coherent whole is of course a well known feature of the architectural discourse. The rebuilding of Berlin, for example, witnessed many disputes arising out of this issue. The narrowness of the planning regulations certainly drove many architects to self doubt, but these same regulations also exemplify the dilemma regarding the question of which cultural ideal should form the basis of such rules. Of course, planning regulations offer the democratic opportunity to determine the individual identity of a location. And it is quite natural that such an identity should relate to the history of a place, which means that the tendency to use planning regulations as a means of establishing connections with earlier, dominant, classical building styles is quite logical. The problem is, simply, that the result is rarely satisfactory because such demands tend to be stylistic rather than qualitative. Here is an extract from the Berlin Planning Regulations referring to Unter den Linden and its surroundings:

"Of particular significance for the appearance of the street 'Unter den Linden' are the form and dimension of the window openings. In reference to historical examples the aim here should be to have vertical openings and a façade which is at least 50% closed."

This is an expression of naked helplessness. How can such regulations lead to an architecture which creates identity? Each façade which investors bravely build respects these rules, but the result does not begin to be a reference to the historic quality of the street, because the chosen type of façade simply reflects standard and widely spread cheap building methods. This really has nothing to do with creating identity.

The real problem with the making and applying of regulations is the lack of a real discussion about the values that could enable a building to express an identity. If the basis of the thought process does not get beyond the mechanical relationships of the openings which façades are permitted to have, then a dynamic discussion of current cultural values is not possible. It is simply impossible to express our relationship with our historical identity in terms of the size of window openings.

The difficulty of addressing this issue of identity can also be seen in the discussions about the rebuilding of the Stadtschloss in Berlin. The extremely sensitive issue here is the orientation of a society which has enormous problems escaping from linear, industrial ways of addressing social processes, even when these social processes are organized in non-linear, parallel structures. Many "truths" are simply no longer valid. And architects, who tend to be trapped at the level of structural and constructional issues, have difficulty in addressing such issues as social change.

This is precisely where media façades can offer more options: they can counter the constructional insularity of façades as a way of bringing them closer together with each other and their urban contexts. To do this, one simply needs to use media techniques to synchronize the projection software of the various buildings. The result can be that the urban space recovers a quality of sensually perceptible cohesion.

It is for this reason that I propose the development of medial design rules. And these rules must have a completely different structure than standard design rules, because their first role must be to describe the mentality of a software. This represents a paradigm change. As ag4 realized the first transparent media façade for T-Mobile in 2004, I was only able to do so by drawing up a building application in consultation with the city authorities which simply regulated the mentality of the media content. This was, in my opinion, an historic moment. Here was a project in which not only the mechanical size and position of the superimposed façade but also the process for playing pictures upon it were regulated in the planning permit. The city was granted the right to shut the façade down if the agreement between it and T-Mobile covering the mentality of the images was broken. Only by handing over such power was it at all possible to gain permission for the façade. And quite rightly so: the city can only become involved in the design process when it has the right to get involved in the culture of content. If a city does not grant permission for a façade then this is largely because it does not know how it will be able to have control over the atmosphere in the city which could be provoked by the contents of that façade. The potential of media façades makes cities nervous because they can completely change their appearance in an instant. When there is no engagement then the result can be chaos. And this is also contrary to any attempt to establish identity! To this extent I can understand the anxiety of many cities with regards to media façades. The problem, however, is that media façades are becoming ever more the subject of hype, with the result that many cities are coming under such enormous pressure that, as in the case of basic billboards, this is once again becoming a simple question of power. And this is clearly no longer a simple European problem. In New York, away from Times Square, bitter disputes are being fought over LED boards.

An agreement over a series of media design principles could form the basis for medial design rules. The establishment of such principles is, of course, completely new territory. The first demand must be that medial content is never used in a linear form. The reason for this is that constantly repeating, closed loops of clips and serial images make impossible any attempt to creatively organize the projection because the cost and complication of changing such complete sequences is simply too great. The content must always be put together in a modular way in order to allow it to be creatively programmed. This creates the potential for interactivity. This could make it possible to create some sort of hands-on link with passers-by but much more interesting is the possibility that individual modules can interact with each other. This would permit the behavioral interaction of several media façades.

Here is a first proposal for a set of media design rules:

Preliminary: *The chapter about media façades shows that there are many ways of designing this new form. Not all work with simple series of images. The following proposal for a set of medial design rules concentrates primarily on such image-based façades because it is images which have the most intense psychological presence in urban space. Such images are also at the heart of advertising. As it is automatic that medial design rules must cover all media, the approach to advertising is naturally a very important theme. Of course such rules could allow any city which chose to do so to simply apply a blanket ban on advertising. It would however be unrealistic to think that such a ban could apply everywhere.*

Aesthetic Regulation of Content

All uses of moving images by electronic media should be regulated - and this applies especially to advertising in public spaces. The use of advertising clips is problematic because they, generally, consist of a closed, narrative dramaturgy. This structure provokes, for example, psychological problems in traffic due to the distraction of car drivers. This means that the individual clips must be carefully produced so that they do not attract the attention of any viewer for too long.

Depending on the size of the media façade (and especially in the case of large façades) care should also be taken that movement in the images does not sensually overwhelm the viewer. A method must be found for defining acceptable levels of such movement.

And yet the evaluation of such medial effects is culturally highly differentiated. A media concept which is considered lurid and stressful in Germany is seen quite differently in Asia or the USA. Even within Germany I have experienced very different evaluations of this question.

A media façade with a high enough resolution is an instrument which offers no inherent control over the mentality of images. Everything is possible. And this is precisely the fact which troubles planning officials - quite understandably, in my view. If everything is visually possible, how can the planning department use their work as a means of stabilizing the individual character of an urban place?

A set of medial design rules can now define basic aesthetic rules for a particular location which take into account its cultural background. If a planning department has the opportunity to influence aspects of these aesthetic rules, then the approval of media façades and urban screens will become a creative act. An example of this is the possibility of individually designing the colors of the content, in that a city develops a palette of colors which can be used in the development of projection software. Of course these rules should not be so draconic. It must remain possible to incorporate corporate logos in content and the projection of autonomous art must also be exempted from the strict color rules. However, all other levels of content such as information and the emotional and artistic backgrounds could be developed using the palette.

This palette of colors could give the city a certain atmosphere by being adapted to the seasons. We can see that there are many possibilities for using medial design rules to creatively influence the aesthetic effect of a medially orchestrated space.

Controlling the Content

Firstly, rules can determine which sorts of images are, generally, permitted. Sexist or violence-inducing images could, for example, be excluded.

Furthermore, the authorities can use tailor-made rules to divide the city into different zones in which images of different mentalities are permitted. Such rules could determine, for example, whether direct product advertising or merely product-related images were permitted. The most reduced form of advertising is the sponsorship of the cultural content of a media façade, in which the sponsor's logo may be incorporated according to an agreed rhythm.

Such design rules can also determine that advertising can only be shown on the media façades of a certain zone at certain times. This regulation would mark a clear difference from conventional advertising hoardings.

Characterization

Medial design rules can also make targeted use of the potential of media façades as a means of creating emotional spaces. This would represent the birth of a completely new sort of urban planning which regulates not just the physical but also the emotional spaces of a city.

What does this mean? A simple example to illustrate this idea is the analysis of the organic development of individual city districts from an emotional point of view. In doing so one must ask how these districts feel, e.g.: lively, chaotic, chic, multicultural, dignified, introspective, expensive, off-beat, etc.

The characterization of urban zones is not difficult and it is also not difficult to incorporate these soft values into planning policy. The challenge to the medial design rules is to transform these characteristics into instructions for media design. Such instructions cannot take the form of simple equations – instead they should resemble an emotional and thematic masterplan. How to achieve this can probably be learnt by looking at the work of creative agencies. These agencies are used to transforming intangible characteristics into structures and principles. They are used to targeting concepts at the mechanism of the psyche. And they know how to develop a "masterplan" which shows the way ahead without concretely prescribing anything.

This work is clearly a decisive area for mediatects, and it could clearly lead to mediatects structuring themselves as creative agencies. The interdisciplinary nature of such an agency is, naturally, very important.

Such emotionally oriented medial design rules naturally also establish standards for the creativity with which media façades are designed. If an urban zone is, for example, seen as "dignified" then the rules will hardly permit the creation of a cheeky Youtube atmosphere.

Such rules make it easier, on the one hand, to counteract a sense of arbitrariness. The entire medial content of a district is now networked and the individual elements should not be in competition.

On the other hand, such emotional standards can never be definitive. In contrast with normal urban design regulations the instrument of medial design rules can be very dynamic. If the characteristics of a district change due to a change in the population structure, then the emotional content of its medial infrastructure can be adapted accordingly at no great cost. Interestingly, this principle can also be reversed. If it is desired that, for urban planning reasons, the development of a district should be adjusted (that, for instance, a boring banking district should develop a lively multicultural life) then carefully targeted instructions to the medial content can contribute to this.

One can reasonably quickly identify what one wants to change in the atmosphere of a district and, if one adapts the real estate offer accordingly, the change in the population structure can be very quick.

The adoption of medial design rules gives cities a powerful design instrument: An instrument with which the internal atmosphere can be controlled and the external effect of the city directed. If a city communicates a recognizable identity to the outside then this, in turn, has an internal effect. This new external image of the city is reflected back onto its citizens and this simply intensifies the process whereby a city gains its identity. And the citizens become more self assured. One actually lives somewhere! And one isn't lost in the utopia described by Bazon Brock: "Utopia describes a situation in which nowhere is special precisely because everywhere is."

It is for precisely this reason that the use of media art makes so much sense – but at the same time it must be differentiated. The common demand of artists and, in particular, of curators that art must be autonomous leads to problems when art is chosen for a location with which it simply has nothing in common. This represents, once again, the danger of arbitrariness which is also not good for art. Public spaces are not anonymous neutral museum spaces which disappear effortlessly into the background of the art displayed within them. A self-referential work of art which requires a special reception will only be irritating in a public space.

It is currently very "modern" to organize media festivals in public spaces in which videos are projected onto façades by night and many cities are keen to erect a media façade in a certain location because there is a hype about such structures. It is important, however, to be clear that, in doing this, a city is only producing a tool. Exactly the same was the case in the 1980s when "planar glazing" made it suddenly possible to build fully glazed façades without frames as a result of which everyone wanted such a façade for their building because they could afford it. But this is no way of defining good architecture. If the design potential of this new medial tool is not recognized, then the hype will soon become a bubble and, like all bubbles, this one will burst.

***Professor Albert Speer** was born in Berlin in 1934 and, after working as an apprentice carpenter, completing his school-leaving certificate at night school and studying architecture in Munich he now specializes in the planning of entire cities and regions worldwide. For more than quarter a century he was Professor of City and Regional Planning at the Technical University of Kaiserslautern. His practice AS&P Albert Speer & Partner GmbH which he founded in 1964 recently presented both a masterplan for the centre of Cologne and the Plan of Action for Frankfurt 2030 and is currently working on preparing Munich's bid for the 2018 Winter Olympics as well as on large urban planning projects in Cairo and the Megacities of China.*
In 1995 he founded the Professor Albert Speer Stiftung. He has been a member of the National Committee for Urban Development Policies since 2008. Albert Speer is a recipient of Goethe Badge of the City of Frankfurt, the Great Architecture Prize and the Cross of Merit with Ribbon.

Urban Planning & Processes of Identity

Is it still possible today to plan cities in line with a regional culture or are the demands of urban planning largely globalized?

I would like to say very clearly that urban planning is very important for a regional culture because the uniqueness and character of a city will play an even larger role in the future and that architecture is hardly in a position to support this role. The demands on architecture are reasonably constant across the globe; ideas about living or about how, for example, a clinic should look are becoming ever more similar.

But at the same time we are living in a world in which the competition between cities and regions is steadily increasing. This can only mean that such soft factors as quality of life and the landscape of a city will play an ever more important role and I am convinced that the debate about such themes in the future will be much more important than the architectural debate itself.

So you can use urban planning to move a city back towards its roots?

We can very clearly encourage cities like Cologne and Cairo to rediscover something of their original history. In future, cities must do much more in this direction. With the exception of a handful of architectural crown jewels these cities are marked by a very generic architecture, most of which was hardly designed by architects at all but which constitutes 95% of the built fabric. And these buildings are so similar that the difference between such cities can only be perceived in terms of urban culture.

Let us take the example of Cologne. Which measures could you implement to reinstate the city's individuality?

We have come to the conclusion that a large part of the history of Cologne is simply no longer part of the collective consciousness of its citizens. An example is the Cologne Ring which was built by the architect Stübben in 1885 and was something of a model for Vienna. The city saw itself as a major city. But if you look today at how the buildings of this ring have been altered, how parking is organized, how the street lamps are being constantly altered and how the green spaces are overtaking the area, then the original experience of the "boulevard in a major city" has been completely lost. This experience must be recreated – and the people of Cologne would be truly astonished by what we could tell them about their city.

Such measures bring results because they encourage both investors to rekindle their interest in a location and owners to invest in their properties. Restaurants and shops appear and a whole region can begin to feel the effects.

What you are describing here is a new way of looking at public space. What is the meaning of public space in Europe in comparison with Asia and America?

That is very varied. Here in Europe we are in the fortunate situation of having public spaces which have developed over centuries – or even over 2,000 years. If you take Freiburg, for example, then the city plan today is still the city plan of the Staufer. That is the strength of European cities and it is something which you do not find in, for example, the USA. And climate of course also plays a decisive role. In the Arab world rooms are always oriented inwards and the only large space is the square in front of the Mosque. In such countries one should not look for public space at the European scale. That is a very different world in which the medialization of public space, in the way in which you do it, must play a completely different role.

That is precisely what interests me. What is the meaning of public space in China? We tend to think of democracy or of such images as "Speakers' Corner".

The Chinese have three traditions. The first is the street of shops which is also a street of houses. Every Chinese city has these. And then there is the idea of the Imperial City with the palace in the centre, and here we think particularly of Peking with its Forbidden City. This centre is always related to a clear North-South axis which, for the Chinese, is a symbol of eternal life and this axis ends with the Pole Star. This is a mythological experience and, in reality, the streets are often not continuous. But for the Chinese it is the reference which matters.

So this is a virtual space that the Chinese are seeing?

Exactly

It is a space which really exists; for every Chinese citizen?

In principle yes or, at least, for every educated Chinese citizen.

I would also like to mention the third tradition. This is the urban park in the centre of the city which is enclosed on all sides as a result of which one has to pay a fee to gain entry. This tradition was around long before the communists.

But urban projects in China today incorporate enormous open spaces. Are these spaces, for the Chinese, nothing more than a means of organizing traffic?

No, this is changing. Globalization and increasingly international networking have enabled the Chinese to see their cities in new ways. We have to differentiate between the traditional and current understandings of public space. But an interesting continuity can be found in, for example, the fact that only the Emperor was allowed to look to the south while the people had to look to the north. This tradition was adopted in its entirety by the communist leadership and determined, for example, the orientation of Mao's coffin. Communism and the Cultural Revolution have been blown away like a gust of history. This is the old China that we are seeing today, the old structures. And the way the people are taught to think goes back to Confucius.

I would like to return to public space. Can an architectural approach shape this in such a way that it reflects the identity of the local society?

Architecture can hardly do that alone. There are naturally such extraordinary buildings as the Olympic Nest in Peking or the Guggenheim Museum in Bilbao, which express a special need and attract visitors. But this is also an expression of the society in question. In general, architecture is hardly in a position to define public space and the question of what is found in these spaces is much more important. You can learn much about a society from how it uses such spaces. Does it dedicate a public space for traffic or are activities encouraged in the space as a means of improving its quality? The question is how a space is used.

Are cities and city regions becoming more active in the area of exploring their identities?

They are all becoming more active: first the larger cities and now the smaller ones. In the light of our success with the masterplan for Cologne several smaller cities are now looking to such a restructuring as a means of becoming more competitive. The strengthening of city tourism is one motivation for this – and a city which receives more visitors is also more interesting for investors. For this reason, culture is also important. And the result is an interconnection between the quality of work and the quality of life.

Another reason for this is that society is aging and that older citizens are less willing to be shifted out into the suburbs. They want to live in the city with which they identify.

This comparison gives me the opportunity to discuss the competition project that we did together for the Media Avenue in Peking. The objective of the competition was to find a concept which would help Peking to attract international media companies. For this reason we employed a very global pictorial language on our media façades. Today, I ask myself very self-critically if this approach was fair to the city - although one could of course counter with the argument that Peking is a city which has very consciously cut its ties with the past and oriented itself to the global world through, for example, the use of a correspondingly global architecture. My vision is that the combination of this global Media Avenue with media façades which speak a more regional pictorial language could lead to a more appropriate hybrid between "global" and "regional". This would mean that, on the one hand, architecture and urban planning would continue to follow international trends but that, on the other hand, the medialization of the façades would permit the creation of an emotional-sensual atmosphere which reflected the mentality of the local people. Do you think that such a concept could have a chance?

First we must differentiate. In that Peking project we had very little space in which to attract the media industry – a task which I think is as important as it was before – and this space would naturally have ended up having a different character than the rest of the city. But our aim wasn't to change the face of Peking – just of this district. We have, incidentally, attempted to sell the same idea to other Chinese cities but so far without success.

The question is also, naturally, how the façades look in combination with such a media projection. The lettering and the specific way of working with symbols will automatically give a local touch to the sensual effect of the urban space. This parameter must be encouraged. The important thing is to explore the individuality of the city – which can be very well served by this superimposition of architecture and media.

Which means that the individuality of the city is a software?

Of course! One can, for example, experience this software wonderfully in Hong Kong at Christmas when traditional symbols are twinkling everywhere. This creates a very special atmosphere in the city. But it is not something which can be done permanently – it must remain something special.

At all costs! But I am also interested in the use of media in normal urban situations. This is why I have been pushing for some time for the creation of a sort of Media Design Charter. It is precisely through working on such a charter that cities will come to recognize their own uniqueness.

I agree with you completely. This would also be an important urban design instrument which has yet to be fully understood.

Of course I also work in the hope that new ideas will not always be initially met by an unconsidered "no". In my work it is always incredibly exhausting to negotiate planning permission for media façades because those involved in the process often don't begin to understand the design potential.

I can give you a good example. Twenty years ago we suggested that the City of Frankfurt could reinforce the potential of the Museum Area on the banks of the River Main by using a lighting concept which would change with the seasons. Only now is the idea moving towards reality. And this difficulty is only partly to do with media. It is extraordinarily difficult to explain that an instrument like new media can improve the quality of urban spaces because people always associate it with such old chestnuts as Times Square. People, unfortunately, only trust what they can see. You are not the only one to suffer from this – we do too! The only solution is to go the extra mile – to take every opportunity to talk, patiently, with everybody involved.

Which is an exact description of my aim in this book! Many thanks for the discussion.

In 2006 the City of Peking invited several architects to participate in an ideas competition. The requirement was to find an urban design concept that would attract the global media industry to develop a presence in Peking. A framework should be developed which would entice major media companies to establish significant operations in the city. This was the period of economic opening and, above all, a period in which the global ambitions of China were advancing on all fronts. It was simply impossible to be too ambitious and, not satisfied with simply watching the extraordinary advance of China, global companies wanted part of the action.

Probably not even the Chinese could explain the detailed economic concept behind the Media Avenue. The important thing was growth! It is likely that the Potsdamer Platz in Berlin had inspired the idea – and the competition - in some way. But here, everything simply had to be bigger.

Peking Media Avenue

An Ideas Competition by Albert Speer & Partner with ag4

With his North-South axis, Albert Speer had already created one important project in Peking on the basis of which he was invited to take part in this competition. In the case of the Media Avenue project his objective was to give this axis a medial quality. And thus we joined forces.

Speer's concept envisaged, from a very early stage, the creation of a large square in the middle of the avenue. This was to be a major transport hub incorporating an underground metro and railway station.

In our work together it soon became very clear that this square offered the best possibility for some sort of medial projection. And yet its dimensions were extraordinary and it simply could not be compared with our European notion of public space. As traffic in Peking is such a major challenge, it was also important that the square did nothing to disturb the flow of cars. For this reason, a project on the scale of pedestrians wandering around the square became quite unthinkable. Hence: the target group for the medial content of the project became the car drivers.

This naturally had effects on the project in that we had to base our work on the perceptive ability of the moving viewer.

The Creation of Identity in China?

First of all we had to address the questions of, on the one hand, the creation of identity and, on the other hand, the cost effectiveness of the project. It was quite clear that the concept must ensure that the installation would bring economic returns.For this reason, we were very driven by the idea of developing a medial language which could articulate a global identity. We assumed that the square would be a meeting place for companies from across the globe and, hence, become a nucleus of the global media industry.

In writing this book I began to doubt if such an expression of globality was appropriate in a city such as Peking or whether this was not actually a form of aggression against the place and its people. Should one not pay more attention to the historical built substance – especially in the content of the medial installation? But there is hardly a city in the world which pays less attention to its historical built substance than Peking. Everything is based on modernity or, at least, something which one understands as modernity. And what, therefore, is the identity of the city now? Establishing the identity of a people that cannot express itself freely is a difficult task. And my understanding of identity is also probably problematic when one considers that in Chinese culture individuality is, in any case, valued completely differently as it is in the West. How can an approach to identity be developed? My normal solution would have been to propose the creation of a working group with representatives from all levels of culture. But that would presuppose the existence of a culture of discourse and, from my own experience of doing business in China, I have yet to see evidence for such a willingness to discuss openly.

I discussed this subject with Albert Speer during our interview. His opinion was that Chinese culture would immediately find a place in the medial content of the square because Chinese characters would doubtlessly play a part. To this end one has to be aware that these characters function like images – which is already an illustration of the extent to which the people in this culture have a completely different way of thinking: a way of thinking which we sometimes find it very difficult to understand.

On the other hand, China has both a huge interest in western know-how and no inhibitions about mixing its own culture with that of the west. For this reason I felt that the concept which we had developed was fully adequate. It offered a framework which, over time, would doubtlessly both create its own internal substance and develop further. And this is of course the great opportunity offered by medial productions – they are always flexible and can always develop further.

The Concept

The circular form of the square offered the opportunity to create a panorama. The adoption of a traffic roundabout with a medially reinforced circular form as the external edge of the square supported the dynamism of the traffic movement. Hence we were able to combine the proportions of the architecture with a media façade which incorporated all the buildings around the edge of the square. An architectural approach was established with a six storey unifying base which precisely followed the edge of the square. Only above this level were the architects really able to do their own thing. That such an accommodation could be arrived at by architecture and medialization is a testament to the constructive cooperation between AS&P and ag4.

The project combined an idea – of creating a panorama with a 360° projection – with a necessity – that this projection must at least be six stories in height if it was not to be lost at the scale of the square. That one could envisage investing the amount of money required for the media technology for a project of such volume may appear to be completely unrealistic. And this could well have been the case if the idea had been to project videos in bright daylight, but for a solution based on the projection of images at night onto Illumesh®, the costs would have been much more credible (See the Bayer Media Sculpture).

In order to intensify the traffic flow around the square it makes sense to develop the movement of the medial content in the same horizontal direction as the cars and to establish rules for this movement. To this end, a dynamic structure was developed using areas of color which move and change at differing rates, overtaking and organizing each other and creating a choreography for the whole square. This basic artistic media structure is easily expandable, allowing for the creation of a wide variety of moods.

Graphic structures of media design

The framework can also be occupied by advertising modules - although such clips are only to be incorporated in line with a strict set of rules and choreographies. The intensity of advertising images is so dominant that, when shown simultaneously, they can cancel each other out. On the other hand, however, the appearance of such intensive images in some coordinated way in various locations around the square could contribute to the development of a clear sense of rhythm.

In the centre of the square we created a media tower (100m high). The key reason for this was our desire to create a visual icon for the Media Avenue which was visible along its entire length.

The tower should have some limited use, for example as a restaurant or exhibition space. Ring-shaped slats with LEDs were designed to run up and down the external construction forming a range of structures which could combine into elements of differing sizes onto which high resolution images could be shown. This was to be synchronized with the highly sophisticated lighting system inside the tower.

Today I believe as much as ever that it is possible to create an appropriate and highly dense medial performance in such a location without having to copy Times Square. It is precisely when technical media installations have such designed settings that the content can gain so much cultural meaning. And is not the choreographic interpretation of an urban space an extraordinary challenge?

www.mediatecture.info/24

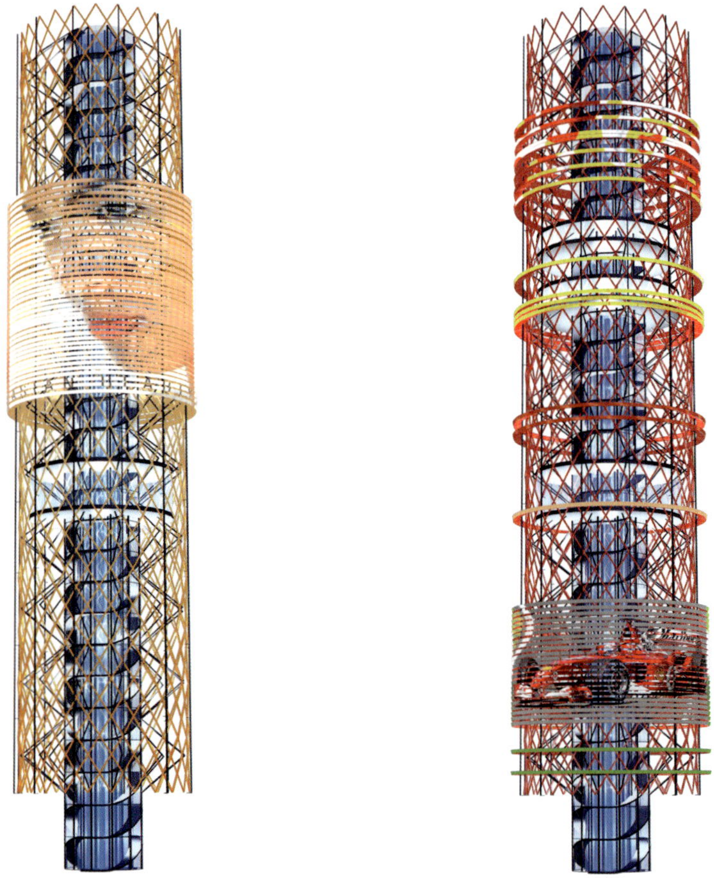

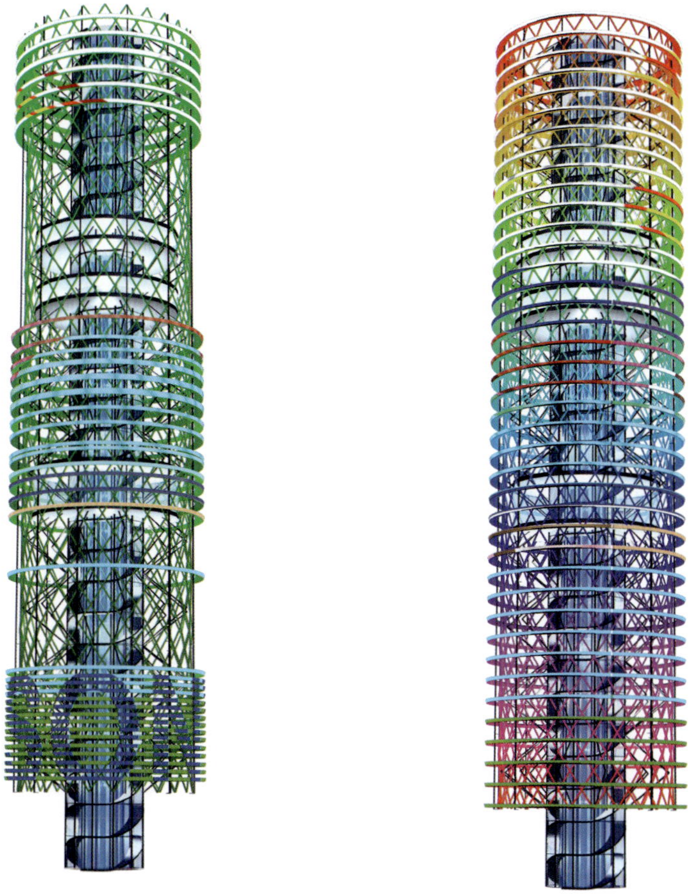

Movement of the annular lammelae, display and illumination synchronized

In the methods and business consultancy company nextpractice, founded by ***Professor Dr. Peter Kruse****, a team of forty psychologists, business economists, computer scientists and designers develops innovative management tools and intervention strategies to support decision-making and harness collective intelligence in companies. nextpractice provides decision-makers with practice-oriented methods and services.*

The applied inventive methods reveal the hidden value systems and complex relationships within businesses and institutions. The objective is to foster discourse within large groups and facilitate concentration on the essentials, and at the same time form a network of the potentials for innovation. nextpractice stimulates creative processes that systematically integrate the knowledge and skills on-hand in organizations.

Making Corporate Values Visible

Departure point:

The Otto Group is an enterprise with a worldwide network and 55,000 employees working in independently run businesses. When there is no feeling of a common ground, no synergies can be generated.

"Paint a stone and send it to Otto in Germany."

This unusual call was received by the roughly 55,000 employees of the Otto Group. With this action, the world's biggest mail order group strives to make its corporate values tangible with the aim of strengthening the network between corporate units. The action seems a bit odd at first sight, but then it becomes clear: It is a personnel development measure of little cost and great inner dynamic.

There are legions of them and no two are alike: Decorated with ladybugs, flowers, landscapes, playing children, parrots, pop icons, shells or abstract forms – they compete in their blaze of colors and artistic effect. We are talking about stones, simple natural stones. Arranged in generous beds, thousands of them have now transformed the foyer of the Otto Group's Hamburg headquarters into an illustrious stone garden. Employees and visitors roam around them in wonder. However, the artistic aspect only plays the second fiddle in this exhibition. It is actually the brilliant final chord of a personnel development project: With these stones, the Otto Group wants to anchor its corporate values in the heads of the employees worldwide.

Flashback: Hamburg, Spring 2004. The responsible people from marketing, corporate communication and personnel development at Otto agree: The corporate alliance of the mail order group must be strengthened. "There are more than 100 Otto companies. For the most part, they are operating independently," explains Jürgen Bock, director of personnel development and the Otto Group Academy, about the status quo. "Through stronger networking, we expect synergy effects such as the exchange of innovations and problem-solving strategies."

Detecting corporate values within the company

The responsibles at Otto are again in agreement: The basis for stronger networking can only be values which are shared by all corporate units. "Common ideals transport the notion of pulling together on one rope, hence creating a 'we' feeling," Bock reasons. However, the director of personnel development also knows: "To develop values within the board and impose them on the individual corporate units from above doesn't make a lot of sense." Such an approach bears the risk that the employees would not adopt the values.

The expert for business culture development and change management, Professor Peter Kruse, works as a scientific advisor for the Otto Group Academy, where he is responsible for conducting executive surveys at Otto. The result of the interview analysis: Passion, innovation, sustainability and working in networks are the values of the Otto Group.

Identification through experience: Values need to be sensed

To identify the values of an organization is one thing; to raise the awareness among all employees is another. Only values present for the employees on a daily basis can generate identity. Bock seeks collaboration with Professor Kruse. The objective is for the employees to experience the values. Based on this premise, the personnel development director and the scientific advisor, together with the Berlin-based artist Ernst Handl and a small group of Otto Group managers, strategically develop a project in January 2005 that should be effective: "Otto Group Meilensteine … gemeinsam mehr erleben" (Otto Group Milestones ... Experience More Together). The heart of the project is a worldwide call to the employees to paint a stone and send it to Otto in Germany.

The project idea: Maximum results with minimum resources

The so-called "avalanche effect" makes it possible to employ only minimal resources – "a small cause triggers a greater effect," as Professor Kruse explains the technique.

In order to encourage the employees even more, the action is combined with a good cause: For each stone sent in, three Euros go to a child aid project. Additionally, there is a 5000-Euro reward for the three best submitted stones. Although the project management draws attention to the project "Milestones" with the folders and posters they distribute in the companies, there is otherwise not an intense promotional campaign. "We trust the avalanche effect," says change management expert Kruse, "the key positions become activated by an inspiring idea, and the dynamic of the system takes care of the rest."

Managers as creative project ambassadors

And it pays off. The management boards of the companies pull out all the stops to activate the employees. The Hamburg mail order company Bon Prix, for example, organizes a stone party with a tombola. Also on board: Asterix and Obelix. The latter carries the oversized symbol of the action in the form of his obligatory menhir. Extra support comes from the canteen: Culinary delights are served up such as "Steinpilze" (yellow boletuses) or "Steinbeißerfilet" (spined loach filets) – dishes that include the German word for stone.

Another action: One morning at Obi@Otto, Hamburg, all employees find a small pot of color on their desk with a message asking them to paint a stone as colorfully as possible. Christina Ziegfeld from the Hamburg-based PR and event agency Mumme + Ziegfeld GmbH summarizes the action: "It is about the network idea." This is the point in the painting project where the PR woman is responsible for the organizational and logistical aspects:

"The task can only be fulfilled when the employees exchange their color pots with one another. In the process, they learn in a playful way what they can accomplish together." In Japan, things get going as well: In the hot phase of the project, the management board of the Japanese branch personally informs each employee about the action – via a personal email from management. "This is necessary as individual appreciation is very valued in Japan," says Ziegfeld.

In other places, such extra motivation by management is not needed at all. "It is no surprise that the employees at Together Design House in Hong Kong, London and Hamburg immediately went full out," reports project leader Bock. Most of the employees are designers, thus creative minds – it is like kicking at open doors with the action there. "Of course, here and there is also resistance," Bock remarks openly. But it evaporates quite quickly given the dynamic the action takes on. Bock experienced that "when the skeptics see the artworks their colleagues were able to produce, the ambition often grabs them too". A remarkable point: Sometimes employees retract already submitted stones in attempt to create better, more beautiful ones.

Professor Kruse finds that all four values of the Otto Group are represented in the project: The passion with which the stones were painted is as impressive as the creativity of the employees. Additionally, there is no object more symbolic of sustainability than a stone. And finally, the artistic diversity that can be admired in the stone garden cannot be accomplished by one individual, rather only by a network. Divided into several exhibitions, the stone garden now departs on its journey through the Otto Group's corporate landscape. "The management boards are literally queuing for the stone exhibitions," Bock tells. Also employees from all over the place want to see the exhibition displayed in their company. "The push effect that marked the beginning of the action has become a pull effect," Kruse draws in conclusion. However, it is still too early for an overall balance: "Due to the exhibitions, the action still has reverberations. It remains to be seen in how far the employee's awareness for the Otto Group's values can be raised," the professor states. Besides reflecting the values of Otto, the project is also an expression of the group's corporate social responsibility. With the action, three child aid projects around the world could be directly supported. Therefore, Professor Kruse also describes the "Milestones" action as a global networking project with a regional effect. "Through an inspiring idea and a little bit of input, strengths inherent in the system could be activated. A personnel development measure has snowballed and become independent," Kruse says.

The "Milestones" project in numbers

» *55,000 employees of the Otto Group from more than 100 companies in 19 countries received the call to paint a stone and send it to Otto Germany.*

» *33,687 stones were submitted. This equates to a participation rate of 65 percent.*

» *For each stone, three Euros went into a donation pot. In the end, a sum of 101,061 Euros was collected and handed over in equal parts to three child aid projects in Indonesia, Austria and USA – the home countries of the winners.*

» *A jury of five members selected the winning stones: Dr. Michael Otto, Director of Personnel Development Jürgen Bock, the gallerist Michael Schultz, an artist and a creative director. The winners received 5000 Euros.*

» *From all the stones, 10 were chosen by lot, and their senders were awarded 500 Euros.*

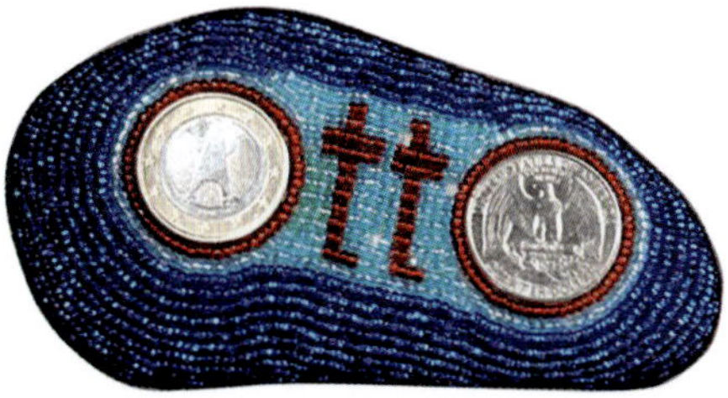

SPACE

SPACE + MEDIA

When mediatecture allows electronic media to play a role in spatial design then the task of defining space is replaced by the much more complex question of what space really is. At the heart of this question lies the confrontation between reality and virtuality. This chapter confronts this intellectual challenge and attempts to suggest ways of navigating this complex area between architecture and media.

SPATIAL CONCEPTS AND AUGMENTED SPACE

It is hard to overlook: the spaces in which we live are becoming ever more complex and hybrid. The physically experienceable space is being overlaid with countless sensual stimuli from a variety of media. These media have an influence on our relationship with reality: What is real and what only appears to be there?

And what does this mean for us? When we run under a bridge our senses make a very quick check that this bridge appears stable and that it is not likely to suddenly collapse upon our heads. Visual media are not part of this combination of sensual observation and experience. Indeed, they seek attention by very consciously detaching themselves from the physical interrelationships of a concrete place. It is to be feared that an uncontrolled meeting of virtual worlds and concrete places could overstretch our ability to orient ourselves sensually in the world in which we live. The removal of all the billboards in São Paolo was obviously a reaction to such a fear of sensual overload.

A common reaction to this phenomenon is the demand that urban spaces should, basically, be kept free of such media. Instead, the focus should be on a return to a more substantial urban culture which sensually communicates that which is real as a means of avoiding irritating illusion. Only physically reliable structured space is capable, so goes the argument, of helping man to continue his social life unhindered.

This, in my opinion, is a completely understandable reaction to the current fully anarchic and unstructured convergence of urban space with both media and signs of all sorts.

And yet this convergence has never really been qualitatively addressed. Advertising and media installations are always on the look-out for opportunities - they take control of areas of façade either because these façades, having lost their urban and architectural value, are no longer under observation, or by offering enough money to a city which, in its own commercial interests, is happy to encourage such developments. Naturally, it is also often the case that aesthetic questions are simply not asked and that everything just develops. But, regardless of the forces driving it, the fact is that this uncontrolled convergence of media and urban space can lead to a cacophony of the sort which would arise if an architectural design simply combined all possible materials without any sort of plan. That the result is regarded as sensual overload is hardly a surprise.

This lack of interest in creating a contextual framework for combining media and city also hinders the development of urban spaces. Without a considered use of media it is very difficult for these to keep up with the development of the globalized world, because globalization is very much to do with hybrid space.

We need a differentiated coordination of architecture and media. And this process of coordination cannot start in line with accepted urban and architectural rules. First of all, it is necessary to answer some key questions regarding perception:

How do we really perceive space and how are the sensual stimuli which affect us processed and combined in such a way that we find ourselves in a space? And is this space really here and now – or is it just in my head?

The relationship between reality and virtuality is (and always was) very complex. I believe that the development of electronic media, with their potent ability to create a virtuality of contagious intensity, both underlines this complexity and puts it in a new light. At the same time, society is developing ever more complex structures and these too will influence our environment. Taken together, these factors offer great challenges to the designer of urban space in the 21st century.

I would like to start with the question of the relationship between architecture and spatial perception. My starting point is that the phenomenon of space as such is not self-evidently comprehensible. Exactly what a space is cannot be clearly defined, as can be seen in the discussions on this subject amongst philosophers (e.g. Plato, Aristotle, Leibniz, Kant, Durkheim), scientists (e.g. Newton, Einstein), sociologists (e.g. Simmel, Bourdieu) system theorists (e.g. Luhmann) and many others.

Each of us clearly has our individual perception of how our living space is put together – because otherwise we would not be able to orient ourselves in our daily lives. And yet, as each person's perception is overlaid with a wide variety of personal experiences, no one can "objectively" look at a space. Yet somehow we manage, in our social processes, to synchronize our spatial perception in such a way that we can understand each other.

The result of this synchronization process is a set of culturally conditioned spatial concepts which function as input for the spatial design process, and architects mostly operate within the limits of these socially accepted spatial concepts. The reaction of the architecture of each epoch is to attempt to establish its current spatial concept as objective – because otherwise it is difficult for architects to sell their approaches and design principles. Architects must be able to communicate to their clients that a particular architectural form is in keeping with a particular social reality. Here lies the problem: This reality is always dependent on the spatial concept which is dominant at any one time. And, for precisely for this reason, this reality is strongly defended as a universally valid reality – especially in periods of new ideas and new technologies which, in turn, exert their own influence on these spatial concepts. There can be no doubt that we are living in precisely such a period today.

What makes the current situation so interesting is the fact that the virtual and the real are frequently presented as clear (and often irreconcilable) opposites. This conflict interests me as it affects me in my daily work.

To which we should add the following aspects:

My own observations through my work as a designer of mediatectonic spaces have shown me that even ideas create their own spiritual spaces. I am not speaking here of mental pictures of a concrete three-dimensional space but rather of more abstract interrelated thoughts which can then lead to more realistic spatial mental models. My impression is that the human brain has a tendency to give spatial structure to complex thought processes – irrespective of the extent to which this process is conscious or unconscious. This is how man organizes his thoughts as a means of making it easier to orient or process complex relationships. He must deal with an extraordinary amount of information which feeds his intellectual processes and, at the same time, compare this information with his reservoir of experience, creating in the process an ever greater "quantity of information". This is only possible because man can filter this information and transform it into an abstract structure which is easier to oversee. The division of an idea into a series of more elementary ideas which he then mentally organizes is one way in which man can simplify how he thinks.

Of course it could be the case that my basic architectural education trained me to think spatially. Perhaps others organize complexly interrelated information in different ways. Whatever the truth, however, I am pretty sure that the human mind has an intuitive tendency to simplify and clarify these complex thought processes in such a way that they become something which no longer corresponds with objective reality. And the dividing up of a whole into its separate parts as a way of achieving clarity is similar to the process of creating an imaginary space. This, for me, is the crux.

Any attempt here on my part to engage in the current neuro-scientific debate would be beyond my ability. My thoughts are, rather, based on very personal observations. And my hope is that the result of these observations is a series of conceptual approaches for the design of the 21st century urban space.

But first of all it appears to me necessary to delve deeper into this conflict between the real and the virtual. And the key question is: what can come out of this connection between the real and the virtual? This essay is based on three basic statements of belief:

Modern architecture encourages the creation of virtual spaces.

Social systems come about when a group of people agree upon certain rules and objectives as a basis for their interactions. The larger the social system, the more it requires signs which can make concepts widely understandable. The rules and objectives for acting together must be filtered in order to simplify them to the point that they can be understood by every member of the social system.

Architecture is the basis of such a cultural communication of ideas due to the way in which it transforms these into spatial structures. Our society is still very strongly attracted to ideas of the "modern" and one way in which this relationship with modernity is articulated is the creation of spaces (e.g. cities) which form platforms for social processes. Such spaces become a collective means of orientation which enables the communication of the notion of modernity via the sensual perception of built substance. Yet the modern spaces which help us in this act of social orientation are only possibilities – they are physically non-existent – because the idea of the modern transmitted by architecture is an idea which can only emerge in the mind of the viewer or user.

Electronic media make it possible to "augment space".

Modern architecture employed the potential of all constructional approaches in order to create modern spatial concepts. Yet the first role of this constructional potential is actually the protection of the user against the weather and, in doing so, it is essential to establish the physical parameters for a construction – the first of which must be the need to obey the laws of gravity. For, even if a structure incorporates many mechanically moving elements, it is intuitively static. Ideas can only be communicated to the extent allowed by constructional reality.

On the other hand, electronic media are only dependent on the availability of a power supply.

The potential of electronic media is that, when they are integrated into architecture, "augmented space" becomes possible. Thus they create interfaces with virtual worlds.

The virtuality of architecture is thus enhanced by the virtuality of electronic media.

This potential to augment space offers society opportunities to integrate its "ideas" into our living space much more intensively than the opportunities offered by architecture alone. As globalization makes social relationships ever more complex, it could be that we shall need to take advantage of this potential of augmented space.

If space is to be augmented, architecture and media must both find new technical forms of expression.

A technical coordination of architecture and media is desperately required because, otherwise, the various sensual stimuli will have an unstructured effect on human perception with the result that the augmented space will not achieve its objectives and that we shall develop the unpleasant sense of drowning under a sea of stimuli.

THE PRESENTATION OF INDIVIDUALITY

In order to try and identify the extent to which architecture must technically adjust itself to the requirements of media, I would first of all like to investigate how architecture itself creates virtual spaces.

How, in particular, were the ideas of the modern implemented in the form of architecture? What were and are the visions of architects and to what extent have these been implemented and to what extent have they failed?

There is little sense in addressing the spatial concepts of the modern without first of all, even if superficially, addressing the nature – and the history - of the modern itself. In comparison with the renaissance or the gothic, the modern cannot be simply defined as a design method. Rather – it should be seen as a way of thinking. The possibility of describing a way of thinking as "modern" already existed long before the beginning of the industrial age. It is this which makes understanding the modern so difficult – and yet so fascinating.

We can start our search for the origins of the modern with a search for the first use of the word "modern": Bernhard of Chartres (11th/12th centuries) used the image of the "Moderni" to picture the dwarves who sit on the shoulders of giants. A reverence for the past is combined with a recognition of typological advance: the "Moderni" as representatives of the present see further ahead than could be seen in the past. Perhaps there is a hint here of one of the first characteristics of the modern: the creation of individuality by means of independent thinking. Bernhard of Chartres was a representative of the "Nominalists", who claimed that the world consisted only of individual things with individual qualities. The "Universals" (general concepts) are a consequence of thought and do not reflect the quality of things. Nominalism holds that general concepts are just names given to things which are really individual and, hence, the nominalists believe that these general concepts only exist as a result of our experience. (Phillex – Lexicon of Philosophy)

Since ancient times, disputes about "proper ways of thinking" have influenced every form of society. This is just as true today. Perhaps one could even say that it has always been "modern" to argue about the meaning of thinking. And it is also important to consider that freedom of thought is not self-evident. It is by thinking that we develop a "self image" and this is something that one must "want".

There is certainly a difference between modern thinking and the beginning of a truly modern age in which society tries very actively to remodel itself in line with this modern thinking. It is not easy to determine when this change occurred. Most likely, the social restructuring required for this new thinking can be connected in some form to the phenomenon of secularization. And one cannot go far wrong if one assumes that the idea of the modern was one which affected society at the beginning of the industrial age.

The central modern value is certainly focused on the "new", which constantly distances itself from that which has just gone before. This idea of always wanting to replace things with something new is a requirement of individual freedom and personal development. Only when one is not stuck to the past can one welcome the new - even when it appears at first unusual and even challenging. This is how the principle of the "avant-garde" came about. The driving forces behind a society create an advanced guard which leaves the masses behind in order to investigate what lies ahead – the new. It makes absolutely no sense when such explorers return with stories of something which is already known. But as soon as these explorers report successfully about something new then society can follow their call – and this is regarded as "modern". It is this that guarantees that society continuously evolves.

"Because the new, modern world differentiates itself from the old in that it opens itself to the future, repeating and driving on the start of the new epoch with each present moment and, itself, giving birth to the new." (Habermas[1])

In order to keep this process alive, the modern must continuously reinvent itself.

"The modern can and will no longer orient itself using examples from the past. It must create its own rules from within. The modern must look after itself and escape is not an option. This explains why one can be so disconcerted by the conviction of the modern that it alone can define the dynamics of the continuous attempts to assert itself." (Habermas[2])

In this regard, art and architecture soon play a very important role. These two disciplines have always offered the perfect stage for self-reinvention. The task now, however, is not to discover the "genius loci" of society in the shape of the glorification of a heavenly truth, individual person or institution, but in the idea of the modern itself. This explains the special relationship between the modern and representation. Representation – especially in the early modern – is seen as a device of the classical.

"In every age, although the forms and emphasis varied, it was always a question of burdens and supports; of representation via a façade; of geometrical proportion and rhythmic harmony based on the classical world (Vitruvius) and of a confirmation of current social relationships. This is why the early modern - with its freedom from the many sets of rules (styles) and from the schematics of social representation as well as with its relationship with the built environment which no longer had to be "conventional" but could be freely determined on a case-by-case basis - represented such a break". (Jürgen Pahl[3])

Representation – just like the orientation towards the past - reduces the space in which the individual can flourish.

Reconciling spaces with this notion that each individual is free to model his own life – together with the related Status Quo which held that the highest possible amount of individuality must be experienced - became central themes of the modern. He who was not constantly working on his individual position was not behaving in a modern way.

It became clear that the rational approach of modern architecture did not have to be functional in the sense of use – but in the sense of it being a representation of the modern. The meaning of representation is "to be represented in a public context" or "having a fitting appearance" – and this is precisely true of the presentation of one's individuality in the public space which, in this case, had to be done in line with the paradigms of the modern. One had to have a fitting modern appearance.

Interestingly, the virtuality of the appearance of modern architecture seems to have been strongly promoted precisely by its inherent contradictions, even when this was not its declared aim. With the result that society is urged to act in line with the ideas of the modern.

Hence the principle of modern architecture is to develop spaces of possibility – as in the very successful example of Mies van der Rohe's Barcelona Pavilion – but this possibility should be seen more in terms of virtuality than in terms of concrete use. The magic which such architecture exudes is extraordinary and this transforms the ideas of the modern into the experience of an immaterial, virtual space. This architecture offers enormous possibilities precisely because each visitor can react to the space in his own way. But this has little to do with the use of the space, which is relatively fixed and inflexible and, hence, quite the opposite of what this architecture wanted to be.

I would like to use this example of the architecture of the Barcelona Pavilion to highlight the individual virtuality which Mies adopted as a means of achieving his objectives. Mies' great technical gift was the way in which he designed his architecture as part of a virtual space. He wanted to hint at a space that would become concrete in the imagination of the visitor. Mies did not communicate his notion of modern space with walls, floors and ceilings but rather he organized the reduced space-defining elements in such a way that an image of his space could only emerge in the mind of the visitor as he moved around it. The space proposed by Mies is partly material and partly immaterial.

In his book "Mental Spaces – a Brief Journey in Time through Virtual Spaces[4]" Gottfried Kerscher addressed the question of how Mies achieved this virtual dynamic. Kerscher observed how one perceived the pavilion and arrived at the conclusion: " …, that Mies had recognized that movement – and not stillness – was the most important motif of his architecture. [...] In the best sense one can describe a space as virtual […], when it is present, but simply not there. In one's head one still has the spatial situation of point X, then one moves to Y and suspects that one is in another space – but in reality it is the same one."

In moving through the pavilion one feels that the space which one is experiencing only comes alive as a result of this movement – and yet, despite this, one still has a sense of calm.

This would suggest that architecture has assumed the role in modern society of mentally establishing modern values. Modern architecture encourages a "modern concept of space" due to the fact that, in modern architecture, it is not a question of the space in which we move but also of the space which we create in our imagination – it is about the architecture of our inner, existential orientation.

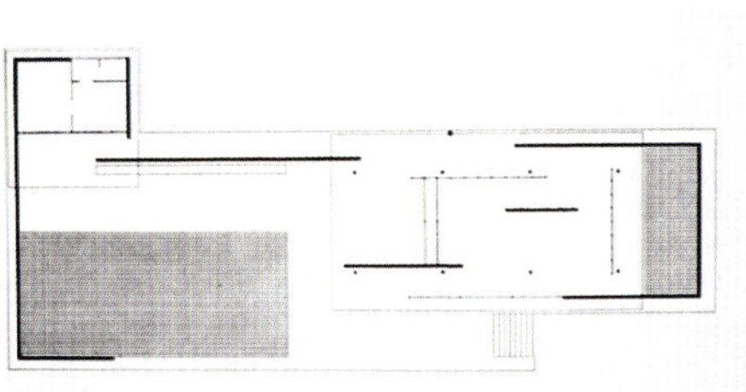

PARADOXICAL SPATIAL PERCEPTION

The dilemma for each architectural work is to establish the point at which the process of creation is over. There is always a moment after which one can no longer reduce, rationalize, add or deconstruct. The "author" must come to terms with the result and say: This is it. The basic construction can now no longer be changed. This is a tragic moment for an architect. On the one hand he is desperate to see his idea implemented but, on the other hand, the finished object will make clear to him the concrete limits of the space which he has designed - in the sense that these limits have been made real by the building materials. Now he will see whether the result is a closed system which forces its users to behave in a certain way or if new possibilities for further development open up. But further development for whom? Should the building be designed in such a way that its structure can always adapt to new demands? Or should the spatial relationships inspire such emotions in the viewer that he himself can develop from within? The dividing line between these two extremes is very flexible. And sometimes they probably interact: if the viewer can simply see that there is a potential for modifying a spatial structure then this can bring him a certain inner peace. When the building is not completely defined this also leaves room for his personal development. This could help to explain the fact that so many buildings are – at great extra expense – built to be highly flexible, but that, in reality, such potential for change is rarely made use of.

There are many examples of contemporary architecture in which spatial relationships are so cleverly designed that they offer the viewer the opportunity to mentally redefine the space. Concrete space has such a complexity that it can hardly be perceived in its entirety. And then there is suddenly a magical moment in which one senses this inner form – in a similar way to music.

This perceptional phenomenon – the notion of developing space in such a way that its deepest truths are not fixed - is part of the program of modern architecture. And as architecture can only achieve such objectives using constructional methods and technical artistry, the principle focus is on the materials. These should be used in such a way that, when future users and observers experience the resulting building, their minds are stimulated to generate new ideas. Of course, as the immediate role of the design process is to optimize the physical quality of the design as a means of creating the required usable space, this virtuality cannot be the prime focus of the architect. But he has only fulfilled his obligations to the modern when this virtual space has been created. This virtuality can affect each individual in a very personal way and lead to an individual experience. But it also has the role of communicating some sense of connectedness between members of society, because it is precisely this virtuality that gives the building its sense of being modern. This phenomenon is something which already connects people in large parts of the world because it makes no attempt to present the potential for individual development and collective behavior as opposites. This simultaneous acknowledgement of both individuality and the notion of a collective idea leads to a mutual convergence of peoples across cultural barriers and demonstrates the level of globalization of design principles – and hence can be seen to be extremely successful!

Hence, the architect should hand over spaces which describe something which is simply not there as a means of offering users the possibility of creating individualized spaces in their minds. The program of good modern architecture is to offer people an open and dynamic living space, despite its concreteness and despite the sense of authority resulting from its structurally required solidity.

Modern architecture always attempts to devise ways of building which offer maximum flexibility and openness (transparency). The role of its virtuality is to encourage the creation of a context in which the individual can flourish and, in doing so, it encompasses true modern values: freedom, openness and individuality. It is this that makes architecture ambiguous.

The virtuality of modern architecture is now confronted with the virtuality of electronic media. These media open up completely different sorts of virtual space. Electronic media have human-human, human-machine and machine-machine interfaces as a means of enabling communication in non-physical spaces.

The virtual objects and spatial structures created in 3D programs are, in comparison, to be seen rather as structures and images which belong to "cyberspace". The question here is, rather, about the types of spaces which will be created by electronic media.

Electronic media must work concretely and unambiguously if they are to fulfill their role as "intermediaries". Electronic media are designed to propose solutions - unambiguous solutions – in every situation. They produce technical solutions in which at least two independent systems are synchronized and enabled to communicate by means of a technical interface. In order for the interface to work, the communication with this interface must be absolutely unambiguous. This process allows absolutely no opportunity for individual interpretation. The solutions enabled by these electronic media are generally short term, because soon a new solution will have to be sought in a new context. Which means that the connections created by electronic media generally have very short stability – and hence are continuously changing (see "The Network (R)evolution").

In both professional and private life, communication increasingly takes place in non-physical spaces. Electronic communication opens up fully self-perpetuating spaces in which many (but certainly not all) social, economic and scientific processes are much easier to organize than in the static medium of the concrete built space. We perceive these complex communicative processes "spatially" because the human brain has a tendency to make pictures of processes as a way of organizing them (it is no coincidence that MySpace.com is called my space). The act of perception of pure facts gives them a different "icon" to that given to direct partners in a conversation and these icons are then to be found in another part of the virtual organizer of stimuli which is located in the brain. But all these icons are interrelated and this means that our spatial consciousness of our living space is very different from our spatial consciousness of concrete space.

Without this synesthetic ability we would immediately lose the overview when working with complex systems. (One can, incidentally, observe the extent to which elderly people need such synesthetic training when they start late in life to work with computers whereas younger people, having already developed such an ability as part of their basic learning processes, have absolutely no such need).

So what is the difference between a virtual space which emerges through the experience of a physical space – let us call this experiential space – and a virtual space which arises from a stream of data?

Experiential space does not really exist; it can only be found in a person's imagination as a result of which it is different from a space resulting from the simple processing of one's sensual perceptions of a concrete place.

And a space which is based on flows of data is, being so ephemeral, very difficult to grasp. Data connections on the other hand – and the resulting communication between humans, machines and places are factually present. There are humans and/or machines which communicate with each other and this communication leads to concrete results.

And in turn, this virtual space arises through the networking together of highly varying spontaneous activities using electronic media – and this space also, incidentally, exists only in our imagination.

So what is the relationship between this spatial concept which is inspired by electronic media and the spatial concepts of modern architecture?

The objective of architecture is to create a stable structure which is able to inspire virtual experiential spaces

The objective of electronic media is to create unstable spaces which, at the moment of their creation, represent an exact reality – on the assumption, that one accepts communication as reality.

And this is a paradox: one medium creates concrete, physically comprehensible spaces while seeking to create virtual spaces and the other medium, although in itself virtual, creates a result which is concrete reality.

Or, putting it another way:

Conditioned by its role of protecting and preserving, architecture is obliged to deliver stable, long-lasting solutions. But these solutions are never unambiguous in order to allow the maximum number of secondary solutions – or applications – to remain possible.

The architectural example of the Barcelona Pavilion shows the extent to which architecture attempts to escape this obligation of stability, because stability always leads to closed systems. It is for precisely this reason that architecture, rather than focusing on clear outputs, seeks to find specific and creative constructional solutions which encourage openness in the mind of the user.

The quality of a demanding architecture is thus measured by how long it remains valid within its particular social system – how long it is able to convey modern values in a social context. Because only when architecture can fulfill this mediatory role over a long period can the cost of creating this architecture be justified. And an architect who can demonstrate such ability can dramatically increase his market value.

In contrast, the success of electronic media is measured in terms of the speed with which they can create a space – a subject which is addressed by Paul Virilio's "The Aesthetics of Disappearance" and other books.

The paradoxical relationship between architecture and electronic media can, exaggeratedly, be put as follows:

Modern architecture must physically develop concrete and stable spaces in order to be sustainably ambiguous.

In contrast, electronic media develop spaces which are as elusive and unstable as possible in order to be as unambiguous as possible for the short moment of their existence.

Or, put another way, the stability of architecture gives it a quality which we had thought belonged to electronic media and the instability of electronic media gives them a quality which we would have rather attributed to architecture.

This observation reveals to us the fundamental difference between the virtuality of modern architecture and the virtuality of electronic media.

And this is a "difference" which interests me in the context of the project of the modern. At the core of the modern program was the freedom of the individual, and it was in response to this program that the modern developed its ideal of open, flexible, content-free space in which the citizen could develop personally as he saw fit without his thoughts being channeled in any particular direction – not that there was any lack of stimuli for such thoughts. This approach is also well known in modern art – as, for example, in the case of monochrome painting.

The ambiguity of ideal modern spatial forms contrasts with our everyday media world and its eternal search for clear communication. This media world is taking over our living space ever more and more and it follows that architecture as a medium of the modern will seek to defend itself if it sees its own ambiguity challenged by the potential for clarity of electronic media. This is why we hear so many architects today emphasizing the meaning of concrete space and their desire to create it using exclusively architectural methods. Only thus can they match the spatial concepts of the modern. But the objective of these spatial concepts is to create not a concrete space but a virtual space: the notion of localization which is central to the idea of the modern. Seeking ambiguity in such a way is hence an unambiguous declaration of an ideology. Architecture can thus not really create an open living space for society as promised in line with the ideology of the modern. It can only thematize it.

Architecture cannot meet the requirements of the modern – it can only communicate them – architecture is a mediator, but the result of its attempts to mediate is precisely the return to the creation of closed systems which it intended to hinder. Modern architecture is not what it thought it was or thought it must be. Habermas noted the extent to which the modern must continuously establish things but " [...] the signs of a crisis in modern architecture which are apparent today have less to do with a crisis in architecture and much more to do with the fact that architecture has simply allowed itself to attempt to do too much."[5]

At least now, if not before, we have the feeling that the relationship between architecture and electronic media is like an Escher painting in which we cannot decide where is the beginning and where is the end. My observation is that we have not yet consciously taken in this irritating reality, as a result of which we constantly confuse the "concrete" and the "virtual": which could be one reason why we so often feel overwhelmed today by real relationships. Unfortunately, as described above, the relationship is very complex and not at all easy to simplify – which is the price of an ever more complex (global) world.

This complexity was never a burden as long as the medium of architecture had a monopoly on the formation of our living space. But now, many areas of this living space are formed much more effectively by electronic media and these work according to completely different rules.

AUGMENTED SPACE

I was surprised by my discovery that Lev Manovich had already invented the term "Augmented Space" in 2002. I had developed the term during the course of my many years of dealing with the theme of spatial concepts and I was fascinated as I recently got to know Manovich's ideas and analyses. It was for this reason that I asked him for permission to republish his text in this book.

From my point of view the incorporation of electronic media into architecture presents the opportunity to create highly emotional rooms which are able to give public space the potential to communicate.

My aim in using the term "Augmented Space" is to use electronic media to work towards the creation of a virtual space which has characteristics different from the virtuality of modern architecture. The virtual space of modern architecture is driven by such values as elegance, individuality and dynamics and it embodies these values for the entire life of the architectural work. In contrast to this, the virtuality of the electronic media can only create spontaneous, momentary emotionality. This becomes particularly clear when electronic media present images. These can stimulate a very precisely targeted emotional and intellectual mental process. Furthermore, electronic media can communicate continuously changing impressions. People register and process this dynamic in a completely different way from the way in which they deal with the effect that architecture has on them.

The sequence in which images are shown can transport moods and emotions of greater or lesser intensity into the public space. For this reason it is necessary to take a sophisticated approach to the subject (see "Media Design in and for Spaces").

The decisive fact, however, is that the spontaneity of these changing impressions can be controlled. This is the basic precondition for communication (see the principle of mimic in the interview with Bazon Brock).

Architecture can also use electronic media as a way of creating a direct connection with the spaces of the virtual world which are created by the internet. The simplest such application is the direct presentation of a webspace on a media façade. I am personally fascinated by the possibilities for the concrete linking of physical space and the web.

One particularly interesting example is the use of a specially created interface with content from the internet to generate a medial presentation. A simple and commonly used example is the creation of a relationship between the media content of a media façade and the weather. Specific design parameters of the media façade react to the coming weather. If this design is so simple that it makes it easy for the inhabitants of the urban space to know what the weather will be, then the space is augmented with a perspective or dimension that the three-dimensional architectural space cannot offer.

This is a very simple example which I have incorporated because it both clarifies the potential of augmented space and demonstrates a decisive challenge in the design of such interfaces. If the weather is simply reported in a textual manner, then this is a trivial addition to the set of such commonly used signs as those which direct the traffic or indicate the empty spaces in a parking garage. If the interface is, however, included more abstractly or as part of the artistic design of the façade then, while the information is undoubtedly present, it can only be regarded as augmented space when the viewers have learnt the artistic code. For this reason, if such projects are to work in the intended way, then they must be accompanied by communication campaigns.

The Bayer Media Sculpture has an application which has a much more pragmatic interface with the weather. This building, particularly in autumn, lies directly below the migration path of many birds. Under specific weather conditions the birds tend to lose their natural ability to orient themselves and begin, at night, to look for lights – which frequently results in them colliding with buildings. In doing this, however, they only react to certain colors. As the entire content of this media façade was created in processing terms, the color base of this content is now automatically synchronized with the weather in order to avoid attracting the birds. The temporarily altered colors of the building are also easy for the locals to see. If Bayer succeeds in effectively communicating the background to this change, there will be two clear effects: Firstly, locals will be able to see, just from observing the building at night, that the migration is underway and, secondly, Bayer will boost its positive image as a company which interacts intelligently with nature - while at the same time widening its communicative spectrum.

The resulting augmented space offers, in this case, not only concrete benefits but also creates new space for the soft values of a corporate culture.

Electronic media in the public space create spontaneous and interactive spaces in the minds of the viewer. This potential for spontaneity and interactivity is precisely the basis upon which the public space can be shaped as a medium of communication. The principle of augmented space which I am describing here is, therefore, always a communicative space. An architectonically designed public space makes communication possible between people whereas a mediatectonically conceived public space becomes part of the communication.

The resulting augmentation of the public space as a medium of communication is, in the cultural context of modern architecture, certainly a surprise and could even represent a conflict between the two. But when considered in the context of the great efforts of various cultures and previous periods to represent the longing and ideas of people in their architecture, this is a very consistent development.

There can be no doubt that, if an open space is to become a medium of communication, the design of the architectonic space and the design of the mediatectonic space must be closely related. Indeed, in this respect, the effect of this paradoxical spatial perception on the design of both mediatectonic installations and architecture itself is certainly an area worthy of future study. The complexity of the psychology of perception as it applies to perception in public spaces is very clearly described in the article by Professor Lukas.

FEELING SECURE IN AUGMENTED SPACE

The inclusion of soft values in the sensual atmosphere of our living spaces can both improve the quality of life of the citizens of a city and create a feeling of security.

If additional emotional content is fed into an urban space and this content appeals to individuals on not only an aesthetic but also a very personal level then the result is a sense of proximity and obligation. An open space whose experienceable edges are composed of glass, steel, stone and concrete etc offers the individual few opportunities to find himself. Architecture creates a relatively anonymous surrounding which is largely closed and communicates little of what is to be found behind the façades.

The augmentation of space with medial content can, however, speak directly to people. This notion of the sensual-communicative addressing of people in a public space can only be achieved by contemporary architecture with a great amount of effort and, in systemic terms, is only possible to a certain degree. This explains why, for instance, the architecture of Herzog & de Meuron is so revered as an exception, because here is a way of addressing the public using signs which was long frowned upon by modern architecture. Signs are used here in such a way that the viewer can discover the logic behind them. This is very reassuring. When someone is making an effort to communicate something in such a way that I can understand it then the likelihood increases that he is not fundamentally my enemy.

In this respect, both electronic media and traditional design methods offer the potential for augmenting space. Graphic elements can also augment public space.

But, naturally, moving pictures have much more potential for directly talking – and reacting - to people.

The augmentation of space using electronic media can, paradoxically, augment architectonic space both inwards and outwards. Space is augmented inwards when information about the user of a building or about a city, city district or region, etc, can be communicated via medial content. In such a case, the viewer receives information about the internal make-up of his cultural surroundings. Light could be shed upon public reaction to particular local events or upon more complex aspects of urban culture. In doing this, the media façade is looking inwards, communicating qualities which have developed independently of the spatial concepts of the modern. The result is a so-called interior space of the city. At the same time this content also has an effect on the outside in that the city presents itself not only to citizens but to visitors to the city, projecting itself far beyond its own borders. Hereby the city widens its spatial presence.

But there are further paradoxes: Urban space is at the same time also charged from the outside. This begins with the electronic medium itself, because this is one of the key symbols of globalization. Only with the emergence of electronic media could data flow intensify to the extent that complex economic processes can be synchronized into a whole new dynamic.

My experience is that companies are attracted to the idea of cloaking themselves with media façades because this is a way of showing the extent to which they are part of these global processes. This is often a decisive motive for such a project (just as, in the Middle Ages, expensive stained glass windows were perfect proof of membership of the Christian community because these were a key feature of cathedrals).

In the same sense, one is attracted to the idea of becoming connected with distant parts of the world through the installation of a media façade which can dissolve the great distances from such places and introduce a sense of simultaneity. In this way any location can be connected to the greatest city. This has psychological importance because it gives one a sense of participation. One belongs to a very large whole.

This potential is increased if the control of a media façade is influenced by internet processes. The result is a new definition of "urbanity". The digital networking of urban space with the rest of the world is becoming a sensually experienceable part of our personal living space. City boundaries are dissolving and this process can be described, wholeheartedly, as augmented space.

Designing this augmented space is a great challenge. Just as many cities as before represent that condition which Alexander Mitscherlich, in the 1960s, referred to as "the inhospitability of cities." The fact that it is media which can play a positive role in attempts to escape from such inhospitability is certainly astonishing for many. Whether this transformation can be successful is, of course, yet to be seen. But at the very least it can only work if one recognizes the potential.

My observation is that contemporary architects are beginning to address the issue of augmented space as a means of incorporating the influence of medial developments into their architecture. Even when the Russian suprematists and constructivists are clearly the inspiration behind the development of deconstructivism, one cannot also avoid noticing the extent to which breakthroughs in cyberspace were and are key moments for this movement. The deconstructivists are protesting against a world which orders everything into static patterns. They say that the world is not made in this way and that architecture should mirror this real world, a world which is in a state of permanent change. In the words of Zaha Hadid:

"The most important thing is movement, the flow of things, a non-Euclidean geometry, in which nothing is repeated: A new way of organizing space."

Making the construction of a building more dynamic by employing jagged edges and contortions as a means of creating the impression that the building could look completely different again the next moment is sculpturally very impressive and emotionally moving, but this is not necessarily the best way of encouraging communicative, augmented space. It would be absolutely no problem to take this architecture further by giving it a medial dimension. As Lev Manovich demanded in 2002, it is simply up to architects to open themselves to the possibility.

Cologne/Bonn, 2001-2010

1 Jürgen Habermas, Der philosophische Diskurs der Moderne (Suhrkamp: Frankfurt, 1985) p.15,

2 ibid., p.16,

3 Jürgen Pahl, Architekturtheorie des 20. Jahrhunderts (Prestel Verlag: Munich, 1999) p.90,

4 Published in German as: Gottfried Kerscher, Kopfräume – Eine kleine Zeitreise durch virtuelle Räume (Ludmig Verlag: Kiel, 2000) The discovery of virtual inner space: p.90

5 Gerd de Bruyn, Stephan Trüby, eds., Architektur_theorie.doc: Texte seit 1960, (Birkhäuser: Basel, 2003), p.160

Space + Media

ETH-WORLD

In 2000 the ETH (The Swiss Federal Institute of Technology) in Zurich organized an international competition. The objective of the competition was to identify a concept for a virtual campus combined with a physical interface with the ETH at its Hönggerberg Campus. ag4 developed a competition entry[1] which merits special attention in this book due to the very clear way in which it embodies the notion of augmented space.

Our concept was not considered by the ETH for further development. Instead, the ETH used the competition process as a basis for developing its own myETH platform. This platform offers students a full range of possibilities for networking and intelligently exchanging data.

In contrast with this, our concept attempted to establish a perspective that, without doubt, exceeded the pragmatic framework established by the ETH. But it is not without a sense of pride that I say that our project of the year 2000 already foresaw methods which, as a result of developments in social networks of the past few years, now seem very familiar. It was clear to us that the filtering and profiling of personally relevant data would be the great challenge of the decade and we suspected that the creation of communities would become hugely important for global society.

Of course the basic requirement of creating a virtual campus with a physical presence corresponded precisely with the mediatectonic orientation of ag4. Our understanding of this task was that the virtual campus required the design of a concrete place which would bring together the social processes inspired by the virtual campus and allow these to express themselves in the form of concrete human interaction. We were convinced that the living space of the future would be of a hybrid nature. This has been confirmed since to the extent that all the dreams of purely virtual cities of the 1990s are no longer relevant.

It was important for me in this project to make use of Bruno Schindler's idea of sensual orientation (see "Showplaces of Power"). I assumed that the design of a connection of a virtual platform with a physical interface could only work through some notion of sensual relationships related to the task in hand. This means that it is the task itself which dominates the process of designing forms in the cases of both architecture and media design. My hope was (and is) that if the functional creation of forms in the virtual world can find a design-based interface with the creation of forms in the real world then these two worlds will grow together and make a sensual orientation possible.

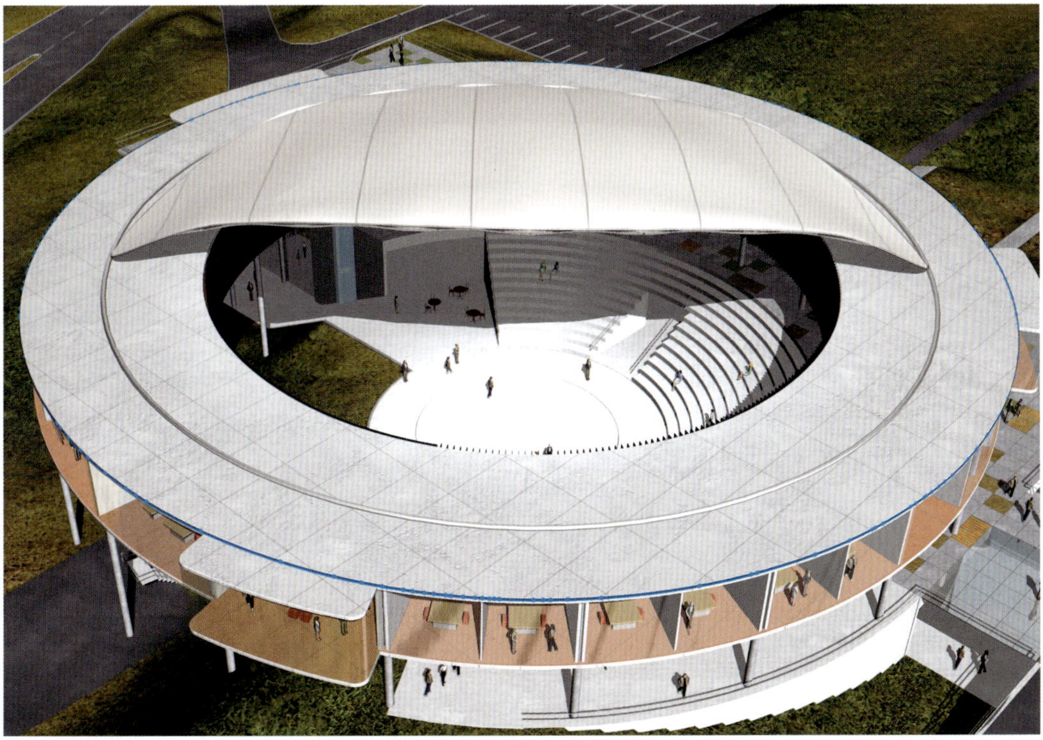

Renderings © Deepartmend/Sascha Selent

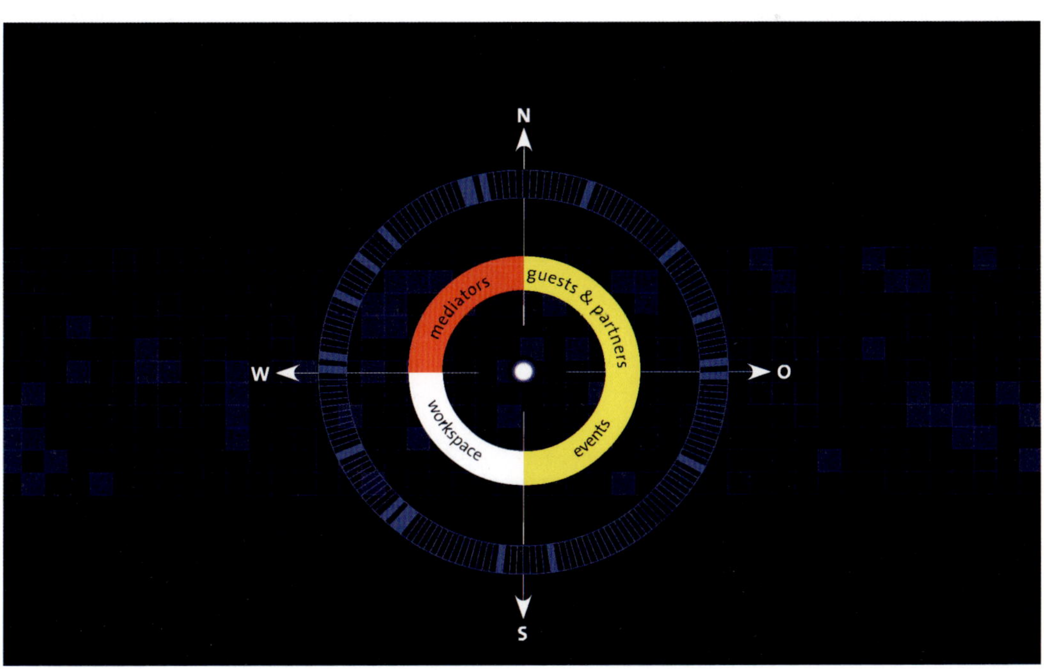

Flash-progrmming © Variovision / Contanze Fries

The highpoints of previous design epochs can be explained precisely in terms of this notion of sensual orientation. Take, for example, the Gothic cathedral in which a dissolving construction and back-lit pictorial windows combine to create the sense of a glowing heavenly space – a perfect convergence of a virtual world with an earthly location. One should underestimate neither the importance of this sense of orientation for the society of any epoch nor the level of social cohesion which is required. While most people in the gothic age required simple images in order to orient themselves due to the fact that they could not read, the demands on the information-processing ability of the individual today are so enormous that we, once again, need simple and rapidly absorbable images.

THE BRIEF

The ETH is a university very dear to the heart of Swiss society – a sort of model project. The facilities are fantastic. But, as with all elite universities, the ETH depends upon a community of people which is also dynamic long after it has moved on from the campus. These alumni are very loyal to their university and offer an opportunity for extremely efficient networking. But the alumni are, today, scattered across the world and no university authorities are in the position to constantly support such a network. One must give the alumni more reasons for keeping close touch with the ETH – they have to want to be linked in! Giving them the chance to support the next generation with generous donations is not enough. A more efficient way of promoting such continued alumni contact would be to find ways of bringing together the creative potential of the university with the scientific and technical needs of private industry. One has to create an offer which makes it attractive for private industry to develop projects on the ETH. The head of research and development of every major company knows that the creative energy of the next generation can trigger surprising innovations - but he also knows that developing projects with inexperienced partners can also bring other, very different sorts of, surprises. The mere availability of the necessary "human resources" is not enough to entice potential industrial partners onto campus. They are much more interested in the principle of interdisciplinary work and such a fantastically equipped university as the ETH can offer much in this respect. But the successful convergence of industry and these various disciplines also requires experienced specialists who can really help to structure this potential cooperation. And these could be the alumni - if one only knew how to coordinate them.

Our concept for the ETH envisaged the creation of a virtual and physical interface which could allow such coordination. The heart of our idea was the development of a social network which was far more than just a way of getting to know - and networking with - people.

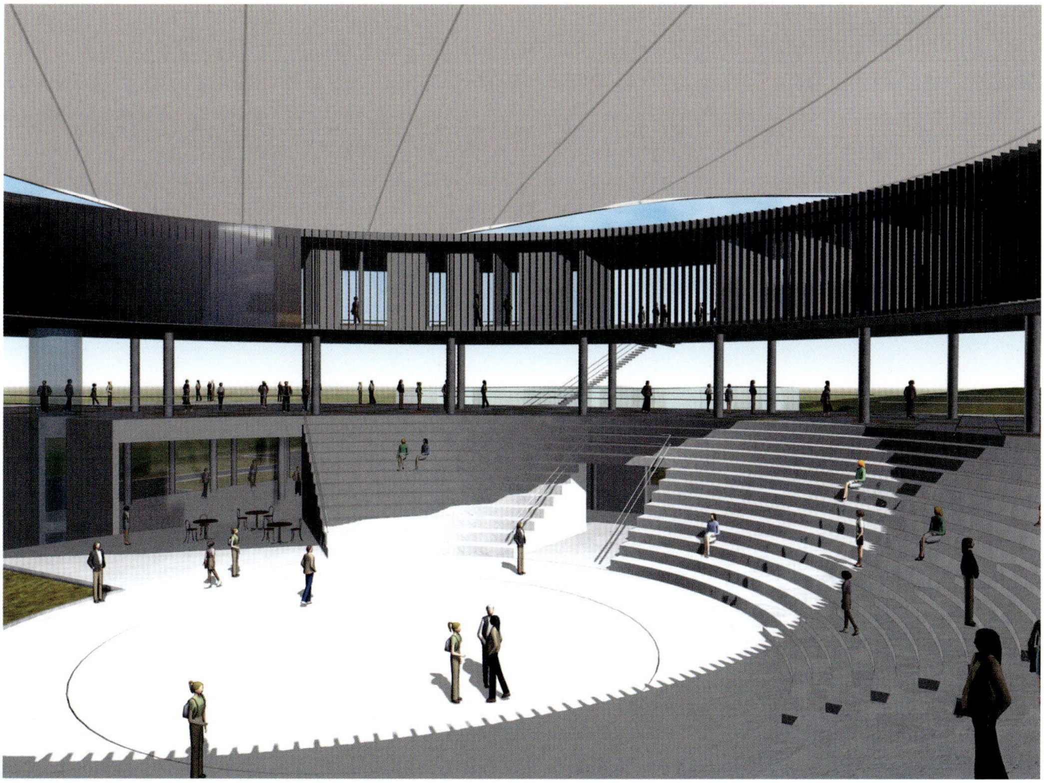

The inner court of the Interfacing Forum serves as a place for events and spontaneous encounters.

In Backbone Park, underground flexible work stations can be booked. They are grouped around the light wells. The work stations are temporarily grouped together into units and can be accessed separately over the light wells.

The Interfacing Forum floats above a conference center.

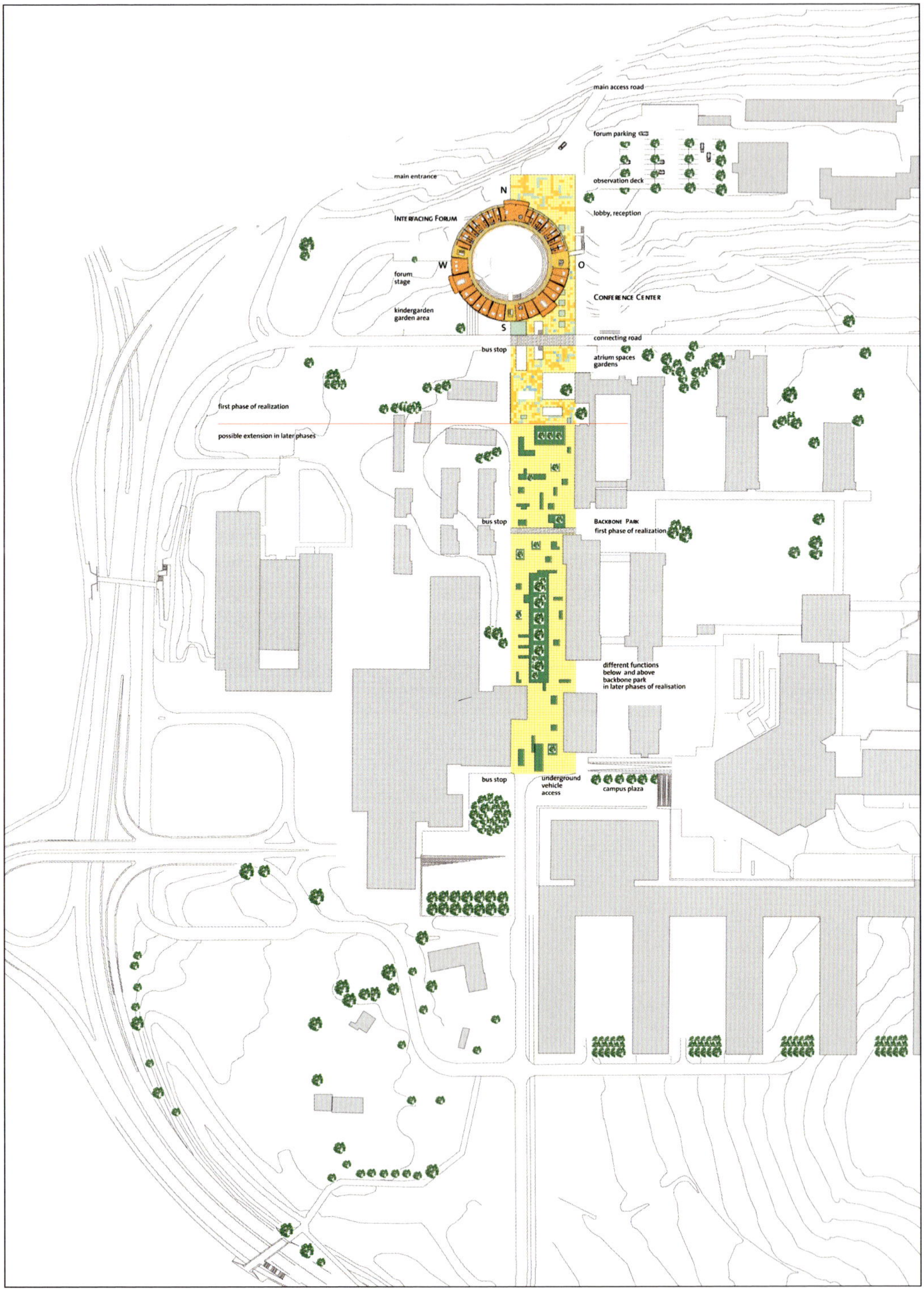
main access road
forum parking
main entrance
observation deck
lobby, reception
N
Interfacing Forum
W
O
forum stage
Conference Center
kindergarden garden area
S
connecting road
bus stop
atrium spaces gardens
first phase of realization
possible extension in later phases
bus stop
Backbone Park
first phase of realization
different functions below and above backbone park in later phases of realisation
bus stop
underground vehicle access
campus plaza

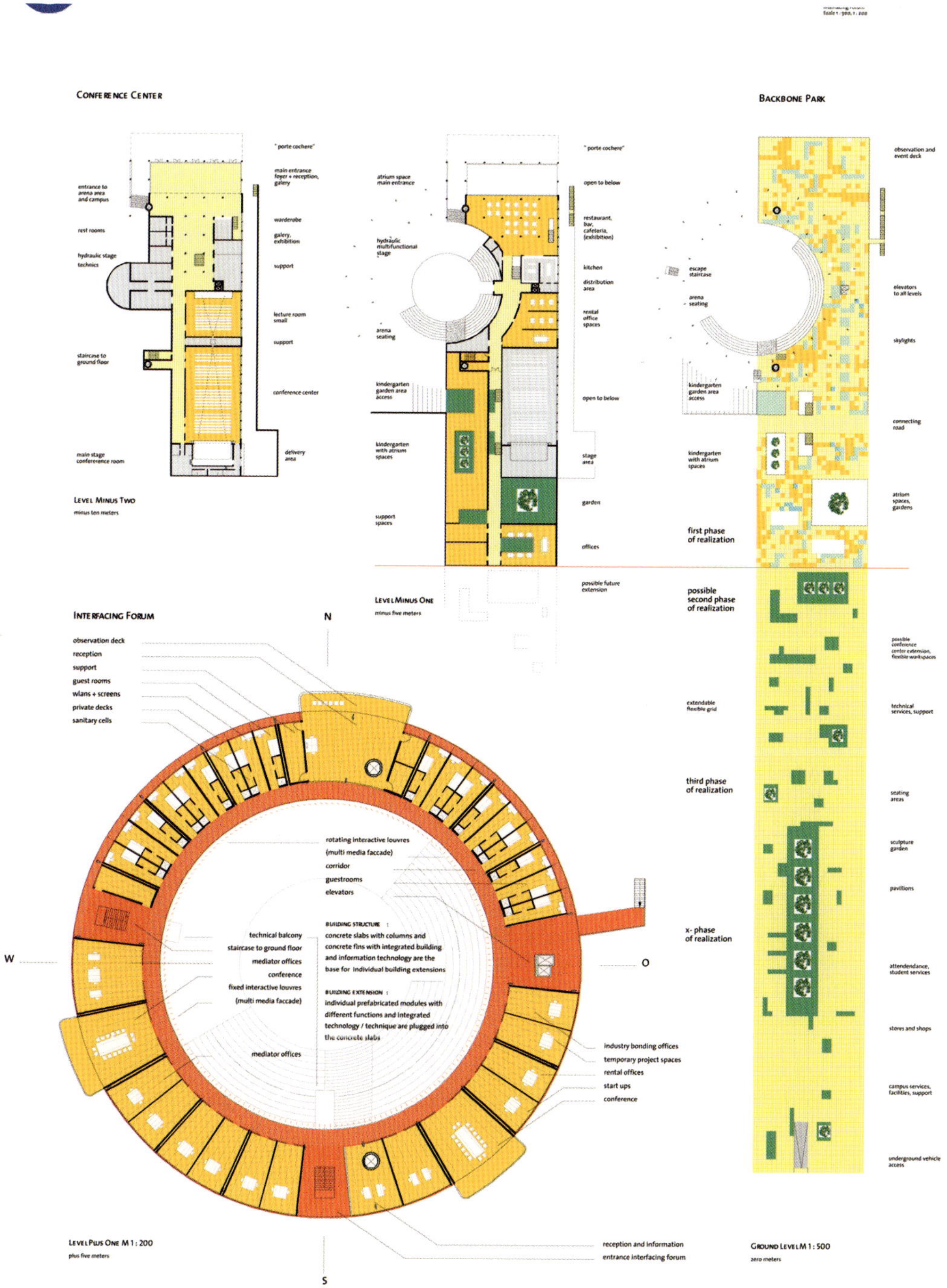
Scale 1 : 500, 1 : 200
Conference Center
"porte cochere"
main entrance foyer + reception, galery
entrance to arena area and campus
warderobe
rest rooms
galery, exhibition
hydraulic stage technics
support
lecture room small
support
staircase to ground floor
conference center
main stage confererence room
delivery area
Level Minus Two
minus ten meters
"porte cochere"
atrium space main entrance
open to below
restaurant, bar, cafeteria, (exhibition)
hydraulic multifunctional stage
kitchen
distribution area
rental office spaces
arena seating
kindergarten garden area access
open to below
kindergarten with atrium spaces
stage area
garden
support spaces
offices
possible future extension
Level Minus One
minus five meters
Backbone Park
observation and event deck
escape staircase
arena seating
elevators to all levels
skylights
kindergarten garden area access
connecting road
kindergarten with atrium spaces
atrium spaces, gardens
first phase of realization
possible second phase of realization
possible conference center extension, flexible workspaces
extendable flexible grid
technical services, support
third phase of realization
seating areas
sculpture garden
pavilions
x- phase of realization
attendendance, student services
stores and shops
campus services, facilities, support
underground vehicle access
Ground Level M 1 : 500
zero meters
Interfacing Forum
N
observation deck
reception
support
guest rooms
wlans + screens
private decks
sanitary cells
rotating interactive louvres
(multi media faccade)
corridor
guestrooms
elevators
Building structure :
concrete slabs with columns and concrete fins with integrated building and information technology are the base for individual building extensions
Building extension :
individual prefabricated modules with different functions and integrated technology / technique are plugged into the concrete slabs
W
technical balcony
staircase to ground floor
mediator offices
conference
fixed interactive louvres
(multi media faccade)
O
mediator offices
industry bonding offices
temporary project spaces
rental offices
start ups
conference
reception and information
entrance interfacing forum
S
Level Plus One M 1 : 200
plus five meters

ESTABLISHING THE CONCEPT

Design

When one wishes to create an organization covering the entire world one requires a simple sign. In the case of ETH World this was easy to find – the world itself. And this sign had the role of helping visitors to navigate through the world of ETH and its alumni. In order to do this - one needs a compass. The world and the compass can be represented as a circle.

And which symbol better represents a community than a ring?

This was how the Community Compass and the Interfacing Compass came about. It was important that the tools which the virtual campus required to organize its network should adopt this ring design.

Architecture

At the same time, it was proposed to develop a building which could become the home of the activities of ETH World. Circular architecture is always suitable as a symbol for community and communication: as a result of which the simple sign of the compass could also be used for the development of the physical interface of ETH World.

Business

No good idea is worth anything when it cannot be financed. It was fully clear that, even for the ETH, financing such a building is not so simple. It was thus essential to develop a business idea which could finance the investment of ETH World over the long term. This idea was that ETH World would organize technical and scientific development projects across the globe. And this led to the development of the concept of the Interfacing Forum.

THE CONCEPT AS STORY

Mr. X is a manager in San Francisco who studied at ETH. He is in charge of a large project but his team is stuck. They know that they need input from completely different areas of expertise and, above all, they need scientific input that has not yet been used together with the standard methodology and, at the same time, is so completely new that they do not know where to start.

The manager is a member of ETH World. He accesses the Community Compass onto which he had earlier entered his personal profile containing details of his professional and technical specialties. The graphic content of the Community Compass is based on a map of the world centered on Zurich. A ring sweeps around the world like a scanner looking for activity and, with its centered grid, blinks at precisely those points where other members of ETH World are online. However, locations only blink when the profile of the online member corresponds with that which Mr. X is searching for. As he is looking for a very specific contact he also fine tunes his profile accordingly. He notices that in a region in which he will soon be attending a congress, many members of the ETH community are currently online. He can now look closer at these alumni and he identifies one former colleague who, although his career has taken a completely different path, is now involved in a field which could well help Mr.X solve his current problem. They arrange to meet during the congress.

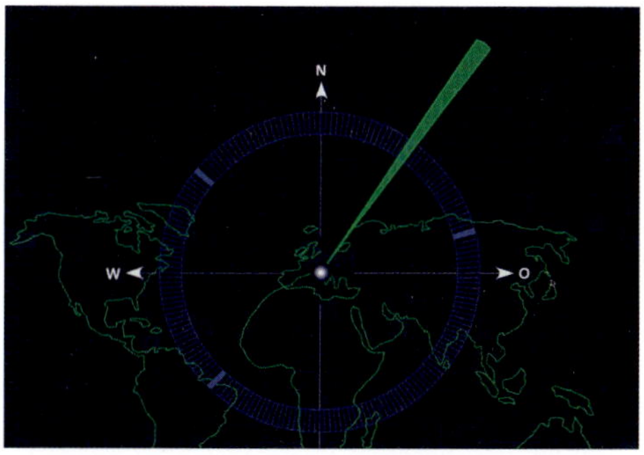

When one clicks on the ring, the world is scanned for active chat partners.

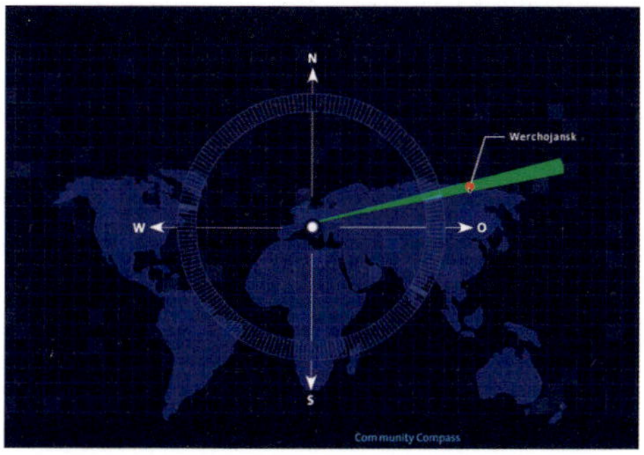

Now only lights that indicate zones where appropriate chat partners are online blink in the ring. When one selects one of these lights, a beam with places where a partner could be found appears.

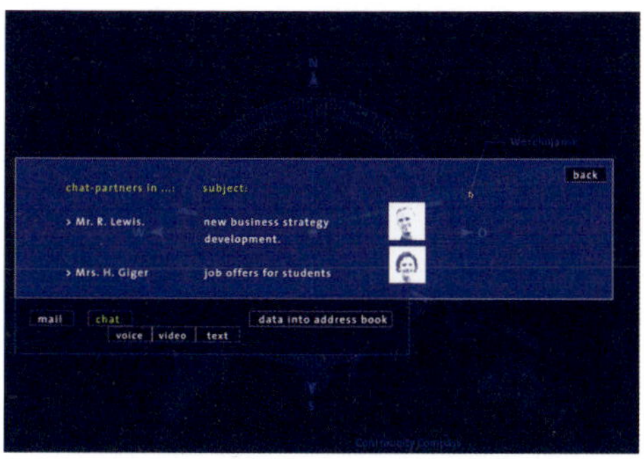

When one clicks on a place, possible chatpartners are listed

www.mediatecture.info/25

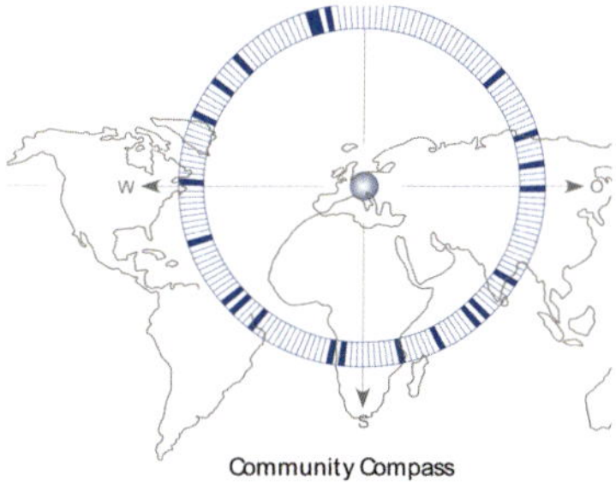

Community Compass

In the segmented ring, lights blink indicating in which zones of the world members of the ETH World are online at the moment.

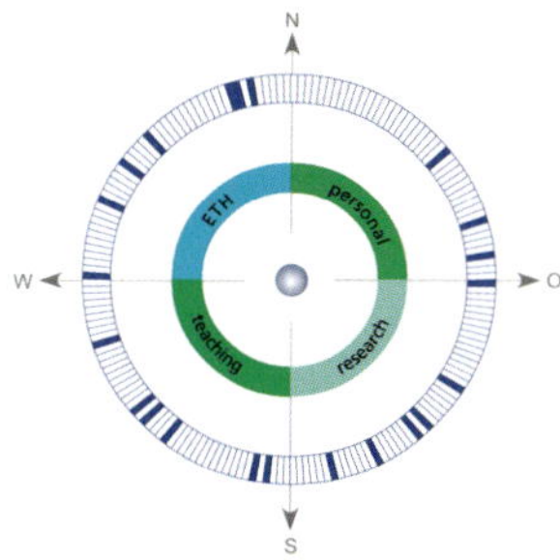

Content Compass

The Content Compass organizes the scientific content of the ETH. This is also the access point for students and for life-long learning.

ETH WORLD Compass

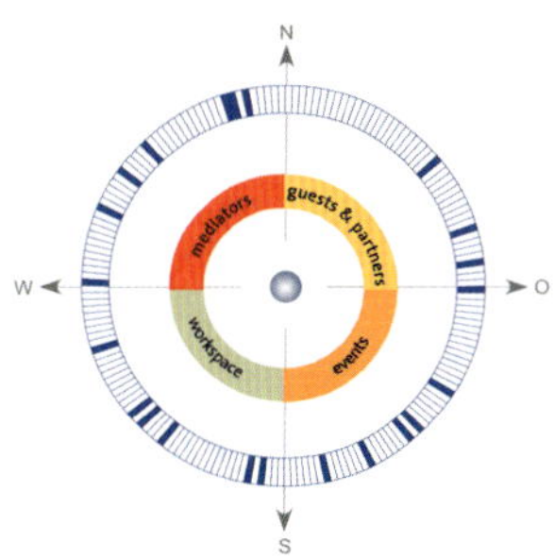

Interfacing Compass

The Interfacing Compass organizes all of the activities of the Interfacing Forum and the Backbone Park.

The meeting brings results because the colleague can suggest the names of other former colleagues who are addressing similar issues but who work in areas with specialist skills which are precisely those which Mr.X is missing. The Community Compass then sets up loose connections between these colleagues before they turn to the Interfacing Compass, which allows them to create working groups. Everything has gone very successfully up to this stage but then, after a while, progress seems to slow down. Interdisciplinary work is not easy because each speaks a different language. How does each specialist communicate to the others the key aspects of his own work? And another difficulty arises out of the simple fact that, due to this interdisciplinary structure, the scale of the challenge has grown. One comes to the conclusion that it is necessary to create a joint working group for about three months and, in addition to this, it is clear that certain specialist scientific input is still missing.

Now one can use the Interfacing Compass to access the Interfacing Forum. Here, Mr. X contacts a mediator, who helps him to book the facilities of the Interfacing Forum for the working group. Here, tailor-made working areas arranged around atria can be created in the flexible basement level and this gives the working group a concrete home on campus. The main Forum structure, a ring-shaped building on pilotis, contains both conference rooms for one-off workshops and hotel rooms which can be booked for those members of the working group who are only working temporarily on the project.

The mediator also helps to establish the necessary contacts with the scientists from ETH who can then direct the work of their students towards various aspects of the project. This work can either be integrated into the ongoing curriculum or treated as work experience. This ensures that, while still studying, students get an insight into project development work and this often leads to contacts which will be important for their later careers.

As the airport is so close, this main Forum building is also suitable for spontaneously organized meetings. Its main role, however, is the support of the community. Events are held in the center of the ring in summer as a means of encouraging the “coming together” of minds. The internal face of the ring is a media façade which can continuously relay information about activities or create an atmosphere appropriate to an event. Mr. X uses this for the party which marks the successful conclusion of the project.

Later follow-up work is once again organized via the Interfacing Compass.

THE ETH WORLD CONCEPT AS AUGMENTED SPACE

Previous chapters have described the notion of “Augmented Space” as set out by both Manovich and myself. However, discovering and writing about such a concept is one thing. The question is: what benefits could this augmented space really bring?

I believe that this concept for ETH shows very well how an augmented space works. Crucial to the understanding of augmented spaces is the fact that such spaces have their decisive presence neither on the internet nor in a building – but in the mind. If one has only ever known something from its virtual appearance then its physical presence appears familiar, and vice versa. One can never experience both realities at the same time but, when one experiences one; the other is always, somehow, present as well. They resonate. Hence, when I experience the Interfacing Forum, the potential of the compass resonates within me because I know that the Forum does not end with its physical form. This is similar to the experience upon entering a Japanese restaurant when one is subconsciously aware of a stream of images from Japanese culture. In the widest of senses, this phenomenon can apply to every designed object.

This is the linkage effect of an object – whether virtual or physical – in augmented space. We all know the importance of such linkage for the perception and processing of complex relationships. This is an accomplishment which we largely owe to electronic media. However, if we add observations from cyberspace to this mechanical understanding of linkage then we see how important it is. The linkage of physical and virtual spaces provokes an individual process, because each person has registered different experiences and images which now begin to take effect. But, whatever the details, the fact is that some sort of linkage definitely occurs and my belief is that one can also understand this linkage as a process of understanding and comprehension.

If we now consciously create augmented space we shall be promoting this understanding and comprehension and this is very important for our inner peace. Each interruption to this process gives us a feeling of unease: one only needs to observe those visitors to an exhibition of modern art who do not understand the artist's code – or the aggressive reaction of the confused train traveler confronted with an interactive terminal. Such reactions are mostly an expression of great insecurity because people tend to assume that a failure to understand can only result from their own inadequacy.

It is for this reason that I believe that the task of mediatects is to use augmented space to promote understanding and comprehension in the world. The task is a paradoxical one: the designer must be able to both take into account and then manage great complexity – with the aim of making the world a simpler place.

1 Together with Jörg Ruegemer (architect and now Professor for Sustainable Architecture in Salt Lake City) and Stephan Mosblech (formerly of ag4 live and now managing director of people interactive GmbH in Cologne)

A light-emitting diode installation on the main building of the ETH transmits the Community Compass ring in the center of Zurich. It continuously shows the social activities of the virtual ETH.

__Lev Manovich__'s books include Software Takes Command (released under CC license, 2008), Soft Cinema: Navigating the Database (The MIT Press, 2005), and The Language of New Media (The MIT Press, 2001) which is hailed as "the most suggestive and broad ranging media history since Marshall McLuhan." He has written 90+ articles which have been reprinted over 300 times in 30+ countries. Manovich is a Professor in Visual Arts Department, University of California -San Diego, a Director of the Software Studies Initiative at California Institute for Telecommunications and Information Technology (Calit2), and a Visiting Research Professor at Godsmith College (University of London), De Montfort University (UK) and College of Fine Arts, University of New South Wales (Sydney). He is much in demand to lecture around the world, having delivered 300+ lectures, seminars and workshops during the last 10 years.

THE POETICS OF AUGMENTED SPACE

How is our experience of a spatial form is affected when the form is filled in with dynamic and rich multimedia information? (The examples of such environments are *particular urban spaces* such as shopping and entertainment areas of Tokyo, Hong Kong, and Seoul where the walls of the buildings are completely covered with electronic screens and signs; convention and trade shows halls; department stores, etc.; and at the same time, *any human-constructed space* where the subject can access various information wirelessly on her cell phone, PDA, or laptop.) Does the form become irrelevant, being reduced to functional and ultimately invisible support for information flows? Or do we end up with a new experience in which the spatial and information layers are equally important? In this case, do these layers add up to a single phenomenological gestalt or are they processed as separate layers?

Although historically built environments were almost always covered with ornament, texts (for instance, shop signs), and images (fresco paintings, icons, sculptures, etc. – think of churches in most cultures), the phenomenon of the dynamic multimedia information in these environments is new. Also new is the delivery of such information to a small personal device such as a cell phone, which a space dweller can carry with her.

Therefore, this essay will discuss how the general dynamic between spatial form and information which has been with us for a long time and which I outlined above functions differently in computer culture of today. Since the kinds of environments I offered, as examples above do not have a recognizable name yet, I will give me a new name - *an augmented space*. The term will be explained in more detail below, but here is the brief definition: augmented space is the physical space overlaid with dynamically changing information. This information is likely to be in multimedia form and it is often localized for each user.

I want to focus on the experience of the human subject in augmented space as opposed to particular electronic, computer and network technologies through which the augmentation is achieved. I also want to re-conceptualize augmentation as an idea and cultural and aesthetic practice rather than as technology. To do this, I will discuss how various practices in professional and vernacular architecture and build environments, cinema, 20th century art, and media art can be understood in terms of augmentation. I hope that this will firmly position the concept of augmented space in historical and cultural as opposed to purely technological sphere.

AUGMENTATION AND MONITORING

The 1990s were about the virtual. We were fascinated by the new virtual spaces made possible by computer technologies. Images of an escape into a virtual space that leaves -physical space useless, and of cyberspace – a virtual world that exists in parallel to our world – dominated the decade. This phenomenon started with the media obsession with Virtual Reality (VR). In the middle of the decade graphical browsers for the World Wide Web made cyberspace a reality for millions of users. During the second part of the 1990s, yet another virtual phenomenon – dot coms – rose to prominence, only to crash in the real-world laws of economics. By the end of the decade, the daily dose of cyberspace (using the Internet to make plane reservations, check e-mail using a Hotmail account, or download MP3 files) became so much the norm that the original wonder of cyberspace so present in the early cyberpunk fiction of the 1980s and still evident in the original manifestos of VRML evangelists of the early 1990s - was almost completely lost.[1] The virtual became domesticated. Filled with advertisements and controlled by big brands, it was rendered harmless. In short, to use Norman Klein's expression , it became an "electronic suburb."

At the begnning of the twenty first century the research agendas, media attention, and practical applications have come to focus on a new agenda – the physical – that is, physical space filled with electronic and visual information. The previous icon of the computer era – a VR user traveling in virtual space – has been replaced by a new image: a person checking her e-mail or making a phone call using her PDA/cell phone combo while at the airport, on the street, in a car, or any other actually existing space. But this is just one example of what I see as a larger trend. Here are a few more examples of the technological applications that *dynamically deliver dynamic data to, or extract data from, physical space* – and which already are widely employed at the time of this writing:[2]

1. *Video surveillance* is becoming ubiquitous. No longer employed only by governments, the military and businesses but also by individuals; cheap, tiny, wireless, and Net-enabled, video cameras can now be placed almost anywhere. (For instance, by 2002, many taxis already had video cameras continuously recording the inside of the cab).

2. If video and other types of surveillance technologies translate the physical space and its dwellers into data, *cellspace technologies* (also refered to as mobile media, wireless media, or location-based media) work in the opposite direction: delivering data to the mobile physical space dwellers. Cellspace is physical space that is "filled" with data, which can be retrieved by a user via a personal communication device.[3] Some data may come from global networks such as the Internet; some may be embedded in objects located in the space around the user. Moreover, while some data may be available regardless of where the user is in the space, it can also be location-specific. The examples of the cellspace applications which are not localised is using GPS to determine your coordinates, or surfing and checkling email using a cell phone. The examples of location specific applications are using a cel phone to check in at the airport, pay for a road toll, or retrieve information about a product in a store.[4]

3. While we can think of cellspace as the invisible layer of information that is laid over physical space and is customized by an individual user, publicly located *computer / video displays* present the same visible information to passersby. These displays are gradually becoming larger and thenner; they are no longer confined to flat surfaces; they no longer require darkness to be visible. In the short term, we may expect large thin displays to become more pervasive in both private and public spaces (perhaps using technology such as e-ink). In the longer term, every object may become a screen connected to the Net with the whole of built space eventually becoming a set of display surfaces.[5] Of course, physical space has long been augmented by images, graphics, and type; but replacing all of these with electronic displays makes it possible to present dynamic images, to mix images, graphics, and type, and to change the content at any time.

If we consider the effect of these three technological applications (surveillance, cellspace, electronic displays) on our concept of space and, consequently, on our lives as far as they are lived in various spaces, I believe that they very much belong together. They make physical space into a dataspace: extracting data from it (surveillance) or augmenting it with data (cellspace, computer displays).

It also makes sense to conceptually connect the surveillance/monitoring of physical space and its dwellers, and the augmentation of this space with additional data, because technologicaly these two applications are in a symbiotic relationship. For instance, if you know the location of a person equipped with a cell phone, you can send them particular information relevant to that specific location via their cell phone. A similar relationship exists in the case of software agents, affective computing, and similar interfaces, which take a more active role in assisting the user than the standard Graphical User Interface (GUI). By tracking the user – her mood, her pattern of work, her focus of attention, her interests, and so on – these interfaces acquire information about the user, which they then use to automatically perform the tasks for her.

The close connection between surveillance/monitoring and assistance/augmentation is one of the key characteristics of the high-tech society. This is how such technologies are made to work, and this is why I am discussing data flows from physical space (surveillance, monitoring, tracking) and into physical space (cellspace applications, computer screens, and other examples below) together.

PANOPTICUM AND INFORMATION THEORY

Let us now add to these three examples of the technologies that are already at work by citing a number of the research paradigms which are being actively conducted in University and industry labs. Note that many of them overlap, mining the same territory but with a somewhat different emphasis:

4. *Ubiquitous Computing*: the shift which away from computing centered in desktop machines and towards smaller multiple devices distributed throughout the space.[6]

5. *Augmented Reality*: a paradigm that originated around the same time as ubiquitous computing (1990)– the laying of dynamic and context-specific information over the visual field of a user (see below for more details).[7]

6. *Tangible Interfaces*: treating the whole of physical space around the user as part of a human-computer interface (HCI) by employing physical objects as carriers of information.[8]

7. *Wearable Computers*: embedding computing and telecommunication devices into clothing.

8. *Intelligent Buildings* (or *Intelligent Architecture*): buildings wired to provide cellspace applications.

9. *Intelligent Spaces*: spaces that monitor user's interaction with them via multiple channels and provide assistance for information retrieval, collaboration, and other tasks (think of Hal in *2001*).[9]

10. *Context-aware Computing*: an umbrella term used to refer to all or some of the developments above, signaling a new paradigm in the computer science and HCI fields.[10]

11. *Ambient Intelligence*: alternative term, which also refers to all or some of the paradigms, summarized above.

12. *Smart Objects*: objects connected to the Net; objects that can sense their users and display "smart" behavior.

13. *Wireless Location Services*: delivery of location-specific data and services to portable wireless devices such as cell phones (i.e., similar to *cellspace*).

14. *Sensor Networks*: networks of small sensors that can be used for surveillance and enviromental monitoring, to create intelligent spaces, and similar applications.

15. *E-paper* (or *e-ink*): a very thin electronic display on a sheet of plastic, which can be flexed in to different shapes and which displays information that is received wirelessly.[11]

While the technologies imagined by these research paradigms accomplish their intentions in a number of different ways, the end result is the same: *overlaying dynamic data over the physical space*. I will use the term "augmented space" to refer to this new kind of physical space. As I have already mentioned, this overlaying is often made possible by the tracking and monitoring of users. In other words, the delivery of information to users in space, and the extraction of information about those users, are closely connected. Thus, *augmented space is also monitored space*.

Augemented space is the physical space which is "data dense," as every point now potentially contains various information which is being delivered to it from elsewhere. At the same time, video surveillance, monitoring, and various sensors can also extract information from any point in space, recording the face movements, gestures and other human activity, temperature, light levels, and so on. Thus we can say that various augmentation and monitoring technologies add new dimensions to a 3D physical space, making it multi-dimensional. As a result, the physical space now contains many more dimensions than before, and while from the phenomenological perspective of the human subject, the "old" geometric dimensions may still have the priority, from the perspective of technology and its social, political, and economic uses, they are no longer more important than any other dimension.

This demise in importance of geometry as seen in augmented spaces can be understood as a part of a larger paradigm shift. If modern society as summed up in Michel Foucault's metaphor of Panopticum was organized around the strait lines of human sight, i.e. the geometry of the visible, this is no longer the case for our society. While some technologies such as video surveillance and infrared communication still require a line of sight, most do not. The examples are cellular and Bluetooth communication, radar, and environmental sensors. Instead of the binary logic of visible/invisible, the new spatial logic can be described using such terms as functions or fields, since from the point of view of these new technologies, every point in space has a particular value on a possible continuum. (Think for instance of a strength of your cellular signal which varies depending how close you are to a cell or whether you outside or inside.) In the case of information delivery into space, these values determine how much, how quickly and how successfully this information can be delivered – in other words, it corresponds to communication bandwidth. In the case of monitoring or surveillance, these values similarly affect how much and how successfully information can be extracted from a point, or region in space. In either case, if the old binary logic of visible/invisible (or present/absent) had still applied in this case, we would either register a signal or not. Instead we witness a new logic, which is described by the key intellectual paradigm of information society –

mathematical theory of communication developed by Claude Shannon and others in the 1940s. According to this theory, communication is always accompanied by noise, and therefore a received signal always has some noise mixed in.[12] In practical terms, this means that any information delivered to or extracted from augmented space always occupies some position on the continuous dimension whose poles is a perfect signal and complete noise. In a typical situation, we are usually somewhere in between: our cell phone conversation is accompanied by some background noise; a surveillance system delivers blurry or low-res images, which needs to be interpreted, i.e. a decision needs to be made by somebody what is the signal being present. Thus along with providing a theoretical framework to describe all electronic communication, mathematical theory of communication by Shannon turns to also perfectly capture the practical reality of our communications, at least up until now. That is, in the majority of cases, the signals we receive are accompanied by noise visible to us.

AUGMENTATION AND IMMERSION

I derived the term "augmented space" from the already established term "augmented reality" (AR).[13] Coined around 1990, the concept of "augmented reality" is normally opposed to "virtual reality" (VR).[14] in the case of VR, the user works on a virtual simulation, in the case of AR, she works on actual things in actual space. Because of this, a typical VR system presents a user with a virtual space that has nothing to do with that user's immediate physical space; while, in contrast, a typical AR system adds information that is directly related to the user's immediate physical space.

But we don't necessarily have to think of immersion in the virtual and augmentation of the physical as opposites. On one level, whether we think of a particular situation as immersion or augmentation is simply a matter of scale - i.e. the relative size of a display. When you are watching a movie in a movie theatre or on big TV monitor, or when you are playing a computer game on a game console that is connected to the TV, you are hardly aware of your physical surroundings. Practically speaking, you are immersed in virtual reality. But when you watch the same movie, or play the same game, on the small display of a cell phone or PDA that fits in your hand, then the experience is different. You are still largely present in physical space, and while the display adds to your overall phenomenological experience, it does not take over. So, whether we should understand a particular situation in terms of immersion or augmentation depends on how we understand the idea of addition: we may add new information to our experience – or we may add an altogether different experience.

"Augmented space" may bring associations with one of the founding ideas of computer culture: Douglas Engelbardt's concept of a computer augmenting human intellect that wasarticulated 40 years ago.[15] The association is appropriate, but we also need to be aware of the differences . For the vision of Engelbardt, and the related visions of Vannevar Bush and J.C.R. Licklider, assumed a stationary user – a scientist or engineer at work in his office. Revolutionary for the time, these ideas anticipated the paradigm of desktop computing. Today, however, we are gradually moving into the next paradigm, one in which computing and telecommunication capacities are delivered to a mobile user.[16] Thus, augmenting the human also comes to mean augmenting the whole space in which she lives, or through which she passes.

AUGMENTATION AS AN IDEA

Having analyzed at some length the concept of augmented space, we are now ready to move to the key questions of this essay. What is the phenomenological experience of being in a new augmented space? What can be the new cultural applications of new computer and network enabled augmented spaces? What are possible poetics and aesthetics of an augmented space?

One way to begin thinking about these questions is to approach the design of augmented space as an architectural problem. Augmented space provides a challenge and an opportunity for many architects to rethink their practice, since architecture will have to take into account the fact that virtual layers of contextual information will overlay the built space.

But is this a completely new challenge for architecture? If we assume that the overlaying of different spaces is a conceptual problem that is not connected to any particular technology, we may start to think about which architects and artists have already been working on this problem. To put it another way, the layering of dynamic and contextual data over physical space is a particular case of a general aesthetic paradigm: how to combine different spaces together. Of course, electronically augmented space is unique - since the information is personalized for every user, it can change dynamically over time, and it is delivered through an interactive multimedia interface, etc. Yet it is crucial to see this as a conceptual rather than just a technological issue – and therefore as something that in part has already - been a part of other architectural and artistic paradigms.

Augmented space research gives us new terms with which to think about previous spatial practices. If before we would think of an architect, a fresco painter, or a display designer working to combine architecture and images, or architecture and text, or to incorporate different symbolic systems in one spatial construction, we can now say that all of them were working on the problem of augmented space. The problem, that is, of how to overlay physical space with layers of data.. Therefore, in order to imagine what can be done culturally with augmented spaces, we may begin by combing cultural history for useful precedents.

To make my argument more accessible, I have chosen two well-known contemporary figures as my examples. Janet Cardiff is a Canadian artist who became famous for her 'audio walks'. She creates her pieces by following a trajectory through a space and narrating an audio track that combines instructions to the user ("go down the stairs"; "look in the window"; "go through the door on the right") with narrative fragments, sound effects, and other aural 'data'. To experience the piece, the user dons earphones connected to a CD player and follows Cardiff's instructions.[17] In my view - even though Cardiff does not use any sophisticated computer, networking, or projection technologies - her 'walks' represent the best realization of the augmented space paradigm so far. . They demonstrate the aesthetic potential of laying new information over a physical space. Their power lies in the interactions between the two spaces - between vision and hearing (what the user is seeing and what she is hearing), and between present and past (the time of the user's walk versus the audio narration, which, like any media recording, belongs to some undefined time in the past).

The Jewish Museum Berlin by Daniel Libeskind can be thought of as another example of augmented space research. For, if Cardiff lays a new dataspace over the existing architecture and/or landscape, then Libeskind uses the existing dataspace to drive the new architecture that he constructs. After putting together a map that showed the addresses of Jews who were living in the neighborhood of the museum site before World War II, the architect connected different points on the map and then projected the resulting net onto the surfaces of the building. The intersections of the projected net and the Museum walls gave rise to multiple irregular windows. Cutting through the walls and the ceilings at different angles, these windows evoke many visual references: the narrow eyepiece of a tank; the windows of a medieval cathedral; the exploded forms of the cubist/abstract/suprematist paintings of the 1910s-1920s. Just as in the case of Cardiff's audio walks, here the virtual becomes a powerful force that re-shapes the physical. In the Jewish Museum Berlin the past literally cuts into the present. Rather than something ephemeral, an immaterial layer over the real space, here dataspace is materialized to become a sort of monumental sculpture.

WHITE CUBE AS CELLSPACE

While we may interpret the practices of selected architects and artists as having particular relevance to thinking about the ways in which augmented space can be used culturally and artistically, there is another way to link the augmented space paradigm with modern culture. Here is how it works.

One trajectory that can be traced in 20th century art runs from the dominance of a two-dimensional object placed on a wall, towards the use of the whole 3-D space of a gallery. (Like all other cultural trajectories in the 20th century, this one is not a linear development; rather, it consists of steps forward and steps back that occur in rhythm with the general cultural and political rhythm of the century: the highest peak of creativity took place in the 1910s-1920s, followed by a second peak in the 1960s). Already in the 1910s, Tatlin's reliefs broke the two-dimensional picture plane and exploded a painting into the third dimension. In the 1920s, Lissitzky, Rodchenko, and other pioneering exhibition designers moved further away from an individual painting or sculpture towards using all surfaces of an exhibition space – yet their exhibitions activate only the walls rather than the whole space.

In the mid-1950s, assemblage legitimized the idea of an art object as a three-dimensional construction (*The Art of Assemblage*, MOMA, 1961). In the 1960s, minimalist sculptors (Carl Andre, Donald Judd, Robert Morris) and other artists (Eva Hesse, Arte Povera) finally started to deal with the whole of the 3-D space of a white cube. Beginning in the 1970s, installation (Dan Graham, Bruce Nauman) grew in importance to become, in the 1980s, the most common form of artistic practice of our times – and the only thing that all installations share is that they engage with 3-D space. Finally, the white cube becomes a *cube* – rather than just a collection of 2-D surfaces.

If we follow this logic, augmented space can be thought of as the next step in the trajectory from a flat wall to a 3-D space which has animated modern art for the last hundred years. For a few decades now, artists have already dealt with the entire space of a gallery: rather than creating an object that a viewer would *look at*, they placed the viewer *inside* the object. Now the artists have a new challenge: placing a user inside a space filled with dynamic, contextual data with which the user can interact. Alternatively, if we want to be more modest, we can say that the arrival of augmented space in the 1980s and 1990s as deployed in urban sphere was paralleled by the development of a similar concept of space by installation artists. If before 3D space was in practice reduced to a set of surfaces – walls in the case of the built environment; flat paintings or gallery walls in an art environment – now it is finally used as 3D space.

WHITE CUBE VERSUS BLACK BOX

Before we rush to conclude that the new technologies do not add anything substantially new to the old aesthetic paradigm of overlaying different spaces together, let me note that — in addition to their ability to deliver dynamic and interactive information — the new technologically implemented augmented spaces also differ in one important aspect from Cardiff's walks, Libeskind's Jewish Museum, and other similar works.. Rather than — laying a new 3-D virtual dataspace over the physical space, Cardiff and Libeskind overlay only a 2-D plane, or a 3-D path, at best. Indeed, Cardiff's walks are new 3-D paths placed over an existing space, rather than complete spaces. Similarly, in the Jewish Museum Berlin, Libeskind projects 2-D maps onto the 3-D shapes of his architecture.[18]

In contrast, GPS, wireless location services, surveillance technologies, and other augmented space technologies all define dataspace – if not in practice, than at least in theory - as a *continuous* field that completely extends over, and fills in, *all of* physical space. Every point in space has a GPS coordinate that can be obtained using a GPS receiver. Similarly, in the cellspace paradigm, every point in physical space can be said to contain some information that can be retrieved using a PDA or similar device. With surveillance;while in practice video cameras, satellites, Echelon (the set of monitoring stations that are used by the U.S. to monitor all kinds of electronic communications globally), and other technologies, can so far only reach some regions and layers of data but not others; the ultimate goal of the modern surveillance paradigm is to able to observe every point at every time. To use the terms of Borges' famous story, all of these technologies want to make the map equal to the territory. And if, in accord with Foucault's famous argument in *Discipline and Punish*, the modern subject internalizes surveillance and thereby removes the need for anybody to be actually present in the center of the Panopticum to watch him/her, modern institutions of surveillance insist that s/he should be watched and tracked everywhere all the time.

It is important, however, that, in practice, dataspaces are almost never continuous: surveillance cameras' look at some spaces but not at others, wireless signals are stronger in some areas and non-existent in others, and so on. As Matt Locke eloquently describes this,

Mobile networks have to negotiate the architecture of spaces that they attempt to inhabit. Although the interfaces have removed themselves from physical architectures, the radio waves that connect cell spaces are refracted and reflected by the same obstacles, creating not a seamless network but a series of ebbs and flows. The supposedly flat space of the network is in fact flat, pulled into troughs and peaks by the gravity of architecture and the users themselves.[19]

The contrast between the continuity of cellspace in theory and its discontinuity in practice should not be dismissed. Rather, it itself can be the source of interesting aesthetics strategies.

My third example of already existing augmented space – electronic displays mounted in shops, streets, lobbies, train stations, and apartments – follows a different logic. Rather than overlaying all of the physical space, here dataspace occupies a well-defined part of the physical space. This is the tradition of Alberti's window, and, consequently, of post-Renaissance painting, the cinema screen, the TV screen, and the computer monitor. However, if the screen has, until recently, most usually acted as a window into a virtual 3-D space; in the last two decades of the 20th century it has turned into a shallow surface in which 3-D images co-exist with 2-D design and typography. Live-action footage shares space with motion graphics (animated type), scrolling data (for instance, stock prices or weather), and 2-D design elements. In short, the Renaissance painting became an animated Medieval illustrated book.

My starting point for the discussion of the poetics of this type of augmented space is the current practice of video installation, which came to dominate the art world in the 1990s. Typically, these installations use video or data projectors. They turn a whole wall or even a whole room into a display or a set of displays, thus previewing and investigating (willingly or not) the soon-to-come future of our apartments and cities when large and thin displays covering most surfaces may become the norm. At the same time, these laboratories of the future are rooted in the past: in the different traditions of "image within a space" of 20th century culture.

What are these traditions? Among the different oppositions that have structured the culture of the 20th century, and which we have inherited, has been the opposition between the art gallery and the movie theatre. One was high-culture; the other was low-culture. One was a white cube; the other was a black box.

Given the economy of art production – one-of-a-kind objects created by individual artists – 20th century artists expended lots of energy experimenting with what could be placed inside the neutral setting of a white cube: by breaking away from a flat and rectangular frame and going into the third dimension; covering a whole floor; suspending objects from the ceiling; and so on. In other words, if we are to make an analogy between an art object and a digital computer, we can say that, in modern art, both the 'physical interface' and the 'software interface' of an art object were not fixed but open for experimentation. Put diffirently, both the physical appearance of an object and the proposed mode of interaction with an object were open for experimentation. Artists also experimented with the identity of a gallery: from a traditional space of aesthetic contemplation to a place for play, performance, public discussion, lectures, and so on.

In contrast, since cinema was an industrial system of mass production and mass distribution, the physical interface of a movie theatre and the software interface of a film itself were pretty much fixed: a 35-mm image of fixed dimensions projected on a screen with the same frame ratio, dark space where viewers were positioned in rows, and the fixed time of a movie itself. Not accidentally, when the experimental filmmakers of the 1960s started to systematically attack the conventions of traditional cinema, these attacks were aimed at both its physical and its software interfaces. Robert Breer, for example, projected his movies on a board that he would hold above his head as he walked through a movie theatre towards the projector; Stan VanderBeck constructed semi-circular tents for the projection of his films; etc.

The gallery was the space of refined high taste while the cinema served to provide entertainment for the masses, and this difference was also signified by what was deemed to be acceptable in the two kinds of spaces. Despite all the experimentation with its "interface," until recently the gallery space was primarily reserved for static images; to see moving images, the public had to go a movie theatre. Thus, until at least the 1980s, moving images in a gallery were indeed an exception (Duchamp's rotoscopes, Acconci's masturbating performance, which can be thought as a kind of animation within the gallery).

Given this history, the 1990s' phenomena of omni-present video installations taking over the gallery space goes against the whole paradigm of modern art – and not only because installations bring moving images into the gallery. Most video installations adopt the same physical interface: a dark enclosed or semi-enclosed rectangular space with a video projector at one end and the projected image appearing on the opposite wall. Therefore, from a space of constant innovation in relation to the physical and software interface of an art object, a gallery space has turned into what was, for almost a century, its ideological enemy – a movie theatre that ischaracterized by the rigidity of its interface.

Since the early days of computer culture in the 1960s, many software designers and software artists – from Ted Nelson and Alan Kay to Perry Hoberman and IOD – have revolted against the hegemony of mainstream computer interfaces, such as the keyboard and mouse, GUI, or commercial Web browsers. Similarly, the best of video or, more generally, moving image installation artists, go beyond the standard video installation interface – a dark room with an image on one wall. Examples of such artists include Diana Thater, Gary Hill, and Doug Aitken, as well as the very first 'video artist' – Nam Juke Paik. The founding moment of what would come to be called 'video art' was Paik's attack on the physical interface of a commercial moving image – his first show consisted of televisions with magnets attached to them, and TV monitors ripped out of their enclosures.

THE ELECTRONIC VERNACULAR

When we look at what visual artists are doing with a moving image in a gallery setting in comparison with other contemporary fields, we can see that the white gallery box still functions as a space of contemplation – quite different from the aggressive, surprising, overwhelming spaces of a boutique, trade show floor, airport, or retail/ entertainment area of a major metropolis.[20] While a number of video artists continue the explorations of the 1960s 'expanded cinema' movement by pushing moving image interfaces in many interesting directions, outside of a gallery space we can find much richer field of experimentation. I can single out four areas. First, contemporary urban architecture - in particular, many proposals of the last decade that incorporate large projection screens into architecture and project the activity inside onto these screens. Example include Rem Koolhaas' unrealized 1992 project for the new ZKM building in Karlsruhe; a number of projects again so far mostly unrealized by Robert Venturi to create what he calls "architecture as communication" (buildings covered with electronic displays); realized architectural/media installations by Diller + Scofilio such as *Jump Cuts* and *Facsimile*[21]; the highly concentrated use of video screens and information displays in certain cities such as Seoul, Hong Kong and Tokyo, or in Times Square, NYC; and, finally, imaginary future architecture as seen in movies from *Blade Runner* (1982) to *Minority Report* (2002), which use electronic screens on a scale that is not yet possible. Second is the use of video displays in certain kind of contemporary spaces where communication of information to public is the key functions: trade show design, such as the annual SIGGRAPH and E3 conventions; company showrooms; airports and train stations. The third is the best of retail environments. These range from small high-end boutiques (I will discuss this type of space in more detail shortly) to mega-size shopping centers / eating/ entertainment complexes which incorporate projection screens, dynamic lighting systems, mirrors, transparent and translucent surfaces to create an experience of an animated and dynamic space. The forth is the multi-media design of music performances, from the concerts of the brand name pop starts, to the numerous VJs performing nightly in clubs in most major cities on earth, to 'hybrid' groups which situate themselves between club and art culture, such as brilliant collective Light Surgeons based in London.

While at this moment they are still imagioned and implemented by the practioners from diffirent fields, we start slowly seeing the diffirent species of augmented spaces being combined into one. A shopping complex leads to an interior shopping street which leads to a multiplex; or an airport complex combines information displays about airline departures and ariival and shopping areas with their own promotions playing on LCD screens, and so on. Although at present the small electronic screens are usually distributed throughout these spaces (for instance, small LCD monitors mounted in elevators of new hi-rize buildings in Hong Kong and China such as CITIC Plaza in Guangzhou), the single larger screen (or other method for large image creation) has a potential to unite them all, offering a a kind of symbolic unity to a typically heterogeneous urban program: a shopping center + entertainment center + hotel + residential units. As as an example, consider Langham Place (Mongkok, Hong Kong, opened November 2004) developed by The Jerde Partnership, the pionners of the ubran version of 'experience design' they refer as 'placemaking.' An entertainment complex with an area of 1.8 million square feet, it combines a 15-storey shopping mall with 300 shops, a 59-level Grade A office tower and the 5-star Langham Place Hotel. The focal point of the complex is Digital Sky which is spanning the entire roof of the mall. Showing continuos visuals, this giant 'screen' is made possible by 200 projectors, PCs, speakers, and special effects lights.[22] No longer a square superimposed on a façade or a wall, here an image envelops the whole space as an ambient "elevator music" sky to shop under.

To discuss the use of electronic images in architecture further, let us turn to Robert Venturi. His projects and theories deserve special consideration here since, for him, an electronic display is not an optional addition but the very center of architecture in the information age. Since the 1960s, Venturi continuously argued that architecture should learn from vernacular and commercial culture (billboards, Las Vegas, strip malls, architecture of the past). Appropriately, his books *Complexity and Contradiction in Architecture* and *Learning from Las Vegas* are often referred to as the founding documents of post-modern aesthetics. Venturi proposed that we should refuse the modernist desire to impose minimalist ornament-free spaces, and instead embrace complexity, contradiction, heterogeneity, and iconography in our built environments.[23] In the 1990s, he articulated the new vision of "architecture as communication for the Information Age (rather than as space for the Industrial Age)."[24] Venturi wants us to think of "architecture as an iconographic representation emitting electronic imagery from its surfaces day and night." Pointing to some of the already mentioned examples of the aggressive incorporation of electronic displays in contemporary environments, such as Times Square in NYC, and arguing that traditional architecture *always* included ornament, iconography, and visual narratives (for instance, a Medieval cathedral with its narrative window mosaics, narrative sculpture covering the façade, and narrative paintings), Venturi proposed that architecture should return to its traditional definition as iconography, i.e. as *information surface*.[25] Of course, if the messages communicated by traditional architecture were static and reflected the dominant ideology, today's electronic dynamic interactive displays make it possible for these messages to change continuously ; making the information surface a potential space of contestation and dialog, which functions as the material manifestation of the often invisible public sphere.

Although this has not been a part of Venturi's core vision, it is relevant to mention here a growing number of projects in which the large publicly mounted screen is open for programming by the public who can send images via Internet or information being displayed via their cell phones. Even more suggestive is the project *Vectorial Elevation, Relational Architecture #4* by artist Raffael Lozano-Hemmer[26]. This project made it possible for people from all over the world to control a mutant electronic architecture made from search lights in Mexico City's Zócalo Square. To quote from the statement of the 2002 Prix Ars Electronica jury, which awarded this project the Golden Nica in the Interactive Art category:

Vectorial Elevation was a large scale interactive installation that transformed Mexico City's historic centre using robotic searchlights controlled over the Internet. Visitors to the project web site at <http://www.alzado.net> could design ephemeral light sculptures over the National Palace, City Hall, the Cathedral and the Templo Mayor Aztec ruins. The sculptures, made by 18 xenon searchlights located around the Zócalo Square, could be seen from a 10-mile radius and were sequentially rendered as they arrived over the Net.

The website feature a 3D-java interface that allowed participants to make a vectorial design over the city and see it virtually from any point of view. When the project server in Mexico received a submission, it was numbered and entered into a queue. Every six seconds the searchlights would orient themselves automatically and three webcams would take pictures to document a participant's design.[27]

Venturi's vision of "architecture as iconographic representation" is not without its problems. If we focus completely on the idea of architecture as information surface, we may forget that traditional architecture communicated messages and narratives not only through flat narrative surfaces but also through the particular articulation of space. To use the same example of a medieval cathedral, it communicated Christian narratives not only through the images covering its surfaces but also through its whole spatial structure. In the case of modernist architecture, it similarly communicated its own narratives (the themes of progress, technology, efficiency, and rationality) through new spaces constructed from simple geometric forms – and also through its bare, industrial-looking surfaces. (Thus, the absence of information from the surface, articulated in the famous "ornament is crime" slogan of Adolf Loos, itself became a powerful communication technique of modern architecture.)

An important design problem of our own time is how to combine the new functioning of a surface as an electronic display with the new kind of spaces and forms being imagioned by contemporary architects.[28] While Venturi fits electronic displays on to his buildings, which closely follow traditional vernacular architecture, this is obviously not the only possible strategy. The well-known Freshwater Pavilion by NOX/Lars Spuybroek (Netherlands, 1996) follows a much more radical approach. To emphasize that the interior of the space constantly mutates, Spuybroek eliminates all straight surfaces and straight angles; he makes the shapes defining the space appear to move; and he introduces computer-controlled lights that change the illumination of the interior.[29] As described by Ineke Schwartz, "There is no distinction between horizontal and vertical, between floors, walls and ceilings. Building and exhibition have fused: mist blows around your ears, a geyser erupts, water gleams and splatters all around you, projections fall directly onto the building and its visitors, the air is filled with waves of electronic sound."[30]

I think that Spuybroek's building is a successful symbol for the Information Age. Its continuously changing surfaces illustrate the key effect of the computer revolution: the substitution of every constant by a variable. In other words, the space that symbolizes the Information Age is not the symmetrical and ornamental space of traditional architecture, the rectangular volumes of modernism, nor the broken and blown up volumes of deconstruction. Rather, it is space whose shapes are inherently mutable and whose soft contours act as a metaphor for the key quality of computer-driven representations and systems: variability.

LEARNING FROM PRADA

Venturi wants to put rich electronic ornamentation and iconography on traditional buildings. In contrast, in his Freshwater Pavilion Lars Spuybroek constructs a new kind of space which he then fills with information – but information reduced to abstract color fields and sound. In other words, in the Freshwater Pavilion, the information surface functions in a very particular way, displaying color fields rather than text, images, or numbers. Where can we find today interesting architectural spaces combined with electronic displays that show the whole range of information, from ambient color fields to figurative images and numerical data?

Beginning in the mid 1990s, the avant-garde wing of the retail industry began to produce rich and intriguing spaces, many of which incorporate moving images. Leading architects and designers such as Droog/NL, Marc Newson, Herzog & de Meuron, Renzo Piano, and Rem Koolhaas created stores for Prada, Mandarina Duck, Hermes, Comme des Garcons, and other high-end brands; while architect Richard Glucksman collaborated with artist Jenny Holzer to create Helmut Lang's stunning New York parfumerie, which incorporates Holzer's signature use of LCD displays. A store featuring dramatic architecture and design, and the mixing of a restaurant, fashion, design, and art gallery became a new paradigm for high-end brands. Otto Riewoldt describes this paradigm using the term "brandscaping" – promoting the brand by creating unique spaces. According to Riewoldt: "Brandscaping is the hot issue. The site at which goods are promoted and sold has to reinvent itself by developing unique and unmistakable qualities."[31]

OMA / Rem Koolhaas' Prada store in New York (2002) pushes brandscaping to a new level. Koolhaus seems to achieve the impossible by creating a flagship store for the Prada brand – and at the same time an ironic statement about the functioning of brands as new religions.[32] The imaginative use of electronic displays designed by Reed Kram of Kramdesign is an important part of this statement. On entering the store, the visitor discovers glass cages hanging from the ceiling throughout the space. Just as a church would present the relics of saints in special displays, here the glass cages contain the new objects of worship – Prada clothes. The special status of Prada clothing is further enhanced by the placement of small flat electronic screens throughout the store on horizontal shelves right alongside the merchandize. The clothes are equated with the ephemeral images playing on the screens, and, vice versa, the images acquire a certain materiality, as though they are themselves objects . By positioning screens showing moving images right next to the clothes, the designers ironically refer to what everybody today already knows: we buy objects not for themselves but in order to emulate the specific images and narratives that are presented by the advertisements of these objects. Finally, on the basement level of the store, you discover a screen displaying the Prada Atlas. Designed by Kram, the Atlas maybe be mistaken for an interactive multimedia presentation of OMA (Office for Metropolitan Architecture, the name of Koolhaus' studio) research for its Prada commission. It looks like the kind of information that brands normally communicate to their investors but not to their consumers. In designing the Atlas, as well as the whole media of the store, Kram's goal was to make "Prada reveal itself, make it completely transparent to the visitors."[33] The Atlas lets you list all of the Prada stores throughout the world by square footage, look at an analysis of optimal locations for store placement, and study other data sets that underlie Prada's brandscaping. This 'unveiling' of Prada does not break our emotional attachment with the brand; on the contrary, it seems to have the opposite result. Koolhaus and Kram masterfully engage the 'I know it is an illusion but nevertheless I believe it' effect: we know that Prada is a business that is governed by economic rationality and yet we still feel that we are not simply in a store but in a modem church.

It is symbolic that Prada NYC has opened in the space that was previously occupied by a branch of the Guggenheim Museum. The strategies of brandscaping are directly relevant to museums and galleries that, like all other physical spaces, now have to compete against that new information, entertainment, and retail space: a computer or a cell phone screen connected to the Net. Although museums in the 1990s have similarly expanded their functionality, often combining galleries, a store, film series, lectures, and concerts, design-wise they can learn from retail design, which, as Riewoldt points out, "has learnt two lessons from the entertainment industry. First: forget the goods, sell thrilling experience to the people. And secondly: beat the computer screen at its own game by staging real objects of desire – and by adding some spice to the space with maybe some audio-visual interactive gadgetry." [34]

In a high-tech society, cultural institutions usually follow the technology industry. A new technology is developed for military, business, or consumer use, and after a while cultural institutions notice that some artists are experimenting with that technology and so they start to incorporate it in their programming. Because they have the function of collecting and preserving artworks, the art museums today often look like historical collections of media technologies from previous decades. Thus one may well mistake a contemporary art museum for a museum of obsolete technology. Today, while outside one finds LCDs and PDAs, data projectors, and HDTV cameras, inside a museum we may expect to find slide projectors, 16-mm film equipment, and 3/4-inch video decks.

Can this situation be reversed? Can cultural institutions play an active, even a leading, role, acting as laboratories where alternative futures are tested? Augmented space – which is slowly becoming a reality – is one opportunity for these institutions to take a more active role. While many video installations already function as laboratories for developing new configurations of images within space, museums and galleries as a whole could use their own unique asset – a physical space – to encourage the development of distinct new spatial forms of art and new spatial forms of the moving image. In this way, they can take a lead in testing outone part of the augmented space future.

Having stepped outside the picture frame into the white cube walls, floor, and the whole space, artists and curators should feel at home taking yet another step: treating this space as layers of data. This does not mean that the physical space becomes irrelevant; on the contrary, as the practice of Cardiff and Libeskind shows, it is through the interaction of the physical space and the data that some of the most amazing art of our time is being created.

Augmented space also represents an important challenge and an opportunity for contemporary architecture. As the examples discussed in this essay demonstrate, while many architects and interior designers have actively embraced electronic media, they typically think of it in a limited way: as a screen, i.e., as something that is attached to the 'real' stuff of architecture, i.e. surfaces defining volumes. Venturi's concept of architecture as 'information surface' is only the most extreme expression of this general paradigm. While Venturi logically connects the idea of surface as electronic screen to the traditional use of ornament in architecture and to such features of vernacular architecture as billboards and window product displays, this historical analogy also limits our visions of how architecture can use new media. For, in this analogy, an electronic screen becomes simply a moving billboard or a moving ornament.

Going beyond the 'surface as electronic screen paradigm', architects now have the opportunity to think of the material architecture that most usually preoccupies them and the new immaterial architecture of information flows within the physical structure as a whole. In short, I suggest that the design of electronically augmented space can be approached as an architectural problem. In other words, architects along with artists can take the next logical step to consider the 'invisible' space of electronic data flows as *substance* rather than just as void – something that needs a structure, a politics, and a poetics.

1 VRML stands for the Virtual Reality Modeling Language. In the first part of the 1990s, the inventors of this language designed it to model and access 3-D interactive virtual worlds over the Internet, and promoted it as the material realization of the idea of cyberspace. (See, for instance, Mark Pesce, „Ontos, Eros, Noos, Logos," the keynote address for ISEA (International Symposium on Electronic Arts) 1995, http://www.xs4all.nl/~mpesce/iseakey.html.) As of this writing (May 2002), Internet-based 3-D virtual worlds have failed to become popular.
2 This text was originally written in early 2002; current edit was done in September 2004.
3 Coined in 1998 by David S. Bennahum, the term "cellspace" originally referred to the then new ability to access e-mail or the Internet wirelessly. Here I am using the term in a broader sense.
4 It is interesting to think of GPS (Global Positioning System) as a particular case of cellspace. Rather than being tied to an object or a building,, here the information is a property of the Earth as a whole. A user equipped with a GPS receiver can retrieve a particular type of information relative to their location – the coordinates of this location. GPS systems are gradually is being integrated into various telecommunication and transportation technologies, from cell phones, to PDAs, to cars.
5 Recall the opening scene of Blade Runner (1982) in which the whole side of a high-rise building acts as a screen.
6 M. Weiser, "The Computer for the Twenty-first Century," Scientific American, 265(3):94–104, September 1991.
7 W. MacKay, G. Velay, K. Carter, C. Ma, and D. Pagani, "Augmenting Reality: Adding Computational Dimensions to Paper," Communications of the ACM, 36(7):96–97, 1993. Kevin Bonsor, "How Augmented Reality Will Work," http://www.howstuffworks.com/augmented-reality.htm.
8 See the 'Tangible Bits' project at the MIT Media Lab, http://tangible.media.mit.edu/projects/Tangible_Bits/projects.htm.
9 Guido Appenzeller, Intelligent Space Project (http://gunpowder.Stanford.EDU/~appenz/ISpace/); Intelligent Room Projects, AI Lab, MIT. (http://www.ai.mit.edu/projects/iroom/projects.shtml).
10 Tom Moran and Paul Dourish, "Introduction to the Special Issue on Context-aware Computing,"Human Computer Interaction, 16:108, 2001.
11 Ivan Noble, "E-paper Moves a Step Nearer," BBC News Online, 23 April, 2001. (http://news.bbc.co.uk/hi/english/sci/tech/newsid_1292000/1292852.stm)
2 If the noise falls below a certain threshold, we are able to reconstruct the send signal perfectly; conversely, if noise is above a particular threshold, the signal disappears. These thresholds are never absolute; they are specific to a particular communication situation, influenced by the bandwidth of a communication channel and also the content of a message.
13 For AR research sites and conferences, see http://www.augmented-reality.org.
14 With a typical VR system, all work is done in a virtual space; physical space becomes unnecessary, and it's the user's visual perception of physical space is completely blocked. In contrast, an AR system helps the user to work in a physical space by augmenting that space with additional information. This end is achieved by laying information over the user's visual field. An early scenario of a possible AR application that was developed at Xerox PARC involved a wearable display for copier repairman, which overlaid a wireframe image of the copier's insides over the actual copier as it was being repaired. Today, additional scenarios for everyday use can be imagined: for instance, AR glasses for a tourist that layer dynamically changing information about the sites in a city over her visual field. Military and artistic applications are also being developed, as presented for instance in the exhibition showcasing AR projects developed by Ars Electronica FutureLab (Ars Electronica Festival 2003). In this new iteration, AR becomes conceptually similar to wireless location services. The idea shared by both is that when the user is in the vicinity of particular objects, buildings, or people, then information about them is delivered to the user.But while this information is displayed, in cellspace, on a cell phone or PDA; in AR the information is laid over the user's visual field. The decrease in the popularity of VR in mass media and a slow but steady rise in AR-related research in the last five years is one example of the ways in which the augmented space paradigm is now overtaking the virtual space paradigm. Interestingly, this reversal can be said to be anticipated in the very origins of VR. In the late 1960s, Ivan Sutherland developed what we came to know as the first VR system. The user of the system saw a simple wireframe cube whose perspectival view would change as the user moved his head. The wireframe cube appeared overlaid over whatever the user was seeing. Because the idea of a 3-D computer graphics display whose perspective changes in real time according to the position of the user became associated with subsequent virtual reality systems, Sutherland is credited with inventing the first VR system. But it can be also argued that this was not a VR but rather an AR system because the virtual display was overlaid over the user's field of vision without blocking it. In other words, in Sutherland's system, new information was added to the physical environment: a virtual cube.
15 Vannevar Bush, "As We May Think" (1945); Douglas Engelbart, "Augmenting Human Intellect: A Conceptual Framework" (1962). Both in Noah Wardrip-Fruin and Nick Montfort, eds., The New Media Reader (MIT Press, forthcoming 2002).
16 And while it may still be more efficient to run, say, CAD, 3-D modeling, or Web design software while sitting comfortably in front of a 30-inch LCD display, there are many other types of computing and telecommunication activities that do not require or encourage stationary use.
17 I only experienced one of her "walks" that she created for P.S. 1 in New York in 2001.
18 For whose readers familiar with these concepts, the artistic augmented spaces I have evoked can be thought of as 2-D texture maps, while technologically augmented spaces can be compared to a solid texture.
19 Matt Locke, in Mobile Minded, eds. Geert Lovink and Mieke Gerritzen (Corte Madera, CA: Ginko Press, 2002), 111.
20 This passive and melancholic quality of video art was brilliantly staged in a recent exhibition design by LO/TEK, Making Time: Considering Time as a Material in Contemporary Video & Film, in the Hammer Museum in Los Angeles (February 4 - April 29, 2001). As Norman Klein pointed out to me, LO/TEK designed a kind of collective tomb - a cemetery for video art.
21 Overview of Diller + Scofilio projects can be found at http://www.labiennaledivenezia.net/it/archi/7mostra/architetti/diller/open.htm.
22 Raymond Wang, "Langham Place offices to roll next month," The Standard (Greater China's Business Newspaper), 19 June 2004 (www.thestandard.com.hk/thestandard/news_detail_frame.cfm?articleid=48588&intcatid=1).
23 Robert Venturi, Complexity and Contradiction in Architecture (New York: Museum of Modern Art, 1966); Robert Venturi, Denise Scott Brown, and Steven Izenour, Learning from Las Vegas (Cambridge, Mass.: MIT Press,1972.)
24 Robert Venturi, Iconography and Electronics upon a Generic Architecture: A View from the Drafting Room (MIT Press, 1996).
25 Robert Venturi in a dialog with George Legrady at the Entertainment and Value Conference, University of California, Santa Barbara, May 4, 2002. The term I 'information surface' is mine.
26 See http://prixars.aec.at/history/interactive/2000/E00int_01.htm.
27 Ibid.
28 See http://www.manovich.net/IA.
29 See Ineke Schwartz, "Testing Ground for Interactivity: The Water Pavilions by Lars Spuybroek and Kas Oosterhuis," http://synworld.t0.or.at/level3/text_archive/testing_ground.htm.
30 Ibid.
31 Otto Riewoldt, qtd. in Mark Hooper, "Sex and Shopping," ID, The DNA Issue (2001), 94.
32 For an insightful analysis of the branding phenomenon, see Naomi Klein, No Logo (New York : Picador, 2000).
33 Reed Kram, personal communication with the author, June 5, 2002. For more Kram projects, see www.kramdesign.com/.
34 Riewoldt, qtd. in Hooper, 2000.

Space + Media

BAZON BROCK, *born 1936 in Stolp/Pommern; 1957 to 1964 study of German language and literature, philosophy, history of art and political science in Zurich, Hamburg and Frankfurt a.M.; at the same time training and activity as dramatic advisor; 1957 first action piece; 1959 first happenings with Hundertwasser, Kaprow, Beuys, Vostell and Paik; since 1968 visitor schools at documenta in Kassel; 1965 to 1976 teaching of aesthetics at the University of Fine Arts, Hamburg; 1977 to 1980 Professor of the University of Applied Arts, Vienna; since 1980 Professor at the University of Wuppertal, focusing on neuronal aesthetics, imaging sciences; 1992 appointment as Doctor of Technical Sciences at the ETH Zurich; Co-founder of the research group Kultur und Strategie; 2004 Award of the Federal Cross of Merit, 1st class.*

INTERVIEW WITH BAZON BROCK

My work is marked by the mixing of physically perceptible space with virtual worlds. This act of mixing soon leads us to ask in which spaces we actually find ourselves. Can you please explain your approach with reference to the subject of media façades?

There are basically three important aspects. The first is architecture parlante, or speaking architecture. In the modern period this was rejected as a form of methodological weakness until it was rediscovered by the post modern in the form of Venturi's book “Learning from Las Vegas” in 1972. Postmodernism regarded the design of the external skin as a means of attracting attention to the contents of the building. In this way, the post-modern swiftly approached the standards of the modern because, when Mies van der Rohe regarded the use of a glass façade to connect inside and outside as democratic and recommended this as a way of articulating ubiquity and presence, then this was exactly what was meant by speaking architecture. Mies' modernism sought to use speaking architecture to express the social and political self-image of its occupants and a glass façade was an ideal means of removing this distinction between public and private.

Hence the media façade belongs to the theoretical tradition of architecture parlante and can, basically, be seen as close to Venturi's idea, with the only difference being that the media façade enables the representation of many different messages. I do not have to move far if I wish to encounter a range of styles, volumes and materials, because the media façade can concentrate all of these in one place. And this can create huge savings when one considers that a change in the ownership of a building can often lead to costly changes to the façade – especially where issues of corporate design are involved.

The disadvantage of this summarizing quality of architecture parlante is, of course, the fact that such buildings are harder to classify. If every company were to adopt such an approach and the façade of every building given the media treatment then there would be an inevitable loss of presence and individuality. It would be wrong to use such technology everywhere. If media architecture wishes to demonstrate its strengths then the use of its reductivist approach should be highly targeted. The maxim "less is more" applies here too and the relentless desire to show that everything is possible is counterproductive.

The second important aspect of the media façade is anthropological and involves, principally, the notion of facial mimicry which holds that any confrontation between two realities must lead to a mutual reaction and that only then does a situation come to life. This means that the media façade must become a means of expressing – of communicating a sense of surprise, disgust, anger and tenderness, etc. It must be equipped with the ability to take into account the mood of the viewer or, put another way, buildings must develop the ability to reflect the viewer's expectations. This principle is indispensable; it is derived from the notion of remote action and is comparable to the way in which two animals use facial expressions to signal if they wish to threaten each other or not. It is precisely this ability which enables animals to resolve disputes at a distance.

Media façades – but not architecture – have such potential and this potential could be realized if they were equipped (through, for example, the use of camera recognition technology) to react and, therefore, set up a dialogue. This principle of facial mimicry is extremely important precisely because it raises completely new possibilities. It is now possible for a building and its viewer to interact and for media façades to engage in the anthropologically important concept of mutual reaction. A company could, for instance, use advertising to evaluate whether its products respond to the needs of its consumers; or set up a dialogue with an instructional character by, for example, using technology to encourage environmental awareness amongst its customers by showing that the attractiveness of a product should depend as much upon the extent to which it is sustainable as upon price.

The third aspect is derived from the fact that the media façade is seen simultaneously by many people, as a result of which it creates – in the democratic sense – a true piece of the public realm.

Which brings us to the second area that I would like to discuss with you. Urban spaces are becoming ever more complex but the aesthetic differences between their buildings are being interminably leveled out.

Stop. It is not a question of being leveled out. The fact that spatial qualities are falling ever more into line is not a sign of leveling out but actually an aspect of the notion of Utopia. "Utopos" is, by its very definition, nowhere. But nowhere can only be made manifest by everywhere. Utopia describes a situation in which nowhere is special precisely because everywhere is. The architecture of the Hilton Group is, for example, the same everywhere which means, for the visitor, that he moves in a sort of utopian space. He can no longer use the architecture as a means of finding out where he is – and this is exactly how he wants it. Rather than complaining that differences are disappearing we should recognize that this is Utopia and that the task is to develop those other more mental abilities which we associate with Utopia; the abilities to imagine, make free associations and lend meaning. For, the more that differences disappear, the more I will have to rely on such abilities.

My suspicion is that this Utopia will make it very difficult for people to develop a sense of security.

Yes, because they lack the spiritual energy. In other words, technology has created the worldwide potential to truly create Utopia – or, to put it another way, the universal nowhere. And the question is whether humanity can survive this situation. How many people have the ability to mentally record differences and locate themselves physically without requiring external signals? This realization of Utopia means that one no longer needs to travel and that the mental work of creating an image of the world can be done at home.

Yes, but on the other hand we long for community, especially in the public space. We long for a sense of belonging.

Well a sense of belonging is naturally no longer possible when everything changes every thirty years. Thirty years is a generation and, if my surroundings don't look any more than they did when I was a child, then how can I feel a sense of belonging? There are no more constants.

But I have the hope that precisely the instrument of medialization will enable us to recreate a sense of regional identity. In Cologne during carnival season, for example, media façades are used to intensify the carnival atmosphere in the city. This was, of course, always done - with flags, etc. - but media offer the potential to intensify such efforts and avoid the danger of architecture creating some sort of universal nowhere. This spontaneous connection with the environment can then lead to a sense of security because a sense of regional difference can be created.

But this can of course also lead to the danger of thinking that media make all this possible with little effort. As soon as one believes that such emotions are available at the flick of a switch they become optional – and as soon as they are optional - they are no longer interesting.

Which means that enough effort and seriousness must be invested in the development of content.

Exactly: the content of the media façade and the design of this content become even more important. And a reductivist approach would certainly help.

And that for me is a perfect last word for our discussion. Many thanks!

***Prof. Dr. phil. Josef Lukas** is head of the Psychology Department of the Martin-Luther-Universität Halle-Wittenberg and chair of General Psychology. His research concentrates on: visual perception (in particular spatial and motion perception); psychoacoustics (in particular pitch perception); mathematical modeling of cognitive processes (e.g. memory, learning, attention); applied cognitive psychology*

What is real? How do you define real? If you're talking about your senses, what you feel, taste, smell, or see, then all you're talking about is how electrical signals are interpreted by your brain.

– Morpheus in the Film The Matrix (1999)

HOW MEDIA CREATE REALITY IN OUR HEADS

INTRODUCTION

In Andy and Larry Wachowski's film *The Matrix*, most humans live in a completely virtual reality. Their actual sensory organs – eyes, ears, nose, oral cavity, skin – are inoperative. The body floats in a sort of nutrient fluid. Only the brain is active and receives signals from the sensory nervous system – but these signals don't come from the sensory organs rather from a computer. Conversely, the "output" of the brain – the efferent motor signals – isn't transmitted to the body's physical periphery but also to this computer, to be precise: a vast computer system, the "Matrix". Sensoric input, motoric output and its feedback to the sensorium are precisely coordinated – not only for each individual brain but also for the interaction with other individuals, and in such a way that one's own perception and actions are not recognized by the brain as being virtual.

The film achieves a consequent, cinematic realization of an idea that is long known in philosophy and the intellectual history, but clearly seems to cause great difficulties for our common sense. For the naive worldview of our thinking, it is simply hard to accept that the world that surrounds us, the everyday objects, trees, buildings, street noise and our conversation partners, as well as all artworks we observe – the Sistine Chapel, the Mozart symphony at the concert hall – that all of this is generated in our head. Whatever we perceive, it is the product of our own brain. This led to centuries-old speculation about whether a certain real existence can or cannot be assigned to the "archetypes" of our perception, the triggers of perception. At this point, however, we don't have to dwell on this still ongoing philosophical debate between realism, constructivism and solipsism. Even for downright realists, the contents of consciousness derived from our perception, that which we see when we see and hear when we listen, are something different than the actual existing things. If the bird we hear singing really exists or not is more a philosophical question. Independent from this, one thing is certain: The song that we hear, the tweeting sound, only exists when we listen.

Virtual realities emerge when our sensory organs receive a stimulus that triggers sensations as similar as possible to those of the "real" object or situation. Thereby, it is not the similarity of the stimulus situation that is decisive, rather the similarity of the resulting sensation. At this point, one has to turn to cognitive psychology, which deals with exactly that: namely, the question of what sensations are, how they come about, which relation they have to the stimulus situation, how they can be defined, etc. In the following, some basic principles and results will be introduced, illustrated by means of examples, and their consequences for media representation discussed. A main focus will be on the question of spatial representations.

PSYCHOPHYSICS: THE RELATIONSHIP BETWEEN STIMULUS INTENSITY AND SENSATION

The strict conceptual separation between our sensoric sensations and the physical intensity of the stimulus triggering the sensation was established in the sciences at the latest with Ernst Heinrich Weber's seminal research on the tactile senses (Weber, 1834, *De tactu*). How loud something sounds, how heavy we find a backpack or how bright lighting appears – these are all sensations that should not be confused with the triggering stimulus which can be physically measured (noises as acoustic pressure, the backpack as mass or weight, for the light source as luminance density). And further: Sensations, per definition subjective occurrences, can equally be measured – hence, quantified just like the physical stimulus intensities. This is the real achievement of the classic psychophysics of Weber, Fechner, Brentano and many others (related details and a broad discussion can be found in Murray, 1993). Classic psychophysics of the first half of the 19th century substantiates that sensation can be quantitatively defined and that the connection between stimulus intensity and sensation is a subject of natural science.

In this context, the *perception threshold* takes on a central role. An absolute threshold is the (physical) stimulus intensity needed to even effect a (minimum) sensation. Stimuli below the absolute threshold cannot be perceived. There is a stimulus but no sensation. This in itself, by the way, already demonstrates the necessity of not confusing stimuli with sensations. The concept of a threshold could not even be formulated without this conceptual distinction. There are numerous illustrative examples: The weight of substances to a few milligrams is easy to determine with a precision scale – however, there is no sensation of weight involved. Also the hands of a bell tower's clock move in a (physically) measurable and continuous way but, in general, too slow for a *perception* of movement. We can observe the hands as long as we can, but there is no movement to detect. And, of course, there are sounds, noises, light stimuli, etc. whose intensity is too weak to be perceived. However, absolute thresholds are not natural constants. Rather, they are highly variable and depend on a multitude of factors: the kind of stimulus, the environmental conditions (the hearing threshold is naturally much higher in a noisy environment than in a quiet environment) or the quality of the sensory organs (there are great individual differences, and the hearing threshold, for example, increases for all of us with age). This, on the other hand, provides perception researchers with a very effective way to measure the sensitivity of a sensory modality: Sensitivity is defined as the reciprocal value of the absolute threshold. The lower the threshold, the higher the sensitivity. That is the simple logic. (Notice what your eye or ear doctor establishes first thing when you visit: It is the different absolute thresholds that are being measured.)

Equally as important as the absolute thresholds are the so-called difference thresholds. A difference threshold is the difference between stimulus intensities that one is still just able to detect. The famous Weber's law, named after the already mentioned Ernst Heinrich Weber, formulates the dependency of the difference threshold on the stimulus intensity. It says: *The difference threshold is directly proportional to the magnitude of the original stimulus intensity.* A tripling of the stimulus intensity also results in a tripling of the difference threshold. In practice this means: With objects of a weight of about 30 g (e.g. letters), differences of 1 g are already perceptible. This is not the case with 30 kg backpacks – here, the difference must be well within the range of about 1 kg. This regularity is relatively universal, also for many sensory modalities and qualities, and also for a relatively large scope of intensities. Occasionally, it is even used as the reason for linear wage increases: A wage increase of 50 Euros is "noticeable" for someone who earns 900 Euros per month; with a monthly income of 12,000 Euros, 50 Euros can hardly be perceived as a wage increase.

The real importance of Weber's law is that it provides a key for measuring our sensations. With a genius idea, Gustav Theodor Fechner laid the cornerstone of modern psychophysics: He defined the increment of sensation triggered by a stimulus increase of one difference threshold – per definition: the smallest possible increment of sensation as a measuring unit for sensations; he called it a "just noticeable difference" (later abbreviated as "jnd"). And with a bit of mathematics, Fechner could convincingly prove that the empirical findings could be quite clearly demonstrated when sensations are presented as a logarithmic function of the physical stimulus intensity. All this is all a long time ago: On 22 October 1850, Fechner had the key idea for his "psychophysical measurement formula", which mathematically describes the connection between the magnitude of stimulus and sensation. Only in our time does the psychophysical perspective – the distinction between objective and subjective situations – become prevalent in everyday thought and language. Meanwhile, the "felt temperature" has a fixed place in the weather report; at least in this context, it is widely accepted that sensations can be measured (even though only a very few people know how), and – as is often the case with new ideas that struggle to break through at first – the new achievement is used quite inflationary (and not always accurately): What it means when someone sighs that a lecture scheduled for 30 minutes lasted "a felt 3 hours" is interpretable in common speech. It isn't a precise statement in the sense of psychophysics (already for the reason that a subjective sense of time cannot be measured in hours). But we all know what is meant when a soccer trainer calls a tie a "felt loss" or when a film about full-figured young ladies is titled "Felt XXL".

COLOR PERCEPTION

Back to "real" psychophysics. Here, terms like "perceived distance" or "felt temperature" are not meant as vague metaphoric descriptions, rather they define new variables with the same scientific-theoretic intentions that are given to physical terms like temperature and length. The fundamental meaning of these measurements of sensation for the design of virtual media can be clearly illustrated by means of color perception.

Color is – as the introduction in the first chapter should have made clear – a *sensation*. Moreover, color is *not* a *physical* concept (even though many physicists today still find this hard to accept). Of course, there is a physical description of stimulus: Visual perception is based on light stimuli, and what we call color is closely connected with the wavelength of light. The basic facts are common knowledge: Our eye (more specifically: light sensitive cells in the so-called retina of the eye) essentially reacts to light, namely electromagnetic waves in the range of 400 to 700 nanometers. Shorter waves (e.g. UV or X-rays) or longer waves (infrared radiation, microwaves, radiowaves, etc.) do *not* trigger visual perception. Thanks to Sir Isaac Newton, it has been proven that all wavelengths of the visible spectrum are quite uniformly contained within bright white light (e.g. sunlight) and that light appears *colored* when one observes thin sections of the spectrum (which Newton produced with a prism that broke the various wavelengths in different ways). This fact is usually graphically demonstrated, where the spectrum from 400 to 700 nanometers is identified with the colors of the rainbow, from blue to green and yellow to red. This representation, however, is misleading at one point at least. Long-wave light in the range just under 700 nanometers actually looks red, short-wave light around 450 nanometers blue, light in the range of 530 nanometers green, etc. Yet, the *inverse* of this proposition is *not* correct. The sensation "red" is not solely triggered by narrow-banded light on the long-wave end of the visible spectrum. On the contrary: In nature, stimuli with a narrow-banded spectrum occur very rarely. A deep red tomato, for example, reflects a very broad spectrum: however, not every wavelength in the same intensity. The seductively simple imagination that each "color" is associated with a specific wavelength is popular – but simply wrong.

The correct relationship between the physical stimulus properties of light and the corresponding sensations (the color we perceive) can be precisely formulated with the methods of psychophysics, as we strictly distinguish between the color stimulus and the color sensation, define each for itself and then try to formulate the rule that connects one with the other. The color *stimulus* (e.g. the light reflected off a tomato and enters the eye) is physically defined by its spectrum, which specifies which wavelengths are contained in what quantity. The definition and description of color *sensations* are more complicated. Over the centuries, painters, artists and scientists have proposed many different color systems – the consensus that quickly emerged was that the description of all colors presumably requires three dimensions, which are usually referred to as color hue, saturation and brightness. Meanwhile, the three-dimensionality of our color perception has been proven through elaborate empiric investigations. These investigations are essentially psychological perception experiments – hence, not physiology and no mention of physics. Test persons generated stimuli, each of the identical color; they were therefore not distinguishable by color. In the second half of the 19th century, H. von Helmholtz could not only predict with these color mixing experiments that there are probably three different receptors in the retina (which were actually found in the first half of the 20th century) – he could also predict their reaction characteristics with surprising precision. Moreover, the mathematical description of the regularity of color mixing experiments by Hermann Günther Grassmann (1853) provided three-dimensionality with an empirically justifiable axiomatic foundation (Mausfeld and Heyer, 2003).

Nowadays, we can precisely articulate the principles underlying the psychophysics of color perception: In mathematical terms, the physical stimulus space is an infinite-dimensional vector space. Each stimulus is described with a real-valued function (its spectrum). The color space, on the other hand, is only a three-dimensional vector space, which means that each color can be clearly determined by specifying three real numbers. And: Three colors are essentially enough to produce all colors we are able to distinguish.

Meanwhile, there are internationally recognized norms for the graphic depiction of the three-dimensional color space. The most known one is likely the so-called CIE diagram, where each point corresponds to exactly one (!) color. In this diagram, complementary colors are positioned exactly opposite from each other; the results of the additive color mixing of the two colors are placed on the straight line connecting the two colors in the diagram. The in-depth knowledge of these principles is the basis for every use of color in the media. To show a red tomato on a monitor (or on photo paper, a cinema screen, a page of a gardening book, etc.) it is not necessary to produce a stimulus that has the same spectrum as a real tomato. On one hand, this would be extremely demanding; on the other, the transmission of the corresponding information would require completely unrealistic bandwidths. It is enough to generate a stimulus that *looks exactly* like a red tomato. And thanks to the psychophysical formulas for color perception, this can be precisely calculated. In other words: For the presentation of color in media, a system of basis vectors in the color space is selected (with monitors or TV sets these are mostly three lights: red-green-blue; with printers with 3 to 4 inks, mostly cyan-magenta-yellow) with which every perceivable color is produced.

That the technology of color television is indeed tuned to the human perception system becomes particularly obvious when compared with other biological perception systems. The range of visible light for various animals differs considerably from our light spectrum. Bees, for instance, also see light stimuli in the UV range; snakes perceive thermal radiation in the infrared range with their fovea centralis. Goldfish (and also many birds) have a four-dimensional color space, thus they can differentiate between color stimuli that appear identical to us. And also five-dimensional color spaces have been verified (e.g. with birds). For all of these animals, our colored media is more than disappointing!

MOTION PERCEPTION

To create an impression of movement through media is relatively simple in comparison to the phenomena of color and space: Displaying static images quickly enough one after the other is sufficient to generate the impression of continuous movement. As our perception system works relatively slow, only the "natural sampling rate" has to be outpaced. Already at a frequency of circa 25 images per second, flicker fusion begins, and the presentation of discrete single frames creates the sensation of continuous movement. With a self-made flip book (or even simpler: a basic drawing program on the computer) one can easily demonstrate that "fluid" motion can be effortlessly achieved even with very roughly sketched single images. It is very obvious that the "movement", or more precisely, the impression of movement is actively generated by the perception system, even when the image sequences contain only little or no hints of real movements, the positions of objects on the individual images are very far apart, the objects change in color and shape, and seem to simultaneously disappear in different directions, etc. (Hoffman, 1998).

Thus, movement as something "non-static" seems to develop very quickly. In the snow on the television screen we see an abundance of movement patterns. When it comes to motion perception, the challenge for media is not the generation of an impression of movement, rather the precision by which we differentiate movement patterns – in particular, when they are so-called biological movements, movements by humans and animals. But, as the focus of this article is on spatial perception, I shall leave the matter with these short references to motion perception. They are best illustrated with the "BML-walker" by Nikolaus Troje (see, for example, http://www.biomotionlab.ca/Demos/BMLwalker.html) that shows both: (a) how little information is needed to generate a rich and differentiated motion perception and (b) how sensitively our perception system reacts to the tiniest changes in motion dynamics and contentually interprets them (such as atmosphere, body shape, psychic or physical properties of that which is moving. Troje, 2002).

SPATIAL PERCEPTION

We now arrive at the main topic of this article, the representation of spatial realities through media: size, depth, position, distance between objects, distance between an object and the observer, etc. Thus, with "space" what is meant here is the perception of the environment around us, together with the objects within it and their geometric properties. We are familiar with the physical description of stimuli from geometry lessons: The *physical space* can be exactly described as a three-dimensional Euclidian space. If one uses a Cartesian coordinate system, for example, the spatial position of each point can precisely be numerically identified with the specification of three numbers (the x, y, and z coordinate). The *perceived space* is the plastic image in front of our eyes when we look around us. We speak of perceived distance, perceived size, perceived direction, etc. when we expressly refer to sensation factors. On first glance, our spatial perception under natural conditions (bright light, open view, etc.) seems to be exceptionally in line with reality. We are firmly convinced that things actually are as they appear: When we touch things, the haptic sense usually confirms our visual impression; we can successfully navigate a space, avoid obstacles, find targets. We find the car in the parking lot, unlock the door, start driving, overtake a truck – our sense of vision apparently provides us with such a precise image of our surroundings that it almost seems obsolete to differentiate between our environment and our perception of it. However, also in the case of spatial perception, it is exactly this distinction that enables us to approach the question of how our perception comes about.

If we, for the sake of simplicity, depart from a "realistic" standpoint and assume a real environment with real objects, whose location, form and size can be physically described, then the question with regard to spatial perception is: How can we have such a plastic, detailed image of this environment – simply by looking at it? The fact that bats master spatial orientation with the reflected sound patterns from their emitted ultrasounds and, at the same time, are able to move quite fast is for us – understandably – astounding and admirable. The performance of our own perception system usually seems far less amazing to us. The likely reason: Despite perception being an incredibly complicated process, it isn't associated with a feeling of effort. Many cognitive processes appear complicated and demanding to us: to argue mathematic evidence, to play chess (against a skilled opponent), to learn Chinese pictograms. Seeing and hearing, on the other hand, seem effortless for the most part – they are automated and function without conscious control. How complicated seeing really is becomes immediately evident through two facts: A major part of our brain is busy with perception in reality, and attempts to remodel perceptual performances on computers has proven to be one of the most persistent problems of Artificial Intelligence for decades. Computers that play chess and validate evidence have already existed for a long time – but the perceptual performance of computers still is comparatively modest.

So how does the "spatial image" come about? Let's first look at the function of the eyes. The light rays hitting the eyes trigger neuronal reactions in the retina, a layer of light-sensitive nerve cells. This process is absolutely comparable with how a video camera functions: An optical apparatus with lenses and light incidence regulation delivers a more or less sharp picture of the three-dimensional surrounding to a two-dimensional arrangement of sensors that react to light. During this process – which complies with the laws of physical optics and is technically described as a "projection" – a lot of information about the spatial extent, position and size of the depicted objects is necessarily lost. How can this information be "reconstructed"? In general, there are different possibilities: The image can be analyzed in terms of its *content* (objects are identified and then information about the space is extrapolated from their position and relationships, in combination with knowledge of projective geometry). Or one can use *several* images, taken either at the same moment from different perspectives or one after another from the same or a changing perspective. Our perception system uses all of these techniques. Due to historical and systematic reasons, we differentiate between two major classes of heuristics and evaluation methods: the *monocular* and the *binocular*.

MONOCULAR DEPTH CUES

Monocular depth cues are spatial information available for each single eye, hence already contained in *one* two-dimensional retinal image. We owe it to them that we have a spatial (although limited) perception even with a one-eyed view. These depth cues also play a vital role in media representations: Wherever a space is displayed onto a two-dimensional surface – on the painter's canvas, a computer monitor, a cinema screen, or on photo paper – in all these cases, only monocular depth stimuli can be realized. The projection takes place before the presentation.

Therefore, to ask about depth cues means: How does an image have to be designed so that it appears as "spatial" as possible? Picture books on the history of painting supply an almost inexhaustible mine of answers to this question. I will limit myself to shortly listing the most important mechanisms – one could write a small cultural history about each one of them:

Linear perspective: Before the invention of the pinhole camera, painters had to artificially create that which happens automatically in photography: They researched the principles of central projection and then could produce images whose geometry corresponded to that of the retinal image. In other words, one omits in paintings exactly that information which anyway gets lost in the projection of reality onto the retina. The picture should match up as close as possible with the retinal image produced through natural observation. Albrecht Dürer impressively illustrated this method in his painting "Man Drawing a Lute" (1525). The mathematical foundations for the procedure were supplied by Leon Battista Alberti, among others, in his text "De pictura" (Florence, around 1435/36).

Occlusion: In central projection, objects (or parts of objects) are typically concealed by other objects. The pattern of mutual occlusion provides the perception system with important hints about the spatial position of the objects. The depth cue occlusion is beautifully demonstrated in, for example, René Magritte's painting "Carte Blanche" (1965). Empirical evidence for the effectivity of occlusion in spatial perception was provided by, for instance, Albert Ames in a series of experiments and demonstrations (Kilpatrick, 1961).

Relative size: The greater the distance to an object, the smaller it is depicted. Therefore, the relative size displayed on the image contains potential information about its distance from the observer. The situation is similar for:

Relative height: Objects above the horizon appear farther away the deeper they are presented in the image. The exact opposite is true for objects below the horizon.

Texture gradient: Regular textures (e.g. the texture of a freshly plowed field) attain an impression of depth by becoming smaller and denser with increased distance.

Light and shadow: The list of monocular depth cues can be almost infinitely extended. However, the relation of light and shadow plays a key role. This criterion is and has been frequently used primarily in media representations on computer monitors as stunning depth effects can be achieved with relatively simple methods. For the visual impression of a rise or fall, such as the display of buttons for graphic user elements of programs, it is sufficient to indicate a narrow dark shadow. Perhaps you remember the first versions of ball games on C64 computers? A small, dark spot that could be interpreted as the shadow of the ball led to a considerable improvement of the spatial impression. A related classic research was conducted by Kersten, Knill, Mamassian and Bülthoff (1996).

Motion: Additional (and rich) spatial information comes from moving images. The motion parallax is one of them (what moves *fast* across the retina – under otherwise same conditions – is perceived as being closer to the observer than what moves slower). The "optical flow" that takes place when we move (we fixate the target of the movement, which causes its image on the retina to remain relatively static; the farther a point lies toward the periphery, the faster it moves) is often used in movies, for example, to induce a pronounced spatial quality. The movement of an object (e.g. rotation around its axis) results in a spatial impression created by the silhouette of the rotating figure.

Adaptation: The lens of the eye automatically adjusts to the distance of an object in order to obtain a sharp image of it on the retina. The resulting curve of the lens could, in principal, be used as information on the object's distance, and adaptation does indeed have an – albeit limited – influence on spatial perception (see e.g. Fisher and Ciuffreda, 1988). This depth cue is difficult to apply in media representations as the eye adapts to the image surface (therefore causing adaptation to usually conflict with other depth cues). The role of adaptation while viewing images is discussed by Ciuffreda and Engber (2002).

BINOCULAR DEPTH CUES

The most important basis for spatial vision is that we see with *two* eyes. Therefore, there are *two* two-dimensional images from slightly different perspectives available for the perception system at all times. It reconstructs the three-dimensional original image from the differences between the two individual images. Only with binocular viewing do we gain the real, plastic impression of a space, which is called three-dimensional vision.

In order to create this impression artificially, two images representing the perspectives of the two eyes have to be produced, and then each eye is only offered the image that matches to it. The technology for this has existed for a long time: Stereo photography uses two cameras mounted on a track. The two individual images are presented in a stereoscope. Wheatstone developed one of the first devices: By means of a mirroring system, the two eyes are presented different images without the observer having to squint.

The so-called anaglyph technique is often used for the stereoscopic presentation of images or films: Both image parts are designed one on top of the other; the image for the right eye in red, the image for the left in cyan. If one views these images wearing glasses equipped with a corresponding color filter on the left and right side, each eye only gets to view *one* image. Hence, with this technique, the function of the "glasses" for viewing stereoscopic images or films is always that of separating the two half-images: Both graphics are being presented (that for the right and that for the left eye), however, the glasses filter out the image for the respective eye. In place of colored images and filters, also polarized light and polarizing filters can be used to present and separate the stereoscopic pair of images. Another possibility is to display both images *interchangeably* in rapid succession and view them with shutter glasses that alternately cover the right and left eye in sync with the presentation. Applications of the anaglyph technique for computer displays are described, for example, in Janssen (2009).

FINAL REMARKS

The available space for this article only allows for brief insights into the processes that cognitive psychologists refer to as "perception". On the one hand, our perception system is the evolutionary result of a phylogenetical adaptation to the environment, but on the other hand, it is also derived from the individual experience of learning. One who has learnt to read does not only "see" the black and white contours of a printed word but the word in its meaning. In reverse, we are illiterates in most foreign writing systems and, consequently, only see forms and patterns in Korean, Arabic or Indian writing.

Whoever strives to create realistic perception with artificial means must know the principles of perceptual processes. The classic textbooks of cognitive psychology, such as Goldstein (2007) or Wolfe, Kluender and Levi (2009) offer a useful introduction. However, they do not answer the question of how "real" the world around us is either. The naive standpoint that there is a real existing world beside our perceived world and that this "reality" is described by physics and chemistry probably can't be upheld, at least not in such simplicity. Also the physicist's idea of the world is informed by his or her perceptions and inner images. The human spirit of invention and technology have expanded our understanding of the world far beyond direct sensual perception. How close they bring us to "reality" and whether reality beyond human consciousness is even a useful category, however, belong to the questions that cognitive psychology cannot answer.

LITERATURE

Ciuffreda, K. J. & Engber, K. (2002). Is one eye better than two when viewing pictorial art? Journal of Optometric Vision Development, 33, 172-177.
Fisher, S. K. & Ciuffreda, K. J. (1988). Accommodation and apparent distance. Perception, 17, 609-621.
Goldstein, B. E. (2007). Wahrnehmungspsychologie: Der Grundkurs, Spektrum Akademischer Verlag.
Grassmann, H. G. (1853). Zur Theorie der Farbenmischung. Poggendorff's Annalen der Physik und Chemie, 89, 69-84.
Hoffman, D. D. (1998). Visual intelligence: how we create what we see, Norton.
Janssen, J. K. (2009). Räumlich am Rechner. Photos, Videos und Direct 3D-Programme stereoskopisch darstellen. c't, 15, 84-89.
Kersten, D., Knill, D. C., Mamassian, P. & Bülthoff, I. (1996). Illusory motion from shadows. Nature, 379, 31.
Kilpatrick, F. P. (1961). The nature of perception: some visual demonstrations. In: F. P. Kilpatrick (Ed.), Explorations in transactional psychology (36-57). New York University Press.
Mausfeld, R. & Heyer, D., eds. (2003). Color perception: mind and the physical world, Oxford University Press.
Murray, D. J. (1993). A perspective for viewing the history of psychophysics. Behavioral and Brain Sciences, 16, 115-186.
Troje, N. F. (2002). Decomposing biological motion: A framework for analysis and synthesis of human gait patterns. Journal of Vision, 2, 371-387.
Wolfe, J. M., Kluender, K. R. & Levi, D. M. (2009). Sensation and Perception, Sinauer Associates.

In my thoughts on "Spatial Concepts and Augmented Space", I describe that, according to my observation, also ideas can take on a spatial aspect in our thinking. I hoped in vain that brain research might offer me some deeper insight. It became clear that science is still far away from being able to provide qualified answers to such questions. A direct request to Prof. Lukas nevertheless provided a sketchy idea of how to approach this challenge, and Prof. Lukas authorized me to publish this sketch as the closing words of this article.

The question of the role of ideas in the construction of our perception spaces is, in my opinion, as justified as it is profound. Above all, philosophers have dealt with this, in particular Kant and Lotze – however, with philosophy a field as well as an approach for these questions is addressed, a realm that I deeply respect but do not navigate all too safely. In any case, it seems evident to me that the three-dimensionality of physical space cannot be reasoned by physics, rather it is a product of the three-dimensionality of our spatial concept. We think three-dimensionally (thinking in the sense of spatial, pictorial visual imagination), which is why we model our surroundings after this pattern. Seen in this light, it is definitely not the sensual stimuli that form our image of the world, but our cognitive equipment that shape a vivid world from neuronal signals… (As you see, I am starting to dabble – but these are naturally the really interesting questions!)

Space + Media

Klaus Wassermann *and* ***Vera Bühlmann*** *have worked together on various projects addressing issues of design theory, mediality and the digital world, and have also been involved in related practical projects looking at, for example, media façades and the possibility of their autonomous "self-curation".*

Klaus Wassermann *is head of the Noosynthesis Laboratory of the Department for Computer Aided Architectural Design of the ETH Zurich and conducts research into the philosophy, theory and implementation of autonomous software systems, with a special focus on their implications for architecture and urban design.*

Vera Bühlmann *researches media theory and philosophy at the Department for Computer Aided Architectural Design of the ETH Zurich. She holds a PhD in Media Philosophy from Basel University. Since 2003 she has been a researcher and lecturer at various Schools of Art and Design in Switzerland.*

STREAMING SPACES –
A Short Expedition in the Land of Active Medial Façades[1]

TO THE MIDDLE, PLEASE

Without doubt, media façades are representatives of that group of hybrids, real cross-breeds, which comprehensive digitalization has given us. They combine two concepts that could not be more dissimilar – or so it seems at first sight. On the one hand is the façade, made of stone, metal and glass, component of a static artefact with a hopefully long lifespan; on the other hand are the media, that fleeting, hardly ever tangible abstractum, roaming between things and events, those ferrying, formerly even hallowed channels[1] whose interference with the relationship between sender and recipient is so easily overlooked. Whereas the façade stands for a partitioning surface, a mineralized demonstrative of compartmentalization and demarcation, what is media-like indeed lives in an intangible interstice, whose discovery happened not very long ago and is intimately intertwined with the propagation of the technology of the immaterial.

The forerunners of the densification of our media-ness, which by now is quite advanced, show up in a likewise hybrid history partaking of various domains. It may appear as a mere curiosity to see this history cross an imaginary axis between Basle and Mannheim, the basin of the Upper Rhine. But it is this region that Gutenberg's printing press, Petri's printing and smuggling house[2], Euler's mathematics in Basle[3] and Braun's famous tube in Strasbourg[4] are native to. Mathematics, printing and electric sciences are at the root of all present digital technologies. Gutenberg's Universe, at that, is far from being obsolete. Printed electronics, *printed physics*[5] as it were, is now at the eve of market introduction, and this new amalgam will probably media-ize our relationship with things as revolutionarily as Gutenberg's press did with letters and books.

Digital technology opens up for us the potential and the pervasive possibility of virtualizing absolutely everything. In parallel, we have been getting familiar for about ten years with the network idea as the mainstay structure of our society, and it is not entirely clear where this collective exploratory trip into immateriality will lead us. The network, which knows neither identifiable edge nor centre, and which explodes the time-honoured notion of causality, develops quite obviously into a form of living[6] as did alphabetization 100 or 150 years ago.

What role might the idea of the media façade play in these circumstances? Is it conceivable for this hybrid concept to become part of our media infrastructure, even a matter of *apps*? Upon this question hangs not least the one regarding form and potential, or indeed the conceivableness of business models.

Media façades are a relatively new media format, surfacing as a possibility probably first in cinema[7]. Hence, no stable level, no maturity of development with concomitant fixing or standardizing of forms can reasonably be expected. For achieving that, media façades are still too infrequent in Western culture, despite Times Square and other circuses. We are probably still wrapped in an exploratory phase and therefore missing at once a sufficiently differentiated conceptuality as well as appropriately honed procedures for discussing what in a specific place might be conceivable in terms of media façade. We are simply lacking experience with façades that outpour their media-ness in a novel rhythmicity over us passers-by.

It may be found that the parentage and the background of the media façade idea prove important for recognizing the resulting potentialities, and using the many fascinating possibilities that emerge in the technological area, from "automatic" generation of content through large-surface tactual sensitivity to three-dimensional representation of visual arrangements within the expanse of a landscape. By parentage, we don't refer primarily to historically datable events. Michel Foucault exposed that perspective as being unproductive and confronted its book-keeperish thrust with his own concepts of cultural disposition and the expressional field, the abstract and time-spanning space of our perceptions, ideas and concepts that intermingles with concrete actional space.[8] In following Foucault, it becomes apparent how profitable deciphering the sedimented layers of the present expressional field and its transmutations may prove. This is what our reference to origin is about. Evolution, be it of ideas, perceptions, concepts or material arrangements, succeeds – quite in line with Foucault's thought – mostly just as a superimposition, as a duplicating and deflectingly incipient alteration within the fabric of functions. Thus we shall, on our expedition, attempt more closely to examine the soil in which some well-known problems are rooted.

On our quest, we shall principally touch upon three notions: media, façade and image. Generally, they are believed to be physical bodies as it were. This is not necessarily a delusion. But here again, conception undoubtedly foreruns things. Hence, we shall concentrate on the structure of these concepts rather than on phenomena.

MEDIUM AND MEDIA-NESS

Media are not, as such, an invention of the modern age, even if Hugo Ball in 1920 first remarks in modern form upon their relevance. He seems to be stupefied, indeed amazed, as he says: *"All the world has turned media: for fear, for fright, for agony, or because there are no more laws – who knows?"*[9]

Media techniques were developed in all cultures and religions, even prehistoric ones. Strangely – and not altogether by happenstance in the wake of deconstructivism – the atavistic notion of angels resurfaces in today's media science.[10] Our present perception and usage of media, seen in the light of cultural history, seems to have the "daimon" as its forerunner, the animated Nature, which quite into the 20th century rather consistently dealt with extrasensory phenomena. There is a true core to this, seeing that every sign, every signal-like situation depends on a medium for *trans*mission, a substrate for embedding. Strangely, the medium as substrate is part of the narrative, of the exchange of signs – and yet it isn't. The very length it took to discover media-ness as an autonomous givenness demonstrates that media-ness is about something non-trivial. Media-ness, the potential of being a medium that befits every medium irrespective of material actualisation, is not reducible to specific media formats. There simply is no prototypical medium.

So what is it that makes us use the medium for things and phenomena so diverse? To begin with, books are just books, television sets just electronic boxes for transforming or decoding oscillations of analogue electric tension into dynamic visual representation, and computers are just another electronic boxes executing binary-digitally encoded comparative operations. Much as water in the oceans, to begin with, is just water, and the gas mix above it just air.

But now, to whales water is no more "just" water, than air to us is "just" air. We are, as are the whales, since time immemorial, embedded in these material environments, and our as well as their communication has grown in them. The vibrations imprinted upon these environments, the – literally – evoked density variations, transform into signals as soon as a potential receiver becomes imaginable who is capable of separating the signal from the concomitant noise. Such separating requires models regarding the medium and a subjectively given number of possible signals; but it does *not* require a model of semantic interpretation. Much rather, the "communicant partners" achieve socialization precisely by evolving, over time and through their own behaviour, the semantics of those signals and the meanings they should be given. Only once signals become interpretable within a – generally multi-layered – cultured regulatory mechanism can they graduate to signs. Mediazation today means, under this aspect, first of all that all possible "signalist effects" have become signs.

Right on the signals level, the environment is assigned a new quality as "substrate", as supporting "milieu". This means media-ness for the air – given its transmitting role of possible density variations – just as for water in the case of whales. Our signs, perceived as semantic carriers, in turn are not being fired by the air medium; much rather, signals are the genuine medium for signs. Evidently, the initially physical characteristics of carrier environments may well impact upon the concomitant media-ness, through constraints – in the case of media façades actual architecture and its neighbourhood. As the layering proceeds to higher levels, however, the importance of these "physical" constraints increasingly diminishes. A word is a word, irrespective of whether it is spoken, written or encoded in binary bits. Quite clearly, it would be a categorial misconception to perceive "air" or "water" as the proper medium for a "word" and to break down media-ness to the materiality of the environment.[11] The same now holds true not just for air, but for the books, indeed for all physical formats as well. *The medium of signs consists of the signs' own stream.*

Every population-like regularity that is susceptible of having, in reiterated fashion, density or probability variations imprinted upon itself, can progress to the role of "carrier milieu" for media events. In a social sense, this particularly includes rituals, traditions, habits, and other behaviours whose expectative horizons are sufficiently established to allow for irritations and irregularities in repetition. As for semiotically fertile signs – which certainly do not include corporate logos with their simple indexicality, nor corporate-identity-related visual "messages" –, their repetition, when paired with continuing non-limitative interpretations, generates the media-ness that on the one hand provides the context for all manner of narratives and, on the other hand, secures durable development of format. Such pairing, together with the descriptive stream, provides what is termed sign situation – that intimate relationship which denotes the affective simultaneity of the jointly traversed space. Charles Peirce equated the sign to the sign situation itself.[12] In other terms, the sign emerges only when originator and interpreter are in synchronous agreement regarding the signal, code, and pragmatics of a situation. This in our view seems to be an important portent of the success of the novel media façade format.

Communication is not a cybernetic, i.e. ultimately mathematical or otherwise formulary undertaking.[13] Sign situations are conglomerations of very heterogeneous things into one irreducible, indissociable unit:[14] i.e. of the signal, its physical origin, its interpretation which lets the signal presently mutate into a sign, the carrying and enwrapping medium, and finally the surface of the communicational community, within which unceasing determining is going on about which signal becomes a sign and which sign gets demoted to a signal. Hence it is clear that manifold conventions, relative to our use of symbols and to our forms of social intercourse, are operative, of which the design of media façades must take account. We can, at least for the field of non-technical communication, identify such notions as emitter and receiver as inadequate, and thereby avoid misunderstandings that are, in the case of media façades, apt to prove costly.

We find the term of "media façade" lacking in pertinence, if it were meant to imply, even before the recurrence of a yet to be established narrative, that there is a link between media-ness and this particular type of façade. This takes up our above finding, that air is not the sole medium for "word". Much rather that façade is a "façade with increased mediatic capacity", a "mediatically active façade", and in that sense a "mediatic façade"[15]. And inversely, there can be no question of tying a particular so-called media format, such as images or film, to a façade as material format, to the exclusion of everything else, even as for us too the affinity between the image and the media façade is sufficiently strong for us to revert a bit later to certain details of the image concept.

Therefore, we may state that there is no specific "façade medium", not any more than a specific "word medium", no prototypical materiality of a media façade nor one single medium for language. Undisputedly, of course, the medium and the signs, as well as media-ness and narrative are mutually conditional and formative.

Thus, mediatic façades are not primarily to be seen as tools for displaying known media, but rather as media in their own right, which indeed generate further media-ness. Mediatic façades are in specific fashion *mediogenic*. All that is being displayed by means of a technical format may, as we have seen, secondarily become a signal, and a new sign, thereby retrogradely functionalizing the exhibited envelopment and whatever was sufficiently often and similarly repeated, into a medium. Conversely, signs and narratives can in their turn also become "technique". So, the display of cartoons on a façade may itself become a medium. In that case, the medium consists of the modulation of a cartoon and its display on a façade, but not the materiality of the media façade or the featured film. *Façons de parler* are an unconditional prerequisite for a "speaking of the façade", a *parler de la façade* as it were, and even more for a *façade du parler*.

As long as it's all about communicating a previously set message with maximum effectiveness, the onlooker cannot help the strange feeling that something might be wrong with the communicative situation. Best regards from São Paulo[16]. We find here, with the messenger or, respectively, the message, not just in the German language, a leftover of late-monarchistic scientism of the late 19th century, a heavy, alien erratic block from the philosophic Miocene.

Screens with purely commercially-shaped content are getting mentally zapped; what there remains in memory, on the screens and in anticipation is just what an interesting story outline, or an appropriate semantic milieu manage to provide towards attractive stories. As Bernhard Waldenfels clearly demonstrates[17], attention is aroused or happens only in so far as the personal space of the subject addressed is being respected, be it thanks to humour or to culturally encapsulated interestingness; but it is totally impervious to being reduced to an automatism, whether neurologically or behaviouristically inspired. Narratively attractive framings might, and probably will, someday turn up in totally new kinds of games or urban competitions in which the role of buildings acquires additional meanings. An initiative of the city of Munich a few years ago is worth mentioning, where over a period of some weeks poetry was displayed on billboards, not as an art but as a media form. The action proved very popular. Relating to media façades, this example may show the wisdom of bewaring of hasty fixations upon certain classes of possible themes.

IMAGING AND SAYING

As humans, we have a special relationship to depictions. Biologically, the primary sense of man is sight, culturally it's reason. On this soil human cultures are generally developing a rich visuality. The relationship between picture and brains is not unproblematic; there seem to exist intrinsic fault lines. Time and again in history there were controversies over the relative weighting of word and image – iconoclashes, as Bruno Latour calls them[18]. The discussion around ornament, or ornament in architecture is part of them. Actually there are renewed allegations about a change of weightings – the future of our culture to be much more determined by images than was the case in the recent past, or so they say.

But how does the typical effect of an image, or moving images, or a film come about?[19] What is this effect made of, beyond the – loosely termed – primary semantic content of an image, which is never adequately drawn into words? For the future of mediatic façades, much may be at stake from the answers to these questions.[20] In order better to understand the mode of action of mediatic façades, we need an at least sufficient understanding of the envisaged visual formats. In contrast to "normal" still or moving images, the format of the media façade requires us to take into account the scaling, too, that the images will impose upon the size of a building. It makes a massive difference whether an image is displayed on a screen 30 by 50 inches or on one 70 by 160 feet. Neither the specific temporality of images nor the sets of representational modes are simply portable across vast dimensional divides.[21]

Picturability or picturization[22] are neither limited to the arrangement or the dynamics of colours, nor to the configuration of forms or indeed subjects, nor to the cases where images at times seem to do service as visual referents. But neither are images simply referents to things external to themselves, no matter how abstract and derivate the nature of the referents. Picturability is not to be likened to signlikeness, even as images happen sometimes to be used as signs and as picturability without semiotic tools proves impervious.[23] All these and some more dimensions are part of the very net of pre-specific relationships in which the individual with his cognition and social collocation is also caught up.[24] For a cognitive person[25] to be able to make something out of an image, a video or some other format, they must (be able to) revert to a structure of decora, general habits of doing, organizing, and recombining things this way or the other, but not differently.[26]

An image seems to claim from the viewer precisely what a notion consistently and completely forbids: maximum approach, indeed immersion into the dynamics of interior never totally graspable cross-references, wildly roving interpretations. Hence the occasional feeling of being taken possession of by images. As soon as we harness images for a purpose, they stop being images and turn into mere indexical pictographs. If on the other hand we turn to interpreting notions by dissecting them like images, we forgo not only the possibility of discourse but lose the very notion as well. Nevertheless, in the case of images this meandering interpretation leads to a pragmatism all its own. The author of a visual arrangement will be able to rely on it that “his” image will trigger that very kind of interpreting, at least as long as the life forms of author and viewer sufficiently overlaps.

The pragmatism of the imaging act may therefore well be seen as a twofold call[27], first to the establishing of a sign relationship in which the viewer must agree to a self-referential role as “emitter” and “receiver”, and subsequently to direct transformation of the so-called image content. This twin challenge ensconced within the pragmatism of the imaging act directly reveals the difference between shown and “merely” seen images, whereby both an imaging act and its production are themselves in a certain way encoded. Not every shown image will be accepted as an imaging act, were these encodings to be incomplete, whereas conversely so-called conscientious viewing may, thanks to said self-referentiality, transform what is being seen into something shown to itself. Displayed images call for an activity proper, which due to social encoding of image and activity is itself subject to decoding, wherefore the viewer may find himself asking, “What am I to do with it?” Media façades too are a cultural technique that is open to learning.

Particularly in view of the interpretative stimulus implicit in every displaying action, the shown image never points (semantically) constitutionally to what it (visually) shows. As part of a Peircean sign and, what’s more, as itself continually generating such signs, it rather points primarily to the non-observable, to instant actualization of subjective referential networks.[28] This implies the possibility and the potential of a matter entirely different, hiding in a sort of virtual crack. Thus it opens up each time, at every viewing, the space of an onto-epistemological virtuality, in which we as viewers surmise the new rules of the henceforth sayable.[29] Thus, Wittgenstein’s distinction between the sayable and the showable applies also within picturability itself. An image is not as much an object of which we might dispose as of e.g. a manual, than an activated and somehow released *anima*. In this sense, *imaging* – as a doing in or through images – is parallel to *saying*.

Here we find in this image-specific dynamics of the showing action the cognitively constructive aspect[30] of displayed images: we humans need images about something only as long as we do not understand that something; but we need them also every time we seek a new understanding of something. And we utilize them when we are absolutely unwilling to understand something or when the exchange of signs is not about understanding in the first place.

This cognitive trinity of the image is irreducible, and the idea that one of the three aspects be separately claimable just a brave delusion. Therefore, the role of an image shown is a priori not limitable to one of the three directions, whence the impossibility of unambiguity in a pragmatic constellation. Hence, the showing of an image to some degree implies right away the putting into question of oneself and of the habitudinal relations with potential viewers. That is why promotional or explanatory visual efforts targeted at the public at large, on behalf of scientific activities or indeed scientists are prone to dizzying uncertainty.

FAÇADE. PURE FAÇADE

Façades, those originally merely materially-oriented delimiting surfaces between within and without, have acquired their mediatic quality not just in our days. No later than Palladio[31], the play of light and statics developed into a medium initially given to representative purposes. Now, nothing can be *presented* as a certain something without at the same time being taken as a sign. Along with the cultivation of the façade, at the beginning of modern age its vast expanses turn into carriers of signs and sign processes that aim far beyond physical function. The façade expressed something, and the general, dense practice of respective interpretative offers or applicable variants, underlain by multifarious social norms and expectations, led to pragmatics of the façade[32], to a full-fledged medium whose structure over time changed very little if at all. Only Semper, centuries after Palladio, developed a style theory[33] in which he traced the design and the ornamentation of the façade genealogically back to vesture and its ornaments. The modernistic breach, however, only shortly after Semper produced the fundamental misconception of the unambiguously identifiable function, that exaggerative and politically so fateful turn to unambiguousness.

Only Venturi and his assistants amended, on behalf of architecture, the concomitant rejection of interpretative openness[34]. According to Venturi, it is the façade and its semiotics that determines the space in front of the building. In that sense, the façade has forever poured forth into the geometrical space before it and thereby created its own mediatic space. This applies evidently even more to mediatic façades, but also to façades that, as in Las Vegas, define themselves primarily through quasi-animated light, or indeed Palladio's playful façades. The light-façade, and the narrative that is made visible on it and by it, represents the building. The function of the façade is neither limited to physical protection nor to indexicalical representation – indeed, we suggest completely to forgo any pretence to functional inclusion. The primary "activity" of the façade to us seems to be the narrative, irrespective of materiality, means or media. Interestingly already in Las Vegas the emphasizing of this activity – at times in function-like condensity – was apt to lead to complete detachment of the façade from the related building. We are therefore not to assume that a façade without a large screen or digital projection be deprived of being part of a mediatic event. We are just not aware of it because we are so used to the usual arrangements and so deeply steeped together with them in our decora. Perceiving strong habitudes as being such from their inside is not self-evident. Remnants of this media-ness of the sheer façade may become perceptible as one remembers the past effect of the notion of skyline or as one tries to delve back into Baroque and its façades. If however the signs and symbols of buildings determine urban space to a higher degree than built space itself, as held by Venturi, then two distinctive sources of media field lines become apparent in the case of a mediatic façade. There is the façade on the one hand, and the showing of images on the other. Consequently, we are in the presence of an intermediatic phenomenon.[35]

It was probably the formerly merely minerally understood materiality that was responsible for the façade's deprecating connotation smelling of frontage and pretence. Exhibition of a façade image coincided with the façade's physical presence, and against the backdrop of the then prevailing decora and schemata any separate sign process was hardly distinguishable. That static relationship between stone and sign – not least responsible for the adoption of Vitruvius' orthography into the scriptural field – is more than just being rocked by digitalization. The capabilities of the modern digital computer, based on ultimate disassembling, turn it into a symbol-spewing, all-powerful (cross-)media centrifuge[36] and virtualize the heretofore neatly definable elements of media formats. The media formats of musical video[37], VJing or scenography[38] show since the mid-nineties that it has become less and less ascertainable where the demarcations between image, moving images, film, video, theatre, writing, dance, or music are running. The reason for the blurring or even dissolving of these demarcations might lie with the fact that the elements of these formats themselves have become constructionally accessible. *Ortho*graphy, as correct and upright drawing, has given, not altogether metaphorically speaking, way to *choreo*graphy – the designing of dance forms and the arrangement of sequences – or indeed to *sceno*graphy.

Mediatic façades now afford, in related fashion, quite an interesting game. The intermediatic-ness of the digitally media-activated façade can be brought into a self-referential relationship with its role as a "normal" architectural façade. Media façades not only display image-like formats but introduce the possibility of retrieving, through "active optics", the concepts of camouflage, mimesis, etc. into building design – and without particular technical effort even three-dimensionally. The latest advertising clips for three-dimensional TV sets in their specific phantasmics already pave the way for this particular perceptional form. We all know Lewis Caroll's Cheshire cat that can disappear at will *"...beginning with the end of the tail, and ending with the grin, which remained some time after the rest of it had gone."*[39] Quite conceivably, before long there may be nothing left of the façade but a jolly façadality.

Ascribing increasingly important media-ness to cities, buildings and indeed inhabitation is nowadays one of the less risky prophesies. Media-active façades and large media-active inside walls will converge, as cross-overs with media-active "goggles"[40] are conceivable. It is easy to picture how totally new spaces hereby open up for experience and design, up to privatization of the public-space experience. Technology that today almost ubiquitously turns into information technology or submits to it, architecture that reorganizes along the lines of digital deployment, and an evolving media-related perception will see to it that we literally populate and inhabit an explicitly medialike-formed and formatted space. Good reason for us to look, in our last section, closer into the specific relationship between media façades and space.

STREAMING SPACES, CONCRETELY ABSTRACT

In media technology, streaming denotes real-time, point-to-point transmission of digitally encoded media content. The communication model therefore is not anonymous anymore as in broadcasting. The classical distinction between emitter and receiver dissolves at least in part, as the client is able to influence the server's operation. The space where such streaming takes place – in the case of mediatic façades simply the space in front of the façade – might be called "streaming *space*". Alternatively the – at this point already rather abstract – space that enables such streaming might be said to be a "*streaming* space". But in conclusion of our investigation we should like most of all to propose another connotation between streaming and space. The media-related situation in front of a media façade or indeed a cluster of such neo-mediatic mega-surfaces is so unique and particular, indeed so different from other medialike phenomena that this uniqueness, given sufficient dissemination, will in its turn convert into a medium. But whence this special position?

The special position of the media façades is the consequence of several factors that overlap and cross-react. Important aspects are related to the sheer dimensions, or to the overlaying of effects of scale by culturally-historically fostered image formats and their affections. We have already seen that. At this point we shall now touch upon the particularity of the relationship between building and space, and upon the ways this relationship changes along with the medium of large-scale images.

Beyond the statics of the normal façade or simple light-façade à la Las Vegas with their rather stereotypical patterns, the contemporary, digital-technology-based mediatic façade allows of more and indeed subtler means of expression. But thinking the Las Vegas phenomenon further, "learning" as it were, in contemporary manner from it (and from Venturi, of course), there appears even the possibility of equipping a building with several façades or indeed complete silhouettes. As has already been said for cubism[41], one of the deeper-reaching potentials of media façades consists in the "...ousting of the absolute point of view through the relative". Dorner's further considerations may likewise be caught up with before long by media façade practice, when he writes, at the place cited: *"Matter will end up being dissolved into pure surfaces and lines that, massless and transparent, interpenetrate. Thus ... space evolves as the crossing of streams of motion and of energy."* Such an adaptation and cross-over of art and architecture[42] might cure not just the "deficits in radicality", recently deplored by the quarterly *Arch+* in reference to Hans Hollein. Dorner's sensible diagnosis regarding the emergence of the relativity of view-points comes down to the generation of new scales and the elimination of hard dichotomies. To be consequent, one ought then at once to concentrate upon the setting up of various scales, so as to make visible various aspects of the World – the empiricist as magician, as it were.

Venturi's insight that the façade be determining the space before it may be adequately reworded in Dorner's terms by defining this "influence" as a dynamics, a continuous change over time. River or stream are appropriate metaphors for this ceaseless dynamic. In short, the media-active façade pours forth or, less euphorically, generates a specific stream that in this form can be produced only by a media façade and be observed only before one. One might say, in a different vein, that mediatic façades themselves form an abstract space in which a specific media stream actualizes and materializes. Of course, media façades are a further step along the densification of informational floods – if one chooses to insist upon adopting a pessimistically-reactive stance. Mediatic façades are a new element in our unceasingly more volatile lives, whereby however this volatility grows on the ground of a (hopefully) stable technical and societal infrastructure. Manuel Castells coined the term "space of flows"[43], so as to capture likewise the resulting fluid social, digital-age behavioural and organizational modes.

Most likely, however, the real meaning of the streaming spaces of media-active façades lies elsewhere. Media façades are not only specific media-like spaces, rather they are capable – given successful construction of their immaterial qualities – of generating a unique and continuously flowing urban experiential space. Streaming spaces are the continuation of socially construed spaces.[44] This, according to Jaques Derrida, arises from the very course of technical development of representational forms, be they scriptorial, textual or imagerial – or, as in our case, media-active façades. *"It is known that the techniques of direct rendering of words and images, to the same extent that they develop, concurrently interpret, select, filter, and in consequence make the event instead of just depicting it."*[45]

Multiple lines and forces cross in this media space in front of the façade, which as it were exudes from the façade as an experiential space. Hans Hollein in his emphatic manifest[46] carries it at once too far and not far enough when he says, *"architecture is the conditioning of a psychological state."* Irrespectively, individual affection remains one of the foremost dimensions of media façades, in particular for the streaming spaces of mediatecture. As media beings, we can hardly escape the so-called psychological effect of an image that enwraps by its sheer size, hardly escape that imagerial milieu. Here, in the middle of the city, shows up not a mere hint of a landscape, though one that as a "media virtual landscape" follows a completely different temporal characteristics. The imagerial-mediatic shroud might also do service as "CogScape" for a more or less targeted designing of cognitive climate zones. The façade mutates to a hyper-façade, a "deep surface"[47], which for us can but mean a media space of volatile dimensionality. We are in the picture, as it were – which quite literally today is technically easily doable –, and thereby at the same time behind the surface of the screen.[48] Unlike cyberspace however, media façades weave their web as socially encoded and experienceable urban reality, as a dense eventfulness, even as this reality in turn is also immaterially grounded. This eventfulness is indeed the source of the kinship between media façades, large projectional installations, and scenography. The mediatic façades' capability of outpouring spaces of media-reality might prove useful as a scale for their classification. Evidently, some additional research is wanted, hardly to be done without, either under a planning, sociological or economical aspect. This scale will be going through further differentiation, whereby multi-criterial optimization through self-organizing maps (SOM) may prove an important tool. In any event will non-reductive evaluation of various types of media façades urgently call for a comprehensive standard, which based on machine learning may well consist of several components. The mediatic potential so measured would be a direct operationalization of lasting interestingness – and what would be dearer to the investor's and the architect's heart? Here as always, success will hinge upon symbolization and quantitative charting, which will be totally unattainable through heterogeneous-heuristic or indeed enigmatic categories such as "interactivity", "environmental", or "organicistic", seeing that they simply do not allow of a comprehensive standard. Exploration of possible design and conceivable formats of and on media façades would be well advised in making use of the experiences garnered with the erstwhile new media, which demonstrated that the means and encodings that are useable for representation, i.e. conveying attractiveness, can never be migrated in direct fashion from one medium to another. That is also why the common term of "urban screens" falls short of mediatic façades, just as the simile of the façade and a textile garb.

Mediatic façades, as we stated, outpour spaces of media-reality. Consequently, it is no longer adequate to speak of viewers, not least because when turning the notorious corner, one is in passing getting soaked by the outpour. Standing this stream will be beyond us if we do not cultivate it, either, metaphorically speaking, through appropriate canalization, or by learning how to swim.

We have tried opening up a vista upon some of the means for conscious cultivation of these new spaces. It is probably the case, as Manfred Fassler puts it, that every mediatic self-enablement is part of man's abstractional history – to which there obviously is no end. The origin of this abstractional history after all has produced our human cognition, and there are serious indications that the cultural and technical history goes on continuously accelerating the development of our cognitive capacities. Hence let us close with another portable quote from Hans Hollein's radical manifesto: "*Therefore, a genuine architecture of our time is about to redefine itself as a medium, as well as to broaden the scope of its means. … All are architects. All is architecture.*"

1 Hoerl, Erich. Die heiligen Kanäle. Über die archaische Illusion der Kommunikation. (Diaphanes: Zurich/Berlin, 2005)

2 The building where Petri installed his printing press in 1502 housed a print shop up to 1990s, remarkably without interruption. Today it holds a free cultural workshop, the Imprimerie Basel, which among other things founded the Journal for Artistic Research. http://unterwegs.imprimerie-basel.ch/pmwiki.php/Imprimerie/Profil (last access: 23 July 2010)

3 Leonhard Euler (1707–1783) put the imaginary numbers on a solid footing and popularized them in mathematics; without this class of numbers, electrical engineering might not have developed as it did.

4 For a detailed overview cf. Siegert, Bernhard. Die Passage des Digitalen. Zeichenpraktiken der neuzeitlichen Wissenschaften 1500–1900. (Brinkmann & Bose Verlag: Berlin, 2003)

5 This term refers to a more recent development in electronic machinery production, which was the cultural-philosophical subject of a workshop organized in May 2010 by the Laboratory for Applied Virtuality of the CAAD ETHZ at the Werner Oechslin foundation in Einsiedeln SZ (Switzerland).

6 Castells, Manuel. The Rise of the Network Society. Vol. 1 of the trilogy The Information Age. (Blackwell: Oxford UK, 1996)

7 Scott, Ridley. Blade Runner (1982)

8 They both deposit in sediment-like manner, wherefore, according to Foucault, the historian ought to proceed as an archaeologist would. Foucault, Michel. The Archaeology of Knowledge. (Routledge: New York, 2002).

9 Ball, Hugo. Die Flucht aus der Zeit. (Duncker & Humblot: Munich, 1927)

10 Krämer, Sybille. "Medien als Mitte und Mittler. Grundlinien einer Medientheorie aus dem ‹Geiste des Botenganges›". Medien im 21. Jahrhundert. Ed. Klaus Siebenhaar. (LIT Verlag: Berlin, 2008) pp.19-32.

11 But it is precisely this reductionist foreshortening that invades, ever since invention of the information concept, the rhetoric surrounding information flows and their steerability and exploitability. Neither is information in general to be taken as entropy, nor is knowledge or its prerequisite susceptible of regulatory "management", i.e. administration or optimization. These are just some belated blossoms of the old modernistic controlling frenzy.

12 At the same time, this is the central difference from that semiotics to which Ferdinand de Saussure refers. In Saussurean semiotics, a sign is a bipolar contrivance, which robs Saussure's theory of lasting fecundity. Cf. Jäger, Ludwig. Ferdinand de Saussure zur Einführung. (Junius: Hamburg, 2006)

13 On which e.g. Niklaus Luhmann's systems theory is based and therefore beset by the same problems.

14 Re introduction to Charles S. Peirce's sign philosophy, cf. Pape, Helmut. Charles Sanders Peirce zur Einführung. (Junius: Hamburg, 2004)

15 Our concept regarding media-related façades which is here being carried further, will hereafter be characterized by the distinctive term "mediatic façade".

16 Just that, i.e. violence through images, was the subject of a convention held there in 2000. By 2008 publicity signboards were dismantled.

17 Waldenfels, Bernhard. Phänomenologie der Aufmerksamkeit. (Suhrkamp: Frankfurt am Main, 2002)

18 Latour, Bruno. Iconoclash. (Merve Verlag: Berlin, 2002)

19 We are convinced that these questions cannot be solved by means of neuro-sciences.

20 Unfortunately, today not even amongst the image-related sciences may there be found as much as a shade of fundamental consensus. Culture-historical or image-historical approaches think of themselves as being in a sort of natural confrontation against semiotics-driven concepts. Of course, we cannot go further here than to give a summary of the problems and outline some proposed ansrs based on our research work.

21 In this context, we find the transition significant that leads to visual music of Oskar Fischinger. According to Cindy Keefer, he is the "...father of Visual Music, the grandfather of music videos, and the great-grandfather of motion graphics". He at least co-invented the so-called "non-objective film". Keefer, Cindy. "Raumlichtmusik – Early 20th Century Abstract Cinema Immersive Environments". Center for Visual Music. Leonardo Electronic Almanach 16,6–7 (2009).

22 Fassler, Manfred. Bildlichkeit. (Böhlau Verlag: Cologne, 2002)

23 Bild und Medium. Kunstgeschichtliche und philosophische Grundlagen der interdisziplinären Bildwissenschaft. Ed. Klaus Sachs-Hombach. Köln: Herbert von Halem, 2006. Bildwissenschaft. Ed. Klaus Sachs-Hombach. Frankfurt am Main: Suhrkamp, 2005. Sachs-Hombach starts from the Peircean semiotics and is one of the founders of the discipline of "computer visualistics".

24 Belting, Hans. Bild-Anthropologie. Entwürfe für eine Bildwissenschaft. (Fink: Munich, 2001)

25 The term "cognitive person" here brings together all entities that command sufficient autonomy and cognitive capacity for modelling reality, individually and in anticipatory and productive fashion. This may refer to single persons, but also to collectives or (future) machine-based epistemes.

26 E.g. Homi Bhaba, who describes culture as a result and an entirety of all processes for developing differences. Bhaba, Homi K. The Location of Culture. (Routledge, New York, 1994)

27 Some years ago and in succession to Ludwig Wittgenstein, Richard Brandom developed, for speech-based communication, a comprehensive theory of verbal communication, to which we owe the formation of "constitutive challenge". In speech though, self-referentiality is not generally present in similar identity to that in images, since we most often speak to or with someone. Brandom, Richard B. Making It Explicit: Reasoning, Representing, and Discursive Commitment. (Harvard University Press: Cambridge MA, 1994)

28 A message is rarely just what it transmits explicitly in the form of symbols, due to the gap created by the impossibility of complete decoding. The necessity of interpretative completion has long ago reverted to own channels, in which we represent very volatile "messages", or indeed the very volatility of the message. What in speech pragmatics was termed illocutionarity or perlocutionarity by Austin, led McLuhan to his famous saying "The medium is the message". Austin, John L. How to do Things with Words: The William James Lectures delivered at Harvard University in 1955. Ed. J. O. Urmson. (Clarendon: Oxford, 1962)

29 This is demonstrated in interesting fashion by the impressive installations and façades of the Studio Roosegaarde (www.studioroosegaarde.net). Their works prepare as it were the showing itself as a behavioural element of responsive façades through variable transparency, in such manner that there results a space of potential symbolic charges. Thus the image-like must not necessarily be seen as being tied to that material substrate which we now habitually call "screen". This heralds, through pronounced physical dynamics, an actual "behavioural turn" (not only) in the mediatic design of façades and other architectural elements, which ultimately points to the primacy on interpretation or, technically speaking, to a "behavioural coating" as the prerequisite to the possibility of "interaction".

30 Stafford, Barbara Maria. Echo Objects. The Cognitive Work of Images. (University of Chicago Press: Chicago, 2007)

31 Oechslin, Werner. Palladianismus. Andrea Palladio – Kontinuität von Werk und Wirkung. (gta Verlag, ETH Zurich: Zurich, 2008)

32 We are formulating here an extension of the term of pragmatics in accordance with Charles S. Peirce's semiotics-related conception or, too, in reference to the pragmatics of the speech act according to Austin. op.cit. [footnote 29].

33 Semper, Gottfried. Style. (Getty Research Institute: Los Angeles, 2004)

34 Venturi, Robert, Denise Scott Brown, Steven Izenour. Learning from Las Vegas: The Forgotten Symbolism of Architectural Form. (MIT Press, Cambridge MA, 1972)

35 There is no room here for going deeper into this important observation. Yvonne Spielmann provides a readable account of the phenomenon. Spielmann, Yvonne. Intermedialität. Das System Peter Greenaway. (Fink: Munich, 1998)

36 Another important aspect of "computer technology", primarily activated by the software-form abstractive layer, is the phenomenon of dia-epistemic simultaneity, which is the structural origin of the present globalizing thrust.

37 The musical video format has in the past proved an innovative frame for experimental extensions of visual forms; on display were e.g. time slicing, topological spaces and hyper-surfaces, cross-overs of image and video, or mixed dimensionality.

38 cf. Lloyd Morgan, Conway. atelier brückner: form follows content (Avedition Rockets). (avedition: Ludwigsburg, 2002)

39 Carroll, Lewis. Alice's Adventures in Wonderland and Through the Looking Glass. (Puffin/Penguin: Harmondsworth, 1982)

40 e.g. available under the brand names of vuzix, Myvu, cinemizer, or eyetop.

41 Alexander Dorner, 1931: "The significant novelty of cubism is the ousting of the absolute point of view through the relative." As is widely known, this relativity was taken to be a simultaneous relativity, for which there are numerous examples not only in Picasso's later works.

42 Kuhnert, Nikolaus, and Anh-Linh Ngo. "Radikalitätsdefizite." Arch+ 186/187 (April 2008). Editorial about Hans Hollein's 1967 manifesto and Radical Architecture.

43 Castells, Manuel. "An Introduction to the Information Age" in: The Information Society Reader, Frank Webster, Raimo Blom, Erkki Karvonen, Harri Melin, Kaarle Nordenstreng, and Ensio Puoskari (ed.). (Routledge, New York, 2004). pp. 138–49.

44 Löw, Martina. Raumsoziologie. (Suhrkamp Verlag: Frankfurt am Main, 2007)

45 Derrida, Jacques. Eine gewisse unmögliche Möglichkeit vom Ereignis zu sprechen. (Merve Verlag: Berlin, 2003). Our quoting Derrida is in no way an indication of any possible attachment of ours to his deconstruction. Even while seemingly standing on the same spot, we are moving as it were in the opposite direction.

46 Hollein, Hans. "Alles ist Architektur." (1967) «Bau» Schrift für Architektur und Städtebau 23.1/2 (1968)

47 Fassler, Manfred. Bildlichkeit. (Böhlau Verlag: Cologne, 2002)

48 William Gibson coined in Neuromancer (1984) the term "cyberspace", meaning the immaterial net behind the screen. Cf. also Pipilotti Rist's video-performance at Times Square, "Open My Glade". (2000)

Networking

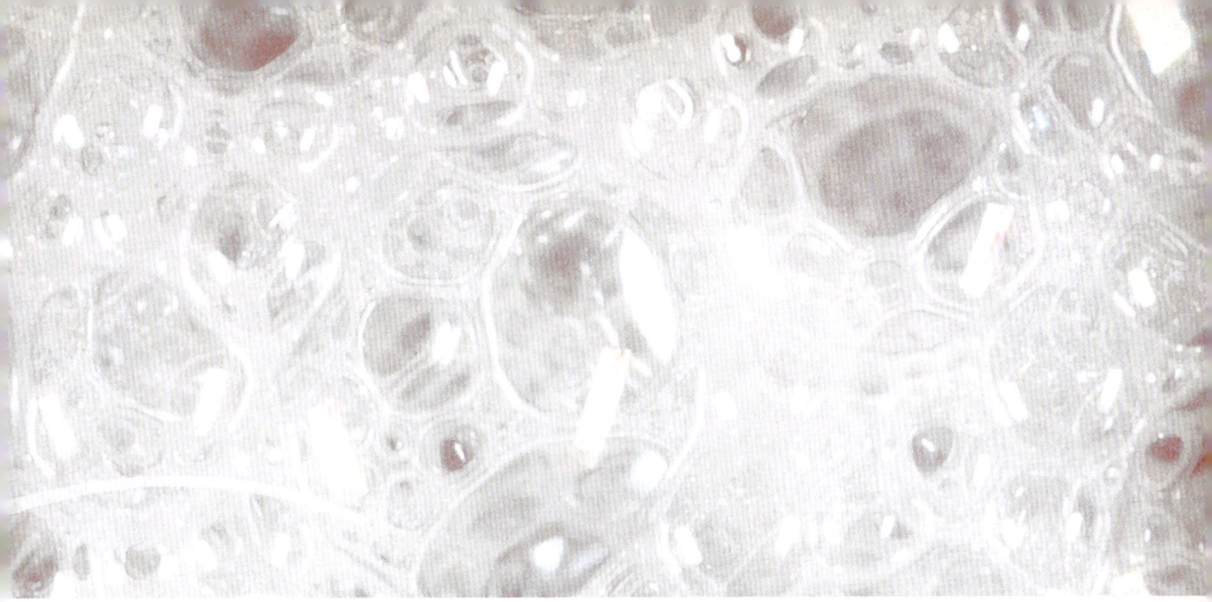

Everything Flows

The origins of the word media-tecture hint at the evolution of a structure into a medium and, hence, at the development of an ability to deal with data. This offers enormous potential for innovation in the fields of industrial production and construction because it suggests that the age of the serial reproduction of built forms is coming to an end. This chapter shows that the mediatectonic way of working is not only relevant for the area of medial content but also for design and production processes.

What is the resulting job description of the mediatect?

The Network (R)evolution

At this point I would like to look at an aspect of mediatecture which, at first glance, has little to do with the media-based orchestration of spaces. Rather, I want to show that, by addressing "the interface between virtual and physical space", we have the potential to fundamentally transform the ways in which we think and work.

This conclusion was triggered by my meetings with Festo, the world market leader in automation technology, and with Ludger Hovestadt, Professor of CAAD (Computer Aided Architectural Design) at the ETH in Zurich. Earlier, I had both worked for Festo and studied with Ludger Hovestadt. Both employ electronic media as a means of transforming design, construction and production processes, which means that they are working very concretely at this interface between virtual and physical space. And I saw my meetings with Festo and Hovestadt as opportunities to develop a plastic description of my idea of mediatecture. There are certainly many other aspects to this subject but I am not a scientist and my aim here is to stimulate discussion.

The question is: what is the role of the "mediatect" in the planning process of a society that is in the grips of a network (r)evolution?

Individuality + Society

Human nature is full of contradictions: we long for stability but hate standing still. There is hardly any area of our culture where we see this as clearly as in modern architecture which seeks to create buildings which should stand as long as possible while being driven by the longing for constant change[1].

One basis of this culture is the controlled design and linear production of building elements. "Linear" because the constructional methods which we most commonly use are generally, for economic reasons, based on the assembly, in a pre-determined sequence, of elements which are, as often as possible, identical. The same process has long governed the production of consumer goods, which is why we are able to produce affordable buildings which offer good standards of living in just the same way that we are able to bring other goods to market at affordable prices.

And yet if we humans were satisfied with the living standards achieved by this process, then this satisfaction would, paradoxically, be a threat to these standards because the result would be a decrease in product development and production which would make it impossible for us to maintain the same standards.

This conscious search for the "new" is a success factor of the modern. Limited space means standing still – but modernity is based on movement (or flowing). "Novelty" is a way of opening up new markets and giving specialists the role of working together to serve these markets.

Luckily, modern culture presents us with another decisive design tool: fashion. The modern ideal is that individuality can flourish.

"Fashion is what one wears oneself. What is unfashionable is what other people wear" (Oscar Wilde)

But, however radically Oscar Wilde saw the situation, the truth is that fashion has long stopped working. At the same time as seeking individuality, people also long for identity and this is invariably linked with the feeling of warmth found in society. One wants to appear in an outfit which is accepted in the circles in which one moves.

www.mediatecture.info/26

The End of the Analog Sequence

Fashion is thus one of the reasons why there have always been new production cycles of new products and why these new products always have one thing in common: the fact that it is repetition that makes them truly attractive. Continuous repetition offers the possibility of lowering costs as a means of capturing market share and guaranteeing commercial success.

This success formula of continuous repetition was also the basis of economical building – but it is also one of the reasons why the forms and methods of global architecture have become so uniform.

It is in this industrialized building process where electronic media have been demonstrating their potential for some time. Electronic media are, essentially, process-related, because they work by integrating highly varied players into their economic, technical and social processes. The result is the swift and spontaneous emergence of a wide range of new systems.

This network (r)evolution is leading to two paradigm changes which will alter life for ever:

1. Products will emerge spontaneously
2. Design will become the art of programming characteristics

This first change refers to a form of product production which does not have to be repetitive in order to be economically successful. Production runs of one item will be possible.

The second change refers to the fact that the design process will no longer need to be a sequence of individual steps in which items progressively become both more precise and more differentiated. This work of differentiation is much better left to the computer. Rather, the designer will divide the design process into a series of small modules - for each of which he determines a basic (but variable) form - and then formulate the characteristics and behavior of these modules in such a way that they can exist, virtually, in the computer. These individual modules can then be combined, adjusted and adapted to each other in the computer in the way which best meets the designer's criteria. The quality of the result will naturally be determined by how well the designer programmed and then accompanied the process in the computer.

But this is just a short introduction to the subject of process-related design which is addressed in more depth later in the book.

The possibility of the spontaneous emergence of products referred to above forces us to examine the words of Oscar Wilde in a completely new light. In such a world, individuality would certainly operate at a completely new level – and everyone could truly wear their very own clothes.

But surely this notion of absolute individuality is an extreme one which is just as unreal and, indeed, just as questionable as any such idealized notion? Do we really want to be individualized, to find ourselves in a situation in which each lives for himself and only feels bound to his own thoughts and wishes?

The truth is that we should not concern ourselves too much with this extreme position, for it is clear that such an approach - in which all our established models of design and production would be swept away - will not prevail. At issue here is much more the identification of development potential. We must assume that we need visions simply to avoid standing still - just as democracy must assume that all men are equal.

And the core of this visionary approach must be that we abandon the notion that design and production are a linear chain of events. It is time to free ourselves from the shackles of mechanical and linear processes.

An Example of a Non-linear Production Process

Electronic media create the potential for the increasing individualization of the production process. I shall now look in more detail at this form of mediatecture, in which media are employed to increase technical and constructional potential.

"Festo is the global leader in automation technology and technical education and training. Our objective: maximum productivity and competitiveness for our clients in the areas of factory and process automation" [2]

ag4 developed both an electronic control system for Festo's new headquarters in Esslingen in 2001 and a forum for presenting the company's philosophy and way of working in 2008. This work enabled me to gain an insight into the future of the automation of production processes.

The creation of production chains through the sequential linear organization of the production process from the manufacturing of constituent parts to the final assembly was the basis of our industrial heritage. Mechanical automation adapted and optimized this process and, for some time now, communication technology has been reshaping the process of optimization. Now, however, this network (r)evolution is leading to a paradigm change because it has the potential to circumvent the linearity of the process by creating parallel processes which are organically developed in real time.

I shall start, however, with a simplified description of the new process-related elements of the production structure. These are based on RFID (Radio Frequency Identification) technology. Electronically encodable labels are fixed to the raw material. This means that each single object which is to be processed is accompanied by data and this data can tell the production process or tool what is to happen to this item. Previously, these tools were either operated by humans or pre-programmed to process objects (repetitively) in a pre-defined way and then hand them on to the next stage in the chain.

Now, however, the organization of the production process has effectively been reversed. This can be explained with a simple image: if Michael Schumacher has driven into the pits then he knows exactly what must be done, as quickly as possible, to his car. He himself is the data carrier and he instructs the tools (the engineers) in the pits to carry out the work.

For product production this technology brings, in itself, no great advantage. This advantage arises, however, when elements of the production process are linked to and communicate with each other. When this is the case, several semi-finished items can arrive at the same point in the process simultaneously and each can pass on different instructions to the machinery. Each item wishes to be processed differently and the various tools (machines/robots) will coordinate amongst themselves the sequence of operations which allows all items to be fully processed in the minimum time.

This process-related coordination would not be necessary if every production station had enough tools to allow each item to have its own dedicated tool for each part of the process, but this would mean the majority of the tools spending much of the time idling and awaiting their moment in the sequence. And this demonstrates the second decisive difference between process-related production and classic production chains. Indeed such "chains" no longer exist. Chains are linearly fixed modules with two articulated joints. In process-related manufacturing, however, the production machinery has become a series of independent objects in the production hall which have no fixed position and which can speed to the item which requires processing.

Festo also intensively analyzes the issue of clustering, although the conclusion of Dr. Post, the head of research, is that process-related manufacturing structures must act "more intelligently" than clusters if the individual elements are to be encouraged to think for themselves.

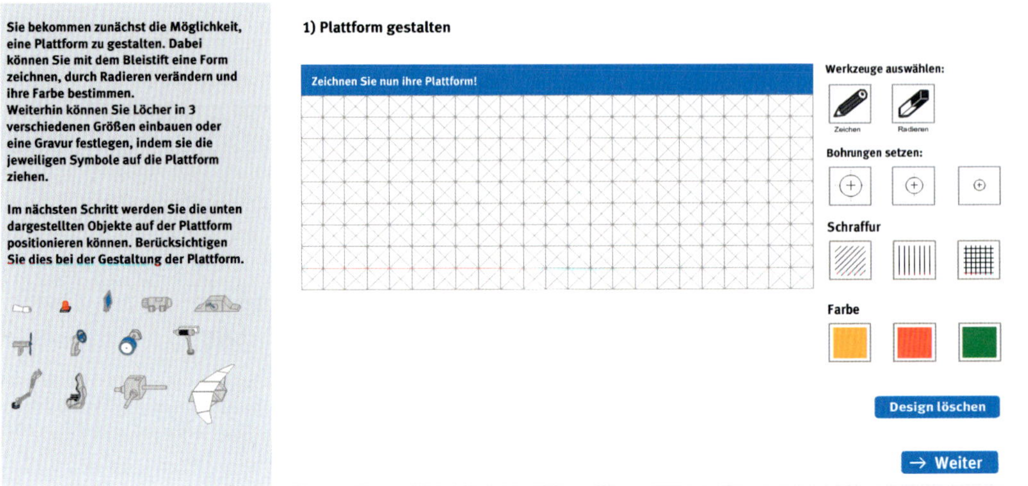

Future Vision: On an interactive display in the Festo forum, the visitor can configure a personalized vehicle.

The objective of this entire development is that products can be manufactured even if they are only sold once. This would mean that products could be 100% personalized: tailored suits – standard in the future and no longer a luxury; the sports shoe with completely individual characteristics - no problem.

Clothes retailers already have access to 3D scanner technology with which they can record the body's individual dimensions and then feed this information directly into the production process.

This connection between data systems and production processes affects our economy in a number of ways. Rapid Prototyping makes it possible to directly realize a form generated in the computer as a functioning part or even as a finished product. There is a vision in which every normal household - just as today it has a microwave - has a Rapid Prototyping Box with which it can directly produce objects for daily use. And, while it is clear that not everyone will ever have the ability to create their own objects on the computer, one can certainly envisage an internet trade in construction information which can be either adapted to the scale of the individual requirements or which, being modular, can be simply combined with other such information. An internet user could, for example, purchase the 3D data which would allow him to model a plastic part for the vacuum cleaner either as a means of replacing a defective part or of adding one with new features. Or he could use a simple computer application to model a coffee cup which meets his precise requirements, allowing each member of the family to create their own cup. That there is such a desire and commitment to actively and publicly live out one's preferences and lifestyle is demonstrated clearly by the success of "youtube" and "myspace".

The fact that self-designed "products" could be either created at home or very simply ordered undoubtedly represents a radical change from the classical industrial society which was based on the creation of standards. The "truth" that one always has to wear in a shoe is both an example and a result of this type of standardization. From the point of view of the shoe-maker, however, this should not be necessary. Modern society has simply developed "truths" to cover its own weaknesses.

www.mediatecture.info/27

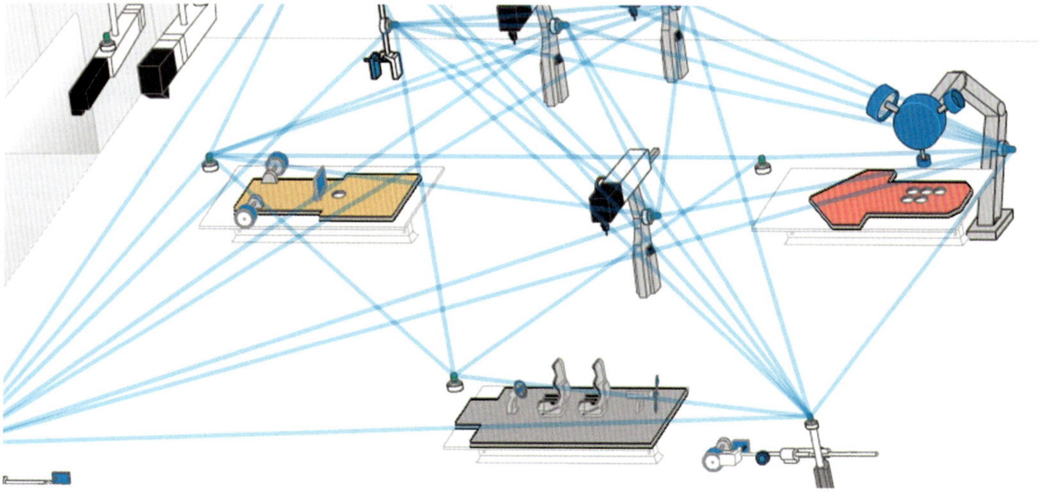

The desired object develops as a batch production simultaneously with other objects in a virtual factory. The programming of the interactive display presents the individualized process in real-time and reveals how the means of production work together in a network.

An Example of Non-linear Building Services

During his studies in the 1980s Ludger Hovestadt first recognized that the use of computers in architecture would bring not only obvious advantages but also great change to architectural design. After completing these studies he carried out research work at the Institute for Industrial Building Production of the Technical University in Karlsruhe under Fritz Haller before establishing the company "Raumcomputer", which created interfaces between building services and the internet. Now he is Professor for CAAD at the ETH Zurich and Chief Technology Officer of Aizo AG, the developer of digitalStrom.

digitalSTROM transforms local electricity networks into communication systems, which connected components use not just as a source of power but also a means of communicating with – and controlling - each other. Aizo has developed a chip which is so small that it fits into a terminal strip, meaning that any piece of electrical equipment can be digitally connected and controlled via the electricity network. Internet connections can also, naturally, be made in the same way. The result is that highly individual user profiles can influence the operations of building services. All electrical equipment is linked and operating performance can be coordinated. This is a very clear implementation of mediatectonic principles: individual user requirements are captured at the virtual level and then have a real effect at the concrete spatial level.

digitalSTROM is, at the same time, an organization which is working to create a uniform standard for digitalSTROM in industrial production, with the aim of integrating this standard interface into as much household electrical equipment and as many building services networks as possible in the near future. The economic advantage of this would be a reduction in electricity consumption due to the fact that equipment would only operate when really required. When not required, equipment would be fully switched off rather than operating in standby mode. And the user can very easily manage the switching of the equipment using electronic interfaces.

But this interconnectedness naturally means more - and the echoes of the dissolution of the linear production chain are obvious. There are many areas of living and working today in which the potential of electronic networks is leading to evolutionary change. Web 2.0 and the success

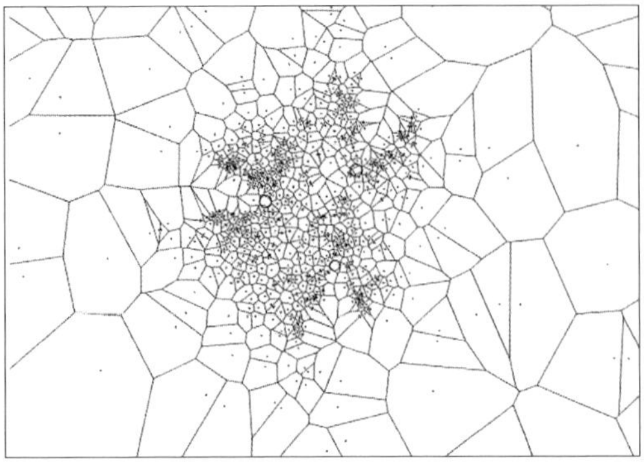

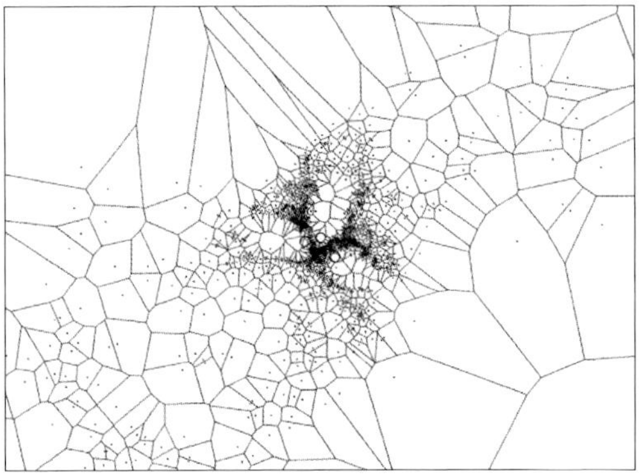

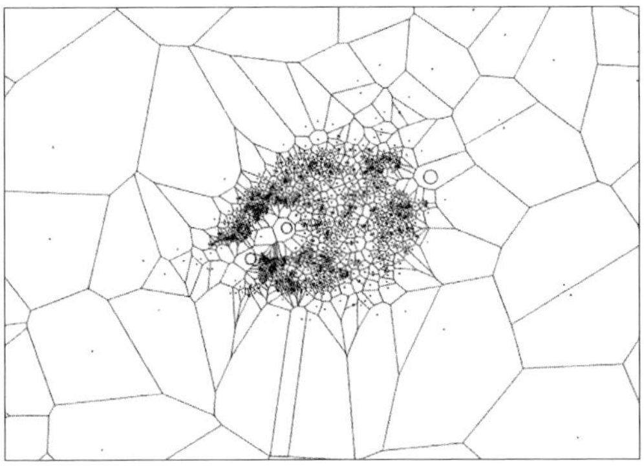

of "youtube" and "myspace" are just some examples of the extent to which social networks can freely emerge without requiring instructions from anybody.

digitalStrom about itself:

"We make use of the phenomenon that collectively organized systems are superior to hierarchical systems. This can be observed in insect populations. A huge number of small organisms of low intelligence use the constant exchange of information to develop a collective intelligence which is superior to any central authority. This intelligence has the ability to continuously develop and adapt to new situations. This phenomenon can be currently seen in the development of Web 2.0 internet applications.

Spaces which are organized in line with the principles of collective intelligence and in which all the electronic equipment can operate as an interconnected team are more efficient, more flexible and more comfortable. The same applies to buildings in which building services are no longer centralized but the responsibility of a network of smaller sensors and appliances distributed amongst the individual spaces."

An Example of a Non-linear Design Process

Closely related to the above is another vision which Ludger Hovestadt is pursuing as part of his work at the ETH:

The structure of foam as the basis for design work

All foams share a specific physical characteristic: when they are moved mechanically, they change their form while the number of cells remains constant. If the size of one cell changes, the neighboring cells automatically adapt accordingly. The same applies to the formal arrangement of the cells.

This phenomenon can of course be described in mathematical algorithms and two-dimensional images. If one pulls at an intersection, then this changes the form and size of not only the cells immediately adjacent to the intersection but, depending upon how hard one pulls, on a much wider area of cells.

But that is just the phenomenological background.

Hovestadt uses this phenomenon to turning the process of designing floor plans on its head: rather than asking how he, as a designer, can produce the ideal floor plan for the user, he starts by simply asking what the needs of the users really are. If the designer really attempts to keep an overview of all the needs of the user it is soon very hard to see the wood for the trees. Hovestadt's solution is to program these user requirements. Each separate use element (or room) is given a profile which describes the parameters of this element. This profile includes, for example, the desired size, relationship with other use elements and lighting requirements, etc. Each use element thus has a unique profile. If one then attempts to use the classic approach of the architect to arrange these elements one can soon be defeated by a complexity which can only be overcome with long experience. Hovestadt's design instrument, however, gives this arrangement work to the computer. Each separate profile is transferred onto the cell of a foam structure and the computer coordinates the relationships between these profiles. Naturally, in doing this, the computer has no "experience" but it makes up for this by quite simply trying out every combination of elements until the best overall solution is found. The use requirements for each element are thus interconnected and the resulting intense exchange of data is an evolutionary process. Nature operates in exactly the same way.

This process is used initially to coordinate the principal user requirements. Further rules are then formulated which, for example, place aesthetic requirements before compliance with functional requirements. The key to this method however is the way in which it reverses the traditional design process. The designer really asks "what do you want?" before asking "and how do you want it?" And in order to be able to ask this how, the designer must intelligently and creatively manage the implementation of the process by the computer. But he is no longer author of the solution – his creative input is, rather, as a programmer.

And Now?

This method of creating spontaneous arrangements naturally establishes completely new benchmarks for the work of the designer. Architects normally work otherwise. They have their own individual preconception of the forms and constructions which could be used to meet the requirements of a project. And, above all, they usually have a special ability to envisage three-dimensional space which gives them the potential to imagine complex relationships – to think space. And now the programming and structuring of space and the creation of form should be left to the computer? The IT experts are to do the real work? What an insult!

I believe that this new way of developing a project will simply free up creative potential to be used in other ways. And of course there will remain many creative tasks which, as before, can only be solved by individually-based intuition. The elegance of a vase cannot be programmed!

What changes, however, is how we think.

There follows the description of a situation which perfectly demonstrates the misunderstandings which can result from the possibilities offered by the use of process-related design:

In 2005 the City of Zurich invited a series of star architects to take part in a limited competition to find an exemplary solution for the development of the waterfront adjacent to the Walchebrücke. The architects were left to propose their own uses.

I was asked by Helmut Jahn to integrate a concept for a media façade into his design. As the ETH was a co-organizer of the competition, it hosted the presentation and, as a representative of the ETH, Ludger Hovestadt also presented a concept. He had developed a normal mix of retail, office and residential uses and then showed how profiles of each of these uses could be employed to generate a spatial concept in the computer. Then he went on to show how the relationship between these uses could change if he altered a principal project requirement (from building over the river to building towers). The computer is able to restructure and optimize spatial relationships – but without taking an architectural position. Nevertheless, the approach permitted the complex spatial relationships – as well as relationships with the transport network and other important elements - to be very quickly established.

Automatic realisation of a spatial program in spatial designs, which can be spontaneously reconfigured according to variable rules. A profile connects each spatial module in a network with all of the other modules and rules. The computer always produces a solution that functionally corresponds to the requirements of the spatial program.

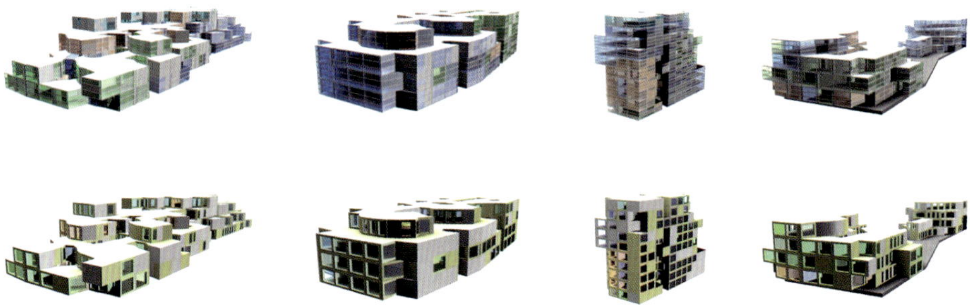

During this presentation I was sitting next to Helmut Jahn and felt his unease very clearly. After the lecture I had the opportunity to introduce Ludger Hovestadt to Helmut Jahn who expressed his opinion that such a design instrument was very dangerous. Using as a metaphor the fact that when a normal human sits behind a wheel of a Formula One car it is pretty certain that he will end up hitting a wall, he said that such an instrument could lead to solutions which were fully inadequate in terms of design. Ludger Hovestadt reacted very elegantly in stating that he would refer precisely to this exchange in every future lecture. He confirmed that this was an instrument which could naturally be very differentially used and that the proposed solutions would always require reworking by a skilled designer. His tool is a design aid and in no way replaces the work of the architect – even when Hovestadt adds that he could program important design monuments by architects into the system modules which could be called up and used at any time!

This idea that even the individual touches of engaged architects should be automated and then reproduced by computers could be regarded as either cynicism or a misunderstanding of the creative process: It is quite natural that the idea automatically engenders resistance. And then there is the quite justifiable question of whether a computer, regardless of how well it is programmed, can really reproduce the creative process of the designer, quite apart from the question of whether we want to accept such a proposal anyway. Should all work really be taken over by computers? Will the art of individual creativity really not be worth anything in the future? In what sort of world would we be living if even the aesthetics of our built environment could be analyzed and controlled by logical processes?

Globusprovisorium, competition entry from the chair of CAAD from the ETH Zurich, Prof. Ludger Hovestadt

One can extend this dilemma to the damage done to the human spirit by brain research. Is there really an individual consciousness? Is there an I?

But it cannot be denied that even this possibility that our individual human environment is theoretically programmable means that we are thrown back on ourselves. What happens when computers can carry out certain tasks so efficiently that it would be madness not to let them do so? The role of the computer is, after all, to carry out a number of minor operations which, in themselves, are quite banal. The computer is never resourceful and never creative - and it is certainly not intuitive.

Ever more complex social relationships are leading to far-leading scientific and technical developments. Or is it not the other way round? Is it not that scientific and technical progress result in society becoming ever more complex? Has the internet made society more complex or do developments in society lead to more efficient means of communication?

Whatever the answer, we find ourselves increasingly involved in a process which we can no longer really ignore. And, while the reality is that this was always the case, it was, until now, always possible to retain the hope that we were at least in control of the process. But the interconnection of countless, minor, modest actions is often more successful than the grand gesture which aims to put everything straight. Not that this is to say that grand gestures – brilliant ideas – which are designed to deal with huge challenges in a completely unexpected way no longer exist. It is rather that it often makes no sense to have such a clear picture of where exactly one is and where exactly one wants to get to when one is dealing with processes based on non-hierarchically organized systems.

Or, more concretely: Successful design means starting a process which should lead to a result which, at the outset, one can only very vaguely envisage. And by starting this process one is accepting, from the very beginning, that, by feeding in the results of countless small interactions in many design areas, things will happen which could not have been imagined at the start but which, by definition, will be good for the overall result. A paradox!

Back to the example of the vase: One can program a three-dimensional outline for the design into a computer. This is a mathematically-based model that can then be made to interact with other mathematically-based structures or influences. The computer carries out a sequence of calculations which it continuously represents in three dimensions until one suddenly notices an interesting development, stops the process and then sets it in motion again with altered parameters. At the end of the process there is a possibility that one has generated a form that, without the computer, one would never possibly have imagined.

This is of course a principle which has now been appropriated by many architects and designers. But it seems to me that the generation of ever more complex building volumes is only the beginning of process-related design. These buildings are a symbolic articulation of the potential of a new complexity. They proudly demonstrate that they are the result of a design process in which the original idea of the architect was integrated into a network of complex mathematical processes. And, in addition to this, there is the fact that without the appropriate computer support the more extreme constructional forms would simply be unfeasible. The building is a conscious result of a network process and, as such, symbolizes a global world which is structured by network processes.

And an additional driver of this architectural movement is its fascination with technical feasibility.

The evidence from the above is that the process of formal optimization is often no longer part of classic design work and that, rather, architects very consciously use technical aids to help them to determine – and to be surprised by - the final form. One of the earliest representatives of this method is undoubtedly Peter Eisenman. His targeted superimposition of basic forms with scientifically generated data about processes related to the location is, in a certain sense, a highly mediatectonic approach: it connects a place with a process. However, for all the resulting fascination with the complexity of the form-making process I am skeptical as to whether the projects which result from this self-generated task of finding targeted connections are adequate, because they are so often so difficult to understand. Naturally one can argue that a cultural object often only "works" when the intellectual background to the object has been explained and when the viewer is open to learning about that background. Of course this is possible – and perhaps it is an approach that one must periodically accept. But it is unsatisfactory when one realizes that a large part of the population is not in a position to address such complex backgrounds. And many have no desire to do so because, such is the insecurity brought on by their initial failure to understand, they switch off immediately. Which is, surely, a natural human reaction?

Process-related Design

There is thus a need to distinguish between a design process which symbolically embodies huge complexity and one which truly masters complex requirements.

The design of architecture naturally leads to the realization of architecture which is complete in itself (leaving aside future alterations). A design instrument such as that created by Hovestadt makes it possible to keep the design process open until very late in the construction process and to give many construction parameters a measure of flexibility which would be quite unthinkable in the normal design process where teams of draftsmen would be required to carry out constant and protracted alterations to the drawings.

Process-related design, on the other hand, allows for output to be produced which, in a linear design process, would be more or less unthinkable. And this output can be shaped directly in the interests of the users.

Process-related Systems

The technical aspects of a building have come to have an enormous significance for the use and functionality of the building. Intelligent air-conditioning systems or flexibility requirements which demand that spaces can adapt to changing uses mean that such buildings are, today, pieces of hardware under the control of ever more efficient IT systems. The natural result is that buildings are becoming open systems which are constantly able to react to a huge range of influences (see digitalStrom).

As a rule, process-related systems depend upon the creation of all sorts of interfaces. And these always have to do with communication: a piece of technical plant has to exchange information with a different piece of plant. These are media-based processes. Media are created in the sense of interfaces in order to allow such interconnectedness to occur.

On the one hand, these media regulate functional processes of which the user of the building is not necessarily aware, acting in the background to optimize the efficiency of the building processes. On the other hand there are processes which have a direct influence on the sensual perception of the user. The differentiated interaction of a lighting control system, for example, which reacts to specific user behavior. The control of the temperature of light alone has enormous impact on performance in an office building. Cold light invigorates whereas warm light helps one to calm down if hectic activity makes it no longer possible to think clearly.

There are many soft aspects of a building which impact upon the emotional intelligence of the user. One the one hand, successful design can appeal to the emotional intelligence as a means of hinting at a social system which would enable the building to fully achieve its potential. But, in order to achieve this potential, the building must also be able to continuously develop and this is also a function of the spatial relationships. Such a development could, for example, be a self-organizing orchestration of the public space in reaction to forms of social interaction. Just as the successful design of a space can encourage the user of the space to relax, such a media-based orchestration can energize social interaction, communicate ideas and free people from their anonymity. But, in contrast to the design of space which always has a concrete structural output, the mediatectonic support of emotional intelligence is merely a process.

www.mediatecture.info/28

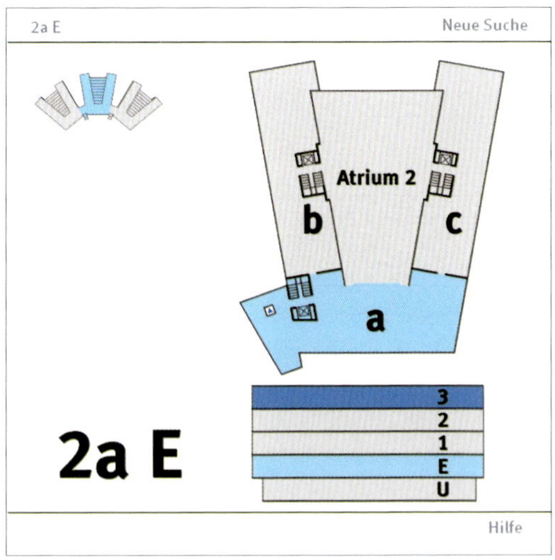

An Example of an Electronic Control System for Festo

Control systems which interconnect people in complex buildings can also be made tangible. In 2001 Professor Thallemer, as Festo's Head of Corporate Design, was responsible for synchronizing the company's new headquarters in Stuttgart with its Corporate Identity. Designed by the Stuttgart architects Jaschek, the newly design headquarters envisaged a highly flexible occupancy of the workplace which would allow design teams to be constantly reformed within the building. The natural result of this was that it could be difficult to localize people in the building and, although this problem could naturally be circumvented if a centralized agency was made responsible for updating and circulating the office layout on a daily basis, it was realized that this would require enormous discipline. Professor Thallemer chose instead to commission ag4 to develop a system which

The Task of the Mediatect

The mediatect creates connections. He organizes a structure of interfaces.

The mediatect does not just create network solutions but is, himself, highly networked into his work. While some architects feel the pressing need to create and defend almost autistically isolated working environments in order to be able to realize their spatial ideas (in a similar way to many artists), the mediatect must position himself in the midst all those involved in the project and set up contacts with each of them. One of the greatest challenges for the mediatect is that he must establish active contacts at all possible levels with the client, the designer and the public authorities, etc.

A typical feature of our culture is specialization. Specialists deliver special services in their area of expertise because they are able to focus on just their area of activity. An automatic result of this is that, within their own circles, specialists develop their own forms of communication - and these can sometimes become private languages. And this means that it can be extremely difficult for specialists from different areas to communicate, a fact made worse by the sociological tendency of each group of specialists to create barriers against others.

Anyone who has tried to implement a holistic and interdisciplinary project for a large organization in which the participation of different departments is essential will recognize the problems to which I am referring.

A result of this tendency of society to shield itself off into small groups is that the mediatect must take a very universal position and develop forms which allow all groups in society to speak.

His work is to do with encouraging processes and this requires him to create interfaces. Here is an example:

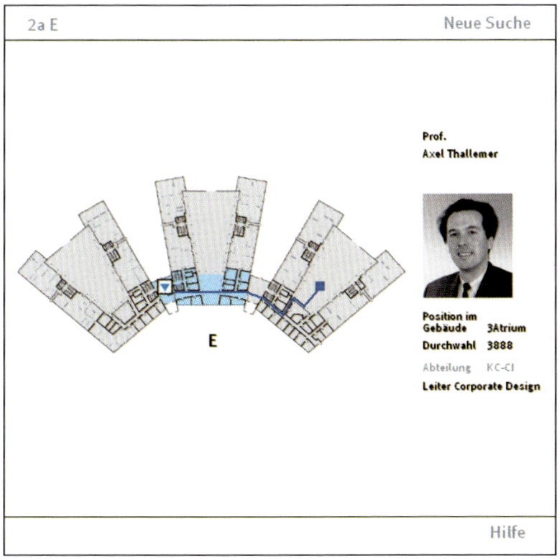

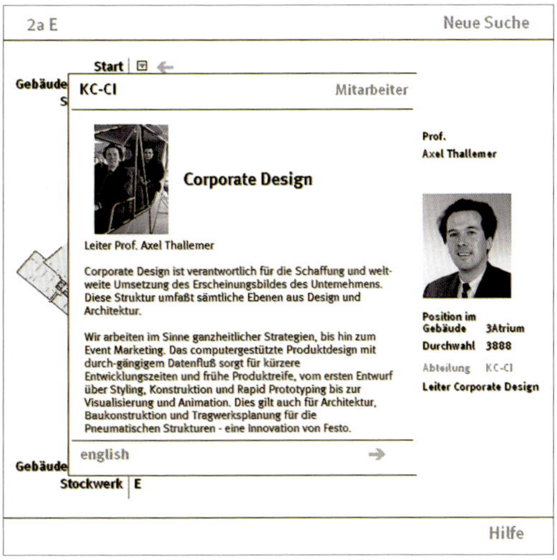

would locate and register the log-in of any employee on a computer anywhere in the building. The central brain would hence know where anyone was working at anytime. An interactive touchscreen was then positioned opposite the lifts in the lobby at every level which would describe online the route from that screen to the target person. Graphically clearly organized floor plans then showed the way in the manner of a GPS system.

At the same time, the database contains an individual profile for each employee and this then appears on the screen. If person A is on the way to meet person B the ability to quickly learn something about the person and what they are working on can be a great help in the social interaction.

In order to manufacture a product which is based on the individual needs of an individual consumer an interface must be developed which is based on as simple a method as possible of registering the characteristics of the desired product and passing this information on to the production process. Such an interface could be a technical installation in the sales outlet which, for example, takes a three-dimensional scan of a foot in order to allow the perfectly adjusted sports shoe to be produced. Or this could be an internet-based interface, such as an automobile configurator.

Experiences of such interfaces in the retail sector have shown that they work best when they are supported by people working directly with the customer. This is, therefore, a new challenge for the salesperson which changes both the sales situation and the requirements for the sales space, which is now an interface between product profile and production. The principle role of the mediatect however is not to restyle the shop but to address the process which occurs between the customer and the production. The result should be a set of rules which can be used by all those involved in the process – whether in the shop design, marketing, training of the sales staff or mediatectonic accompaniment of the entire process.

The fully automatic production of individual products (see Festo) may encourage the idea that human work will be less needed in future. In the implementation of the process, however, it could actually be that direct human interaction could take on a much greater role than previously. At the very minimum, the linear process (I go into a shop, buy something and take it home) will in many cases be replaced. This intensification of social processes can already be seen in the automotive industry when one considers the extraordinary orchestration which often accompanies the collection of an ordered vehicle. The 1:1 process - customer, purchase, finish - is disappearing. Similar far-reaching changes can be seen in classical retailing. The Metro Future Store Initiative shows how RFID technology will lead to the replacement of the traditional payment procedure by direct payment (ag4 has accompanied this project over many years and realized the first concrete applications of many of the concepts involved). And when there are no more cash-tills, then other typical boundaries within the linear retail process could follow.

It is difficult to predict the effect that automated production could have on society. Classic workplace structures will certainly change dramatically and this change will be accompanied by a move away from the linear relationship between production and consumption. Whether the resulting changes in social structures will compensate the inevitable disappearance of jobs cannot be foreseen. What is clear, however, is that the basic paradigm of our industrial culture is changing and that it makes sense to react creatively to such a change. History shows us that neither denial nor ignorance can succeed in preventing social change which is driven by technological progress

In this situation, the role of the mediatect is to test the possibilities for the new processes linking sales and production. If he wishes he can undertake any resulting design work himself - but he is equally free to create a multidisciplinary team of designers to carry out such work. There are designers who have developed such a characteristic way of working that they represent a very clear set of values and emotions and, as such, have almost become a brand in their own right. The mediatect can devise a general concept that already envisages that a particular designer should be given a particular task. This design will then become a medium, playing a very specific role in the consumption process. The job of the mediatect is subsequently to coordinate this design with other media – in precisely the same way that the conductor has to create harmony between the soloist and all other members of the orchestra.

Naturally there exists the danger that this metaphor raises questions of relative power – but so does the original image: Who is more important – the conductor or the soloist?

I have chosen to focus in this way on the processes linking consumption and production in order to create a simple image which can illustrate the mutual dependency of process and design. This image can naturally be transferred to many areas of life: to urban space, for example. When Albert Speer said that architecture alone was scarcely able to define the identity of public space and that the question of what happened in that space was much more important, then he was touching precisely on the issue of process and on the interconnection of spatial structure and use potential. If the job of the urban planner is the targeted design of spatial structures which will support specific social processes, then the task of the mediatect is to ask himself with which measures he can concretely support such processes. He develops means of communication which are integrated into or, better, interconnected with, the urban space. And in order to create these means of communication he must first of all establish rules. What should be the role of the street, square, façade, street furniture, advertising, art, events? He develops a superior model which reveals a particular process-related objective. This can almost be described as composition.

But this composition initially remains somewhat hazy because the various elements need time to establish their mutual relationships. It is not a question of launching a spatial design which can then develop itself further but rather of establishing a concept which creates design possibilities which are not all apparent at the beginning.

Everything flows! In a similar manner to Hovestadt's design instrument, this is a process from which a built structure will emerge that could not have been foreseen at the start.

The central creative challenge for the mediatect is the individual development of a set of rules which are based on the need for social interaction.

This description of a network process involving the greatest possible range of mutually influential design, media and planning inputs is a vision and, for precisely this reason, still far removed from reality. There are however indications that the world is moving in this direction.

1 www.festo.com/cms/de_de/Ueber_Festo.htm

Interview with Prof. Dr. Piller

***Prof. Dr. Frank T. Piller** has been carrying on research about the configuration of customer-focused innovation and value added processes, the successful management of radical innovations, and the use of external expertise for the innovation process since many years. He holds a chair in technology and innovation management at RWTH Aachen and is co-director of the Smart Customization Group at Massachusetts Institute of Technology (MIT), where he worked at the MIT Sloan School of Management from end of 2004 to early 2007.*

In this conversation, I would like to explore with you the influence of "mass customization" on the creation of our living environments. In which fields has mass customization already fully unfolded today, and into which realms, in your opinion, could the transformation of our habitats through mass customization expand?

Already in the 1970s, the vision emerged that clothing, for example, could become completely individualized. There was the imagination that in the year 2000, one would design a garment on the computer at home, and then the fabrics would be cut with laser cutters, etc. This did not come to pass. It has to be said that, in general, the amount of mass customization in the consumer industry is far smaller than in the capital goods sector. The working methods to produce capital goods are not classified as mass customization; they simply have already been applied without any further discussion for a long time. It is just being done. If you don't have it yet, then it just gets implemented today. For this we need stable production processes and modular product architectures. Choice Navigation, meaning assistance for the client to make their choice, has to be newly developed. Previously, this was the task of the traditional production engineer; today, there are configuration systems. However, mass customization remains a niche in consumer goods production, although certain services have been developed in the clothing sector. This doesn't even reach a small percentage of the entire mass customization market. On the other hand, there is nothing that hasn't been touched by mass customization. With regard to living environments, one has to differentiate between the architect house – which isn't mass customization, rather batch production – and the classic prefabricated house, which again has to be seen as mass production. Currently, there are some developments, primarily from Japan, which are something in-between. There are many new ideas in the fields of single-family homes and office equipment.

Among other things, the vision of modernity was to create flexible environments where people can realize their needs individually. Especially in architecture, it was about providing "spaces for possibilities". A foundation of modernity is industrial production, which then led to mass production in order to offer the goods at an affordable price. This automatically coincided with a limitation of possibilities. As compensation, the idea of consumerism was generated to connect products with special purchasing incentives. Our living environments are also accordingly structured. The row house became a symbol of modernity for an emerging middle class, but turned out to be everything else but flexible and individual. Is the potential of mass customization only a possibility to compensate the flaws of industrial production, or will our way of thinking and acting change fundamentally in the context of such transformations that also accompany "social media" and "community build"?

Firstly, one has to differentiate especially in the case of single-family homes. In the USA, they might look all the same from the outside, but inside, they are totally different, as the timber construction enables modifications by the individual. Thus, it very much depends on the building style if there is individualization or not. In comparison, the typical row house here is definitely more standardized, however, the outside doors and windows are often individualized.

To return to your question: I don't think one can speak of flaws in industrial production. For many years, it was not so important for us to offer intense heterogeneity in consumer goods. We were quite content with compromises. Only because today we are on a very high level of consumerism is it suddenly important to get exactly what we want.

I have to be more specific with my question. I had an interview with Larry Weber, a marketing specialist, on the subject of "social media". He explained how much consumer behavior changes due to social networks, because people communicate with each other about what they really need. Now I suspect that this has an influence on the production of products.

You point out an important aspect with social networks, namely, that in many fields today we don't even need individual production as the exchange within the community helps us to find a better standard product. Currently, it can be observed that the so-called "matching agents" are becoming better and better. Therefore, I don't always need mass customization on the level of products. Today, we can say that the clients get what they want. There are plenty of products; the range is broad enough. The only problem is to find the suitable product. That's why it is sometimes easier to configure a fitting shirt, for example, than to look for it in the existing product range. Through social networks and offers on the Internet, I have the chance to find the right product – thus, it falls within the definition of mass customization as it becomes one and the same. Thereby, mass customization creates an individualization that hasn't been possible before.

The designs by some star architects with eccentric formal languages are indeed already examples of mass customization, as many of the building components have to be produced in unique dimensions and forms. In your experience, do you see some other developments that could influence our physical environment?

First of all, it has to be stated that the star architect, of course, does not stand for mass customization at all as he or she pursues classical batch production. Nevertheless, star architects do need the potentials of mass customization to be able to realize their complex forms, that is correct. This is happening on the level of components. But on the level of products, the star architect is a craftsman. It would be something else, if the architect would create a system whose application would result in thousands of different houses.

Where do you see the limits of mass customization? Which realms of our industrial culture will hardly change?

There are three limits to the development of mass customization. The first one concerns the complexity of individualization from the perspective of the demanding party. For a long time, the extent to which the client wants to determine his or her product has been overestimated. There was the belief that if you offer individualization, the clients will just come, but that is not the case as clients often are not interested in this. Moreover, many configurators are so complex that the client feels left alone. The second limit might reside in production technology for some areas, but in perspective this should be manageable. The third limit has to do with Change Management. It is incredibly difficult to convince a management oriented upon mass production in the direction of mass customization. I myself am surprised how strong the force of habit is. There is no example of a large brand manufacturer who seriously set up a mass customization service. The corporations would have to fundamentally modify a number of aspects, and managers don't like that.

The really exciting projects come from startups, because they are not stuck in the familiar structures. A prominent example is the success of Dell Computers. It was a startup company that individually configured computers to the desires of the client, and no big supplier was capable of copying this model. Today, Dell is losing ground since the hardware can already do everything you need.

The limits of mass customization, however, generally reside more in the difference between the aesthetic design and form-fit design. In the realm of consumer goods, mass customization has mostly been applied in aesthetic design and that doesn't have such a great future. It simply is too complicated and annoying for the client to determine the design.

The future of mass customization lies in form-fit design, for example in the business-to-business field, where machines are adapted to the individual situation. But all mass customization efforts by sportswear manufacturers, for example, relate to aesthetic design, and I don't think that this is sustainable.

Something that's never taken into consideration enough is functionality; the muesli, for example, that not only suits my taste, but also my body values. Exactly in sportswear, there would be a need for individual adaptation of many functions, but this has yet to be grasped by the brands.

The profession of the architect in classical modernity has hitherto been characterized by a generalist approach, which integrates every detail of a building into a closed system. In the meanwhile, construction has become so complex in technological terms that the architect might coordinate the various experts and synchronize their skills, but whether he or she is still really able to guide them, is the point at issue. New processual methods of planning and production will only enhance the architects' insecurity in their claim of design. But even if the social and technical networks have increasing influence on the design of living environments, it needs the power of a conductor to maintain a red thread in the designs, to prevent that everything drifts toward randomness. Which properties and qualities does leadership in design need for this in the future?

A very important question. From the perspective of the architects I don't know enough, but the same debate is held in product design. The classic product designer hates mass customization as he or she has developed an integrated product where everything is tuned with each other. Therefore, it is very important not to develop the product or the house through mass customization, rather to create a solution space. This solution space defines what one can change individually and what one must not combine with each other. I believe that there is a great chance in this model, and that also means that mass customization shouldn't be associated with everything goes, rather it is a process. The design task of the future, in my opinion, is neither to think in terms of integrated products nor in modular building blocks, but far more in terms of those systems of rules that connect these building blocks and also clearly state what is not possible. Thus, the mentality of designers should become more open and holistic, since they ought not think in terms of integrated products anymore, where every detail is fixed. Instead, the designer should use his or her creativity for the solution space.

On the other hand, a designer who guides and advises the client is needed, just like a good tailor does it. In the future, the planner has to listen and absorb much more to be able to interpret concisely. This is also the problem with all these configurators, who are extremely bad in consulting the client.

For me, this is a very exciting conclusion to the interview, as you explained exactly that which I tried to cover in the essay "Network (R)evolution" as being the task of the mediatect.

How Bionic Innovations Redefine Design Tasks

__Prof. Axel Thallemer__ has been the head of the Department of Industrial Design scionic® [SCIence&biONICs] and a professor in Linz since 2004. He has been a guest professor in Houston, TX, USA, in Guangzhou, Canton and Beijing in China, as well as in Taipei, Taiwan, and Singapore. Thallemer worked at Porsche for five years, initiating the Computer Aided Styling Process there, and founded and was the head of Festo Corporate Design for 10 years. For more than five years, he has been a freelance innovation consultant in industrial contexts. In 2002, he was called to the Royal Society of Arts, London. Thallemer has been presented in eight museums worldwide and received many international awards, most recently 1st place "Books" at the German Printing Industry Innovation Awards 2009..

After the "form-giver" became the "designer" in the mid-1970s – also for brands and identities[1] – a comprehensive understanding of the profession was established latest at the start of the 21st century, distancing itself from mere "decoration"[2]. "Design" was then understood not only as stylistics, but also as scientific troubleshooting and problem solving. The analytic look at natural systems and processes as a source of sustainable inspiration serves as a methodological approach to develop better solutions. Nature is not directly copied; these examples rather serve as stimuli to obtain innovative concepts. Furthermore, different fields of knowledge are actively connected with each other in the development of a new product – and also of brand and identity – a priori through the interplay of multiple disciplines. This is a way of acting against the restrictive idea of specialization. If the self-conception of design before was "(good) form", today a more holistic understanding of material, production and environmental technologies, of goal orientation and bionic trends in the advancement of this profession is prevalent in order to achieve a – also ecological – functional form. An art term has been coined for this approach, and communications design, inspired by examples from nature, has been developed. This university brand strategy revolves around serial, automated production processes and not, as 20th century design dictates, the artistically inspired original design, which perhaps also takes on the form of a technically or manufacturally unique item with an excessive sculptural approach. Here, freehand sketches also serve as a first refinement of design ideas that subsequently, however, are tested with the help of computer aided stimulation processes in considerably more diversified examinations of variants.

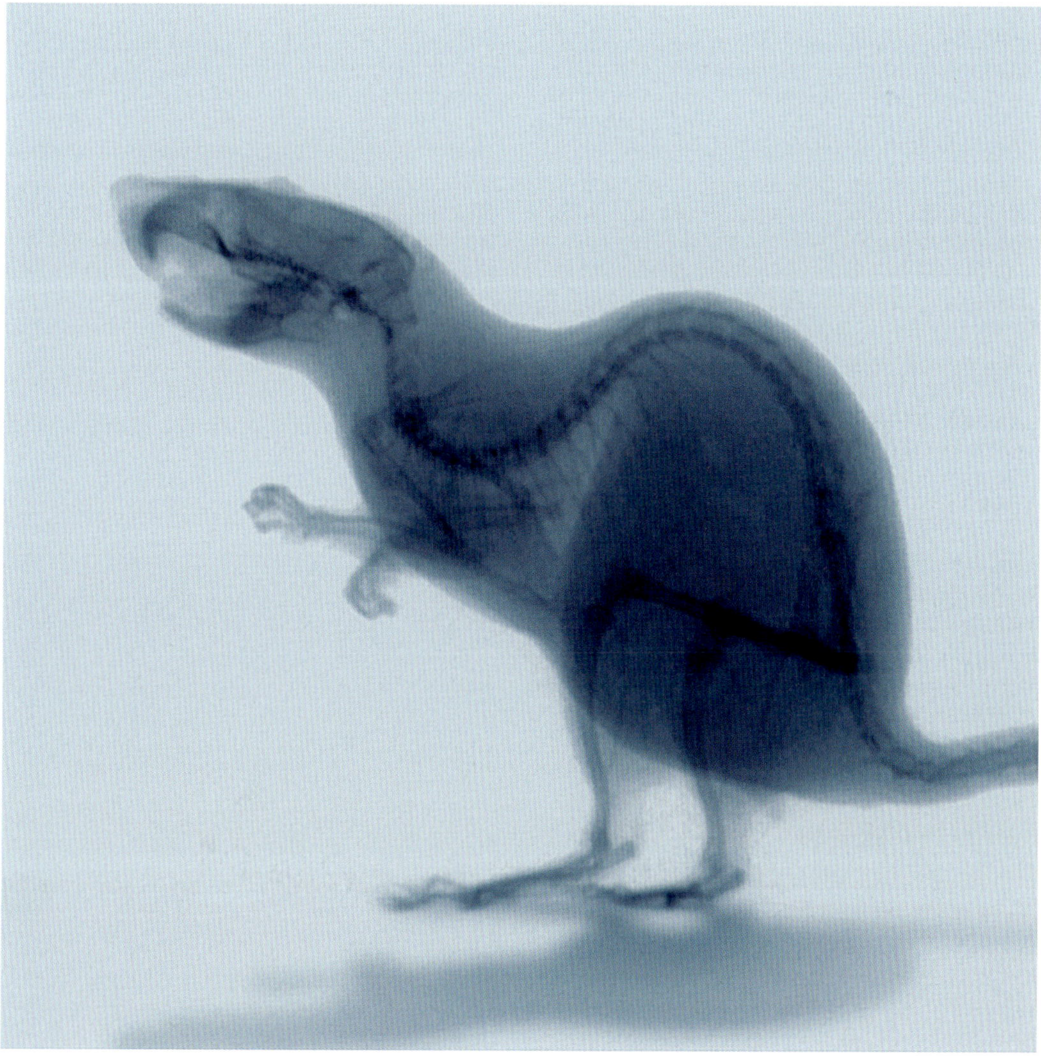

Since "designer" is not a protected job title, the university (not college!) title Degreed Engineer in Industrial Design is awarded for this intellectually more demanding methodology[3]. The leitmotifs of bionically inspired design could be rooted in the idea of ascribing multiple functions to the materials used, instead of requiring one material for every desired function; or, instead of developing unique, albeit exceptional properties, offering the broadest possible spectrum for the intended applications. Thus, an efficient optimization of material diversity and number of elements could be sensibly realized in the future. The respective adjustment in terms of intended function would be achieved through the formal design. Energy-optimized product development addresses the primary expenditure of energy required for the material and serial production as well as for possible re-use concepts after the end of the anticipated usage cycle has been reached. An energy-efficient and material circular economy, in close communion with nature, will play an important role in mankind's further development. The central media for this are air and water[4]. If these two media become material for art and buildings, then this is also mediatecture; however, these media continue to primarily be only used and polluted, which poses the question particularly for the product world of what is really essential to life. What use is another chair design to society? Where could one make out problem solving strategies for the current questions of the total system of Planet Earth? How are the challenges of the future, especially as a result of the industrialization, met? The natural principle of trial and error, by means of computer simulation with a larger number of variants, can better yield prognoses within a shorter period of time and contribute to greater parallel development than consecutive, strictly separated individual steps. None of nature's products can be seen as without having a direct dependency on the purpose of applied form, color and materials as well as their integration in the habitat over the course of the entire "product cycle". If, however, humans are part of this space and cycle, they will have to integrate these principles into further industrial development for the sake of a sustainable future. Fun-design, like light switches decorated in the Hundertwasser-style, and design-art, such as basically useless citrus juicers, would then rather be abundance-objects of a hedonistic "late-society".

Unfortunately, design usually has something to do with consumer goods, brand fetishes, status symbols; investment goods and industrial processes, or even innovations, still are only rarely designed, and even less so in the sense of a corporate identity. Environmental protection or sustainable product development are also far from enjoying the status they deserve. The aforementioned sharpening of industrial design in research and teaching could serve to form a stand-alone feature to counteract these tendencies[5], where, in collaboration with the industry, effective solution strategies for the problems of tomorrow are already worked out today. Neither the basic research typical of universities nor classic teaching would be neglected in the process, and without getting stuck in matters of pure practical use as in colleges. Still, design is rather associated with furniture, living accessories or things we don't really need to live. Accordingly, unique artistic design is more typical of the genre than the orientation on industrial processes.

Example Robotic Arm

As an example of this direction, a robotic arm was developed with the involvement of three universities, which – at the intersection between evolutionary biology, biomechatronics, functional form and information systems – can perform anthropoid movements. For this, we coined the term "anthropofunctionality" in order to also propagate a linguistic differentiation from anthropomorphous machines with robot-like behavior. This robotic arm with antagonistically operating, artificial muscles can roughly reach and grab as far as the size-wise comparable human arm. Its inherent flexibility and light weight predestine it to be a humanoid assistance system in a domestic or work environment. Other development strategies included a minimum of complexity and building parts for a maximum payload-deadload ratio. This leads to a high level of agility – above all acceleration and speed – under similar workloads at a considerably lower own weight in comparison to that of conventional electrically operated industrial robots. The result is not only lower energy consumption; due to the volume-proportionate contraction capacity of the pneumatic muscles, no additional energy – in contrast to electric engines and mechanisms – is required to maintain a position. The economization of resources, from primary energy and material optimization to less moving mass and operating energy, goes in the direction of enhanced sustainability in the sector of industrial automation.

However, design should not draw too much inspiration from nature but merely analytically examine and abstract possible problem solving approaches in nature in order to potentially conceive derivatives, but never copies. Meanwhile, it has become fashionable to purely superficially present "bionics" as a solution approach in the form of a formally phenomenological, however, not contentual equivalent. One example would be the "innovation" of a car manufacturer from Württemberg who styled a concept car in the form of a trunkfish down to the dashboard – a modern, technical fairytale nicely depicting a number of decorative features of "design". However, one can also find such "biomorphic" ad-hoc or retrograde justifications in architecture, such as highrise buildings later communicated by the media – arguably for PR purposes – as structurally derived from "flora and fauna". One should never forget that manmade objects will always remain technical objects – dream and reality! The liberal arts (artes liberales) have since antiquity been distinguished from the illiberal ones (artes mechanicae) according to the Aristotelian categorization.

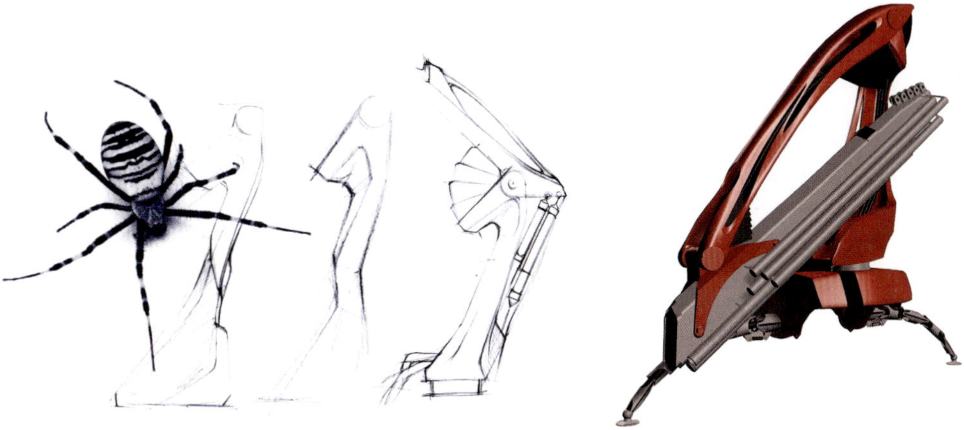

© Project team, scionic® I.D.E.A.L.

Natural example of a spider leg as reference for the joint design of a lightweight concept for a telescopic crane extension.

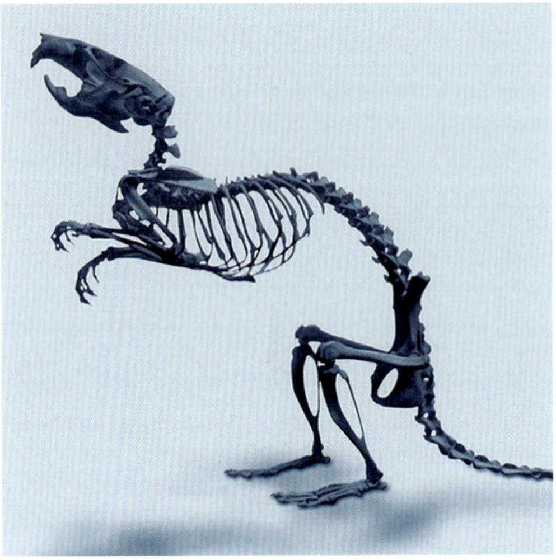

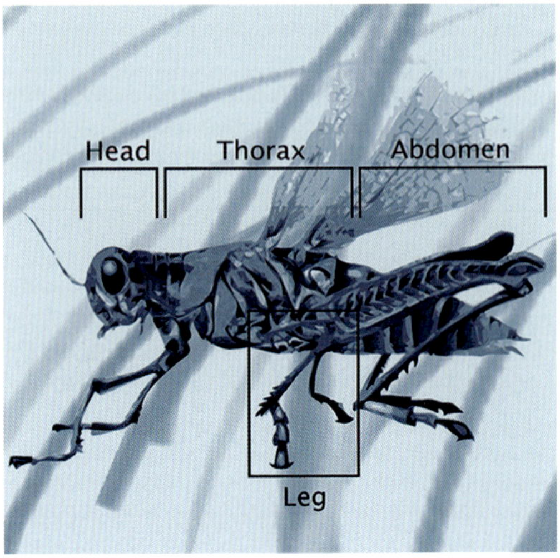

In light of economic pressure and increased speed (also regarding the planning and building of office buildings), design today should be innovative; instead, "fun-design" and "design-art" as artwork have come into fashion. A slave to the marketing and/or advertising industry, design in the classical sense has sold out and contentually annihilated itself anyway. It is absurd to design new fashionable casings for, say, coffee makers, toasters or cell phones in increasingly shorter cycles just to trick prospective customers into believing they would be buying a new product. In reality, this merely serves to prolong the sale of old technical achievements without so much as even taking real innovation into consideration. The economic pressure today is arguably more based on simulating apparent innovations than it is on real improvements. It follows that this pressure is artificial and doesn't work that well anymore; instead of acceleration, it is all about caution. Better to think long and hard before anything is done. If design had not increasingly become a stooge, it would have to be truly innovative. Regarding the design of buildings, designers seem to be advancing into the esta-

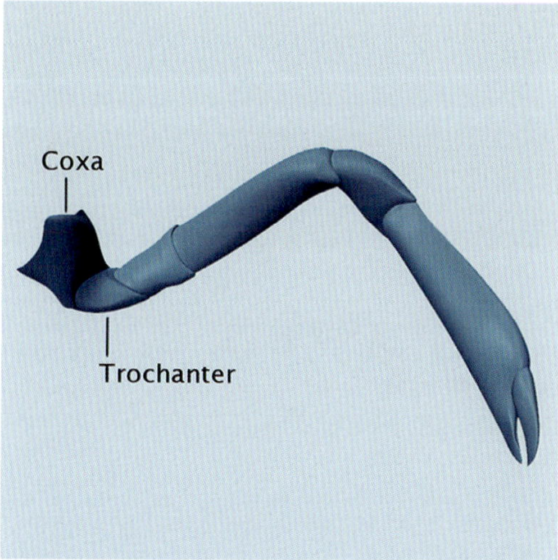

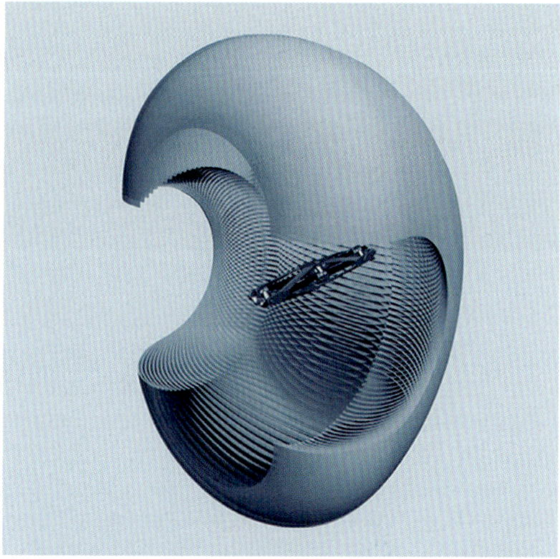

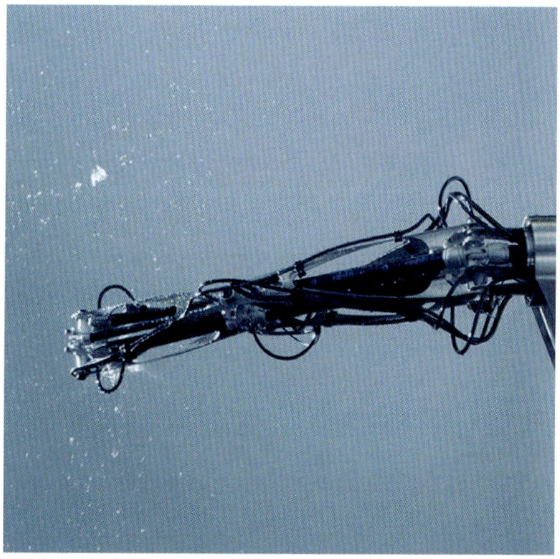

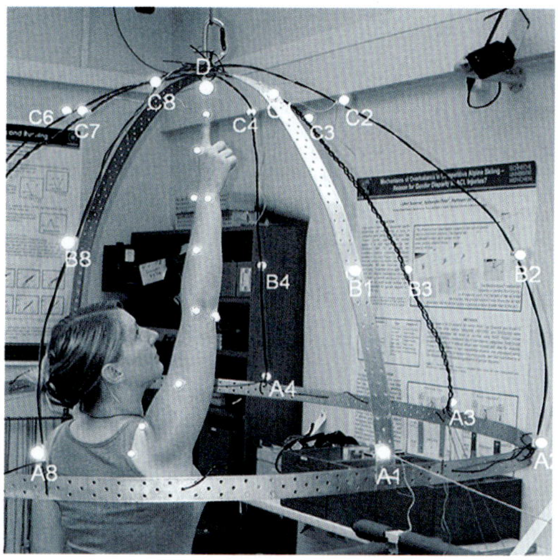

© Professors Drs. Ament, Witte, Ilmenau University of Technology; Dr. Fischer, Friedrich-Schiller-University Jena; Danzer, Thallemer, University of Art and Industrial Design, Austria, scionic® I.D.E.A.L.

blished realms of architecture at the expense of the "holed up" conceptualizers. How many innovations can be conceived in the context of office buildings? Minimal running costs per square meter with optimized income return and tax deductibility, this is usually the reality! Here, mediatecture can transdisciplinarily provide excellent approaches of multimediality.

The conflicting area (or more positive: the work area) between industrial design and technology has changed. The term "technics" – in contrast to technology – has the ability to raise goose bumps and make one think of lubricating grease and screws. Famous industrial architects like Schupp & Kremmer recognized this intellectual problem already in the first third of the last century and described it in their book "Architekt und/oder Ingenieur" (Architect and/or Engineer), published in Berlin, as did the industrial designer Jens Reese in 2005 in "Der Ingenieur und seine Designer" (The Engineer and his Designers; as quoted at the beginning[1]), also published in Berlin. Since nothing has really changed, there is nothing more to add except that "engineering" still

Anthropofunctional movement of a robotic arm as Man-Machine-Pas-de-Deuxl © Bodmer/Hempel & scionic® I.D.E.A.L.

considers itself to be incredibly important, despite de facto being meanwhile located on the periphery of the novelty production process in the realm of basic innovations. This is not least due to educational concepts. While good design is taught multidisciplinarily, preparing students for the anisotropies and non-linearities in the professional practice, typically the classical engineering education is structurally stuck in 19th century conceptual constructs. Mediatecture has the potential to break these social conventions. In order to position multimedia industrial design – especially when inspired by examples from nature – interdisciplinarily as an interface between the applied, fine and performing arts, scionic® teaching embraces the contextual dissolution of purely conceptual boundaries.

In light of globalization, the question arises whether there is a design that can be applied worldwide, or whether regional reference points are always necessary. In the case of industrial components, a global design is possible because the laws of nature are the same everywhere on Earth. However, this is not the case with user interfaces where language and cultural differences already require differentiation in the design. It is crucial to synergetically consider regionality in the design process of products that are directly operated by end-users.

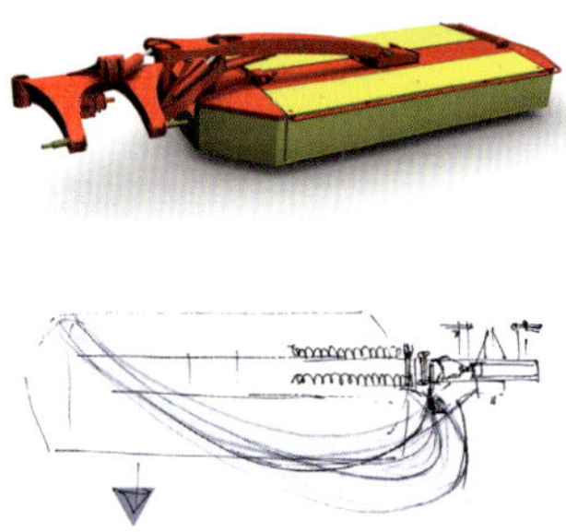

Lightweight concept for a mower extension in the forage grass industry

Natural example of a banana's negative geotropism

Technical Textiles and their Significance for Design and Architecture

Technical textiles – also woven metals – are preferably employed wherever at least a temporary mobility of space-creating structures is required. Lighter in weight and packing volume during transport as compared to their operating condition, composite membranes are predestined for use in space. Given my specific experience on the subject, Mr. Ulrich Kübler, Head of Life Science at Astrium Space Transportation, commissioned a company of the EADS (European Aeronautic Defence and Space Company) to draw up a conceptual design for a planetary greenhouse together with my scionic® industrial design students in Linz.

It will be interesting to see to what extent the future development of building shells will move toward flexible multifunctional media membranes. Generally, it can be observed that the focus in the development of buildings is currently moving more and more toward highlighting energy and material efficiency potentials. Parallel to this, multimedia concepts are emerging on the building surface.

Priority is naturally given to aspects of lightweight design. Should the idea of using examples from nature as a source of inspiration for finding new ideas become increasingly popular also within architecture, then a grand morphological building kit would be available for entirely new modular design approaches from individual elements. Next to the material and function-appropriate design of shells, it would also be important to think ahead about a minimum of energy expenditure in the production process, over the entire period of use, up to the decomposition and recycling of the materials. Moreover, it will also become important to consider how much complexity, how many different functions, can already be integrated in one building part. Today, it is predominantly the case with technical solutions that every single function is ascribed to one individual material or one individual building part. In the future, solutions will become conceivable which, analogous to nature, can be realized multifunctionally in one building part. To what extent it will be

© Hofer, scionic® I.D.E.A.L.

Salad rotator for artificial gravity and natural solar illumination despite pneumatic protective enclosure against cosmic radiation.

possible to copy nature's perfect material and energy-efficient circular economy remains to be (skeptically) seen.

The constructive principles of the exo and endoskeleton have not yet been exhaustively applied in the constructional development of human artifacts. Traditionally, from an evolutionary point of view, the building and construction industry has its origin between the poles of existing caves and wooden sticks found in nature, which were assembled into mobile frameworks and infilled with surface materials, whether of animal or botanical origin. Next to the building material stone, the natural composite material of cellulose and lignin – in the form of wood – antipodally played an important role. Ever since, the majority of all material applications in the building and construction industry have been derived from them: from natural and artificial stone to fiber-reinforced cement materials and concrete components on the one hand, from the girder, just like the tree has grown in nature, to curved wood shell frameworks on the other. With the systematic use of mineral ores and fire, first metals, then ceramics and, eventually, glass were added as building materials. Earlier, animal or plant "skins" likely served as cladding materials; later, they were separated into individual solid fibers and then reassembled into semi-finished products for surfaces. Only recently have new weaving techniques also made three-dimensional textiles possible. Through innovative processes in chemistry, membranes today can also be produced as geomembranes without fiber-bound reinforcing material. Special attention is paid here to the surfaces because the use of coatings makes it possible to influence a number of properties, which up till now could not be done. However, since the historically passed on building materials provided the basis for the relevant rules and regulations, it is accordingly difficult to meet these specific laws today with entirely new and different materials. Without adjusted building regulations that have been coordinated with the new materials and their properties, it will be impossible to achieve sustainable innovations in the building sector.

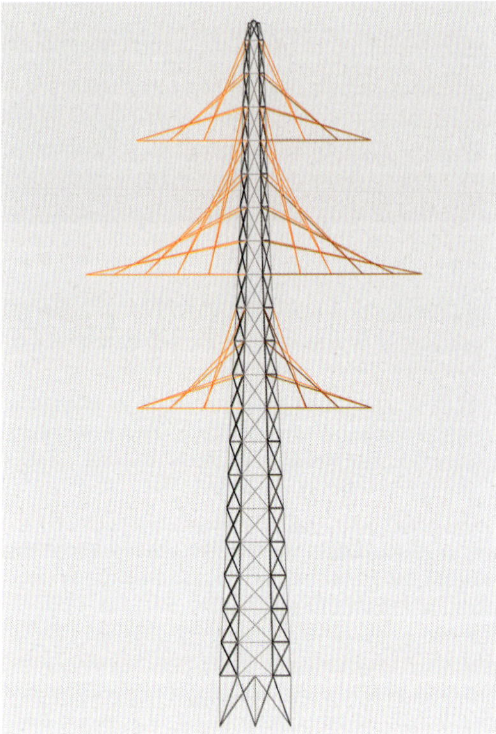

© Frittajon, scionic® I.D.E.A.L.

Mathematical modeling and computer simulations presently enable far more precise predictions on the behavior of building components than was the case in the past. Before, experience as well as trial and error played a more important part than is the case today, exhausting the range of technical possibilities. In addition to new materials, production and joining processes, as well as optimized calculation methods, cross-industry networks of individual trades will become increasingly important in the future. This means that a continuous integration of sensor technology into individual building parts and an active dynamization of building structures are to be expected in contrast to the – to date mostly – static construction. This holds especially true with regard to a future function of integrated multimediality. A biomembrane developed by EADS Astrium Space Transportation today already has the ability to autonomously regulate the gas exchange of oxygen, carbon dioxide and monoxide, thus enabling planetary plant growth as a

© Frittajon, scionic® I.D.E.A.L.

The tensile triangles of the "tree pope" Prof. Dr. C. Mattheck inspired the new conicoidal cantilevered suspension system.

fresh food supplement for manned space flights without health risks. This makes self-regulating and even self-organizing building structures conceivable, which automatically adapt to the respective conditions and do not necessarily have to be made of traditional building materials but rather are composed of fluidically prestressed, flexible membranes[6]. These theoretical approaches can lead to entirely new solutions in the realm of spaces for the continuous stay of human beings, whether on Earth, on the moon or on Mars. But new proposals can also be expected in the realm of air and space travel as well as personal mobility.

Our global society is predominantly based on fossil energy sources. Without these, all technological achievements are useless, not only in the context of mobility. Electricity is the primary energy supply without which public, local or long-distance transport, fridges and most private kitchens could not function anymore. The whole telecommunications system as well as entertainment and information electronics systems run on this energy form. Without electricity, there is also no mediatecture. Out in the open landscape, whose "naturalness" we sometimes enjoy, it is necessary to use power pylons to distribute electricity over a network, even if it eventually comes out of a socket. In my experience, it is easier to topologically estimate a landscape by the lines of pylons. Apart from spatial orientation, these pylons, as contrast points between nature and manmade technology, also give the landscape a special appeal. Entire regions, individual places and various energy supply companies have recently come to realize that special pylon designs serve to better differentiate their identities in the local and transnational competition. Pylons with specific differences in shape from the tee square framework and consisting of various material combinations, such as concrete-and-steel structures, wood glue girders, stainless steel constructions, and as yet unrealized attempts at fiber-reinforced synthetic materials, aim, above all, to draw attention to their location or the energy supplier. In addition to unusual color choices, specially designed models consciously toying with metaphoric or zoo and biomorphic forms are also generally associated with (industrial) design. Due to my scientific approach to innovations, I was commissioned by an energy supply company to redesign a 380-kV suspension tower together with my scionic® students. Inspired by folding patterns in nature, we designed a triangulated framework in such a way that the functional form was not heavier than the mathematical ideal of the pure engineered construction. In addition to pylons, more examples of already developed, innovative designs for the industry[7] that – with regard to the use of energy and material, and their structural optimization and design – lend themselves to functional forms without being heavier or "worse" than the pure engineering solution can be found in the book "Scionic: Purpose-driven Gestalt. The End of Design?", published by Arnoldsche Art Publishers. Published by the same house under the title "Visual Permutations – Creativity in the Here and Now", this is an approach to a system theory of aesthetics where the conceptual boundaries of applied, fine and performing arts are dissolved. We often perceive as aesthetic that which better serves the purpose, regardless of color and materiality.

The increasing fragmentation of specialist knowledge in the building and construction industry can only be brought together again by cross-sectional communication between diverse disciplines. The dissolution of specialist boundaries will play an essential role in the development of future building structures. It will be interesting to see which emergent positions architecture will seek out next to mediatecture and other evolving disciplines. Or will that just leave us with formal aesthetics or "architect-art"?

1. See Axel Thallemer, "Technologiedesign und Marke über einen bionischen Ansatz," in Der Ingenieur und seine Designer, ed. Jens Reese (Springer, Berlin, Heidelberg, New York, 2005) pp. 111-127.

2. See Axel Thallemer, "Die Zukunft des Design," in Im Designerpark, eds. K. Buchholz and K. Wolbert, 1st edition (Institute Mathildenhöhe, Darmstadt, 2004) pp. 206-210.

3. See Axel Thallemer, "Keyword Engineering Design," in Wörterbuch Design: Begriffliche Perspektiven des Design, eds. Michael Erlhoff and Tim Marshall (Birkhäuser, Basel, 2008) pp. 124-127.

4. See Oliver Herwig and Axel Thallemer, Luft – Einheit von Kunst und Wissenschaft (Arnoldsche Art Publishers, Stuttgart, 2005) and Oliver Herwig and Axel Thallemer, Wasser – Einheit von Kunst und Wissenschaft (Arnoldsche Art Publishers, Stuttgart, 2007).

5. See Axel Thallemer, "Raumgestalt und Designstrategien in der universitären Lehre für das 21. Jahrhundert," in ArchitekturRausch – Eine Position zum Entwerfen, eds. T. Arnold et al., 1st edition (Jovis, Berlin, 2005) pp. 36-47.

6. See Peter Sloterdijk, Spheres III – Bubbles. (Suhrkamp, Frankfurt am Main, 2004) pp. 474-476.

7. See Axel Thallemer, "Keyword 'The Future of Innovation – in industrial context,'" in The Future of Innovation, eds. Bettina von Stamm and Anna Trifilova (Gower Publishing, London, 2009) pp. 366-367.

© Berger scionic® I.D.E.A.L.

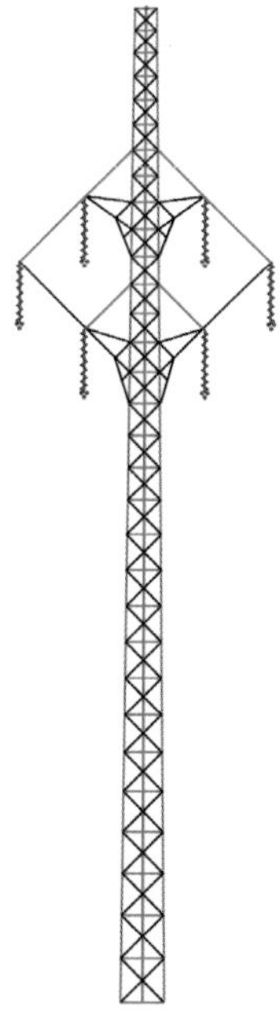

© Berger scionic® I.D.E.A.L.

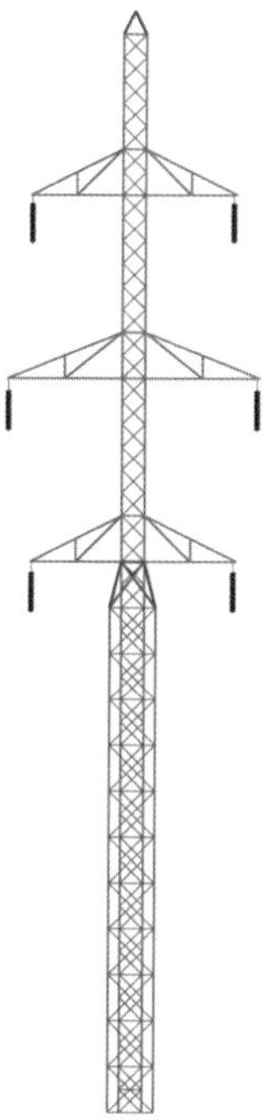

© Dai Yali scionic® I.D.E.A.L.

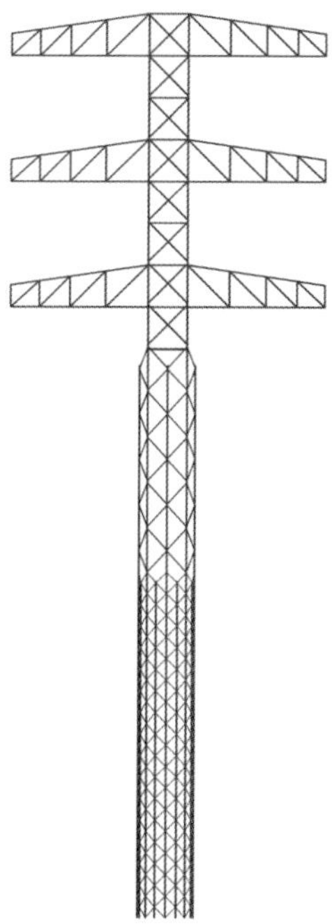

© Colceriu scionic® I.D.E.A.L.

design

MEDIA DESIGN IN AND FOR SPACES

As architecture and media converge, media design will become an ever more important element in the definition of the quality of urban spaces. This will lead to more direct sensual contact between architecture and people. What are the consequences of this for media design?

MEDIAL DESIGN FOR SECOND DEGREE ARCHITECTURE

When I was studying architecture I already had a problem: for some inexplicable reason I was simply not moved by the architecture which was presented to me as contemporary. Of course I could tell when a building was successful and the chosen design approach had been employed with full effect. And of course there are countless works of architecture which inspire – when examined simply in terms of volume or materiality, rhythm, construction or use of light, etc. But somehow I often failed to experience any sensation of closeness – and closeness with what? No matter how good a modern building was supposed to be, it often left me with a sense of being distant – it did not speak to me. I simply did not know what I was looking for.

These questions led me to attempt to understand architecture as a means of communication. But how can architecture communicate? And this brings me to the meta-structures of architecture - to typologies. My observation, that skyscrapers are always a statement of power and that they adopt a phallic form as part of their articulation of architecture as a representation of social relationships, was not so well received among my university colleagues. Apparently this was at odds with their understanding of architecture as a self-referential cultural form. The resistance to my opinion was so aggressive (at least in the 1980s), that my suspicion was merely reinforced. Such resistance can only come from those who feel that they have been unmasked.

Turning to electronic media was the logical next step. If architectural forms are not in the position to respond to the communicative challenges of society then the dimension of time has to be applied to architecture as a means of creating this connection with the surrounding space.

Rem Koolhaas' concept for the ZKM in Karlsruhe in 1989 inspired everyone who was open to ideas of using media in architecture. And Toyo Ito's Tower of the Winds is still, today, a beacon for the growing community of architects and media artists, who, in their very different ways, are working on the medialization of architecture and urban space.

Questioned in an interview about how and why the theme of the media façade had become such an important one, Jan Edler from Realities United suggested that the answer was to be found in architecture itself. The phenomenon of the media façade could be traced back to the weakness of current architecture and the to fact that much of the design process – especially of façades – had been removed from the responsibility of architects and become subject to the technical demands of the building industry.

And this is why the surface – the external skin of a building – presents itself as a projecting surface for anyone who still desires to create a vibrant, directly experienceable architecture. The approach of Realities United to this task is to regard the media technology for each project as a kind of architectural expression whose pictorial points (pixels) substitute the façade of the building.

Media artists such as Mader Stublic Wiermann from Berlin transform façades into complex abstract structures whereas Lab(au) from Brussels underlay their façades and media sculptures

The 1989 concept for the Center for Media and Art Karlsruhe by Rem Koolhaas

with interactive processes, which aim to abstractly reflect such intellectual constructs as, for example, time.

Across the globe there is a huge variety of further approaches to using electronic media in façades. Many play with the dynamic transformation of colored light into a façade whereas others look for ways of getting inside the building as a means of allowing the medial effect to work from the inside to the outside.

Taken together, these projects raise the question of the effect that the medial power of the media façade in the urban space can have on people. Can it be that this conjuring with pixels is a mere fashion item which, despite its undisputable short-term entertainment value, is actually nothing more than an excited, colorful but yet empty show? To what extent do the technical possibilities of this new approach offer us the opportunity to develop a deeper relationship with the viewer than that offered by, for example, a plate glass façade with a simple grid?

It seems obvious that any form of dynamic change in a visually stimulating object – whether a swelling sea or skimming clouds – can provoke a sense of fascination. And it is certainly the case that much movement can be very calming – in comparison with the fear and anxiety which can result from that absolute stillness which in nature is virtually unknown.

A media façade always creates a virtual world. In the chapter "Imagining and Enlarging Spaces", I describe the relationship between reality and the processing of sensual stimuli into a virtual reality. The notion of media design, on the other hand, addresses the issues of how the use of media façades can redress the unattachedness and unrelatedness of public spaces and of which technical methods can help in this. This question of how is a question of content – a question of what one perceives. The evaluation of media façades and their cultural relevance must urgently get beyond the technical medium itself and address the issue of content – and not, as so often, confuse the two. And if the content is created in such a way that the viewer cannot understand it, then the relationship between sender and receiver can be seriously damaged. The establishment of communication between a façade and the public moving in front of it is a relatively new und unresearched area – but there are two important basic rules:

No narrative structures and the avoidance of repetitive loops

If these rules are not respected then the medium has been misunderstood.

Narrative structures lead the viewer to search for editorial content of the sort which, in a television channel, continuously brings new information. And this fact - that such content directly awakes our expectations of something which we already know – means that the inclusion of narrative structure in the content of a media façade merely transforms it into a television. Furthermore; as it is so difficult to take in linear stories in a public space, the viewer tends to instinctively reject such content.

Repetitive loops strongly reduce the attractiveness of media façades. And frequent repetition of the same content quickly leads to the rejection of the façade by its neighbors – it simply becomes annoying. Such content also creates a sense of disappointment. Medially moving images hint at the ability to escape the stiffness and stasis of an architectural façade and it is frustrating when they prove to be just another sort of closed box. In places like Times Square, however, this is perhaps not such a problem, not only because there are so many repetitive loops that one cannot distinguish between them anyway but also as a simple result of the cult status of this Times Square effect.

Of course the repetition of certain forms of medial content is unavoidable – but efforts should be made to ensure that the sequence of this repetition changes. However, it is naturally preferable that changes in content are as organic as possible and that the façade, consequently, gives the impression of being alive

The link with cybernetics is obvious. Heinz von Foerster puts a value on the difference of **"second degree cybernetics"**[1]. This notion of the reflective self organization of complex systems can also lead to **second degree architecture**.

Tower of Winds by Toyo Ito, first order cybernetics

A Vision

I dream of a city whose architecture reflects the emotions and ideas of its citizens. I dream that the design of public space develops a new form of intimacy which gives people an experience of society. I dream that façades can become the interface between outside and inside – the interface between status and existential orientation.

The word façade is derived from the word face and this suggests that there must be some connection here with the concept of face and, more specifically, the face of a building. Bazon Brock uses this starting point to suggest that the medialization of a façade can allow it to develop a sense of mimic which can react to the people around it.

How would it be if the electronic monitoring of movement in the public space enabled the public space to develop a direct relationship with the "mimic" of the façade? If only this electronic monitoring were capable of observing people's behavior. If the character of the content could adapt – or react - to the fact that people were hectically rushing around or calmly meeting in groups. If, for instance, the public space could disseminate calm when people were agitated, or warmly greet the groups of people who had come together. Or if active areas of the public space could be defined which were able to communicate with the façade in line with an established set of rules or interfaces.

Or if the public space could react to the time of day or the quality of light in order to establish a mood.

People's longing for close contact can be seen in web 2.0 and in such fora as myspace. And even when it seems that the quality of the ensuing personal exchange is somewhat shallow, this does at least show that people want to overcome this distance.

Finding a place for this longing in the media façade requires intensive work on ways of creating content. Introducing such transparency to modern building is a technical challenge and this is just as true for media design. And, in just the same way that the technical ability to create huge panes of glass doesn't guarantee that we can build fully glazed façades, the ability to create interactivity does not alone fulfill the medial desire for closeness.

Which means that we need applications which allow façades to react with great sensitivity to their surroundings but must at the same time be careful to ensure that these have more than the sort of purely playful character which can become a one-off and, ultimately, disappointing experience.

The psychologist, internet researcher and management consultant Professor Peter Kruse describes the internet as a space of social resonance. Would it not be wonderful if the façades of our urban spaces could not, similarly, resonate to the rhythms of public life?

1 siehe „Kybernetisches Gestalten oder die Architektur zweiter Ordnung“ Ein Gespräch mit Heinz von Foerster

The Façade as a Space of **Social** Resonance

Living. Distinctive. Unpredictable

Share, participate - joy, sadness, humor .....

And the same from inside to outside, the façade, which shows to the outside how the people on the inside are feeling. Showing a face. Becoming perceptible.

Dissolving borders. Is this not what modern architecture promised us? Allowing spaces to flow together?

SELFGENERATING **LIBRARIES OF IMAGES**

Communication is Process-based

The management of content can be based on purely artistic principles. I, personally, search for uses which also give one the feeling that the project makes sense, which is why I am interested in the use of complex media techniques in order to create realistic presentations. This is precisely why I have invested so many years in the development of media façade technology (Mediamesh®, Illumesh® etc.). A particularly important aspect of this development work has been the principle that the technology which is developed can be relatively transparently integrated into the façade in such a way that the content appears to swim in a transparent layer over the façade. Only in this way can content and architecture become one.

I attempt in other chapters of this book to shed some light on the question of which content is most suited to being communicated via a media façade. Here, however, my aim is to present a method for controlling communicative content in such a way that it becomes part of the social processes surrounding the façade. Mediatectonic flow and second degree architecture are central to this method.

As early as 2004 we had to develop a process-based content management solution for the Serono project in Geneva. The special construction of the wax wall meant that it was impossible to create a uniform image, although the adopted media technology would have theoretically made perfect video resolution possible. The content was therefore divided up between the individual modules of the installation. The result of this process was IMPP – Interactive Media Pool Platform – which ag4 and I are still developing to this day. IMPP was also the basis for the projection onto the Bayer Media Sculpture.

Put simply, this program contains a set of rules for variably connecting individual media modules – that is to say, images or short video clips - and the result is a sequence of combinations which continues to develop without end. The challenge of this method is to draw up a set of rules under which the media modules combine in a logical way. Such rules tend naturally to be associative. The communication of content which can only lead to effective communication with the help of a disciplined logical structure – that is to say content based on a linear chain of information – is not possible with such a media façade. But this would any way make little sense, because such a use of a media façade reduces it to being a monitor.

The emotional and aesthetically successful transformation of complex content into an interactive and self-generating structure is a much more complex concept in the public realm. Why? because the aim here is to encourage the viewer to think about and then contribute to the debate in order to try and ensure that the resulting message has a much more long term effect than if it had been formulated in a short and direct way.

The process-based media façade creates cascades of pictorial impressions. The brain combines these impressions into information and experiences them in a very different way to that in which the brain deals with linear chains of information, in which each piece of information stands in a direct relationship with its neighbor.

The organizational psychologist Professor Peter Kruse has identified a basic alteration in how we process information within our medialized culture. Earlier, we took in each piece of information in its original context as a result of which we could logically put these individual pieces of information together – in a linear process. Now, however, search machines present us with all the information about a particular matter immediately and unfiltered. These pieces of information have no obvious contextual relationship with each other – which causes frustration. But if one is prepared to tolerate a certain amount of frustration one will slowly begin to recognize new information patterns – and these patterns lead to completely new insights. The extent to which this principle has successfully established itself can be seen in the way in which such services as google have taken over our work and, indeed, our lives.

A method of generating libraries of images for media façades developed out of a process-based method of combining images has exactly the same potential for activating our innate ability to search for and recognize patterns.

The Programming of Content

Groups of images with very different origins and mentalities are separated into various pools. Unique characteristics are ascribed to each individual media module, with each being given a profile which can be recognized by the control mechanism, enabling it to be arranged with other modules in line with clearly laid down rules. Modules with similar profiles can then be combined, overlaid over each other or animated by the use of some artistically desired effect. Everything happens in real time with the design process occurring in the following phases:

1. Agreement of the desired atmospheres or communicative intentions which are to be articulated by the images
2. Search for and creation of the appropriate media modules (images, animations, video clips)
3. Drawing up of a masterplan, which establishes the rules for combining the media modules
4. Profiling of the media modules
5. Development of an interactive surface for the individual management of the projection (regulating elements which make it possible, at any time, to adjust the method of combination)

In order to be able to work with such an instrument, designers must be willing to engage with a new way of thinking. One is no longer in control of the library of images which one is creating but one is, rather, creating a sort of genetic code – a mediacode for the development of this library of images. The design process is no longer top down but bottom up. Or – in order to better explain the idea of Professor Kruse – it has an inherent element of destabilization. Only unstable systems can constantly find new solutions or – with reference to the projection onto a media façade - an unstable system for playing images can produce surprising moments.

But the crucial thing is that the viewer will eventually begin to recognize the images as a pattern. And I am convinced that we humans, in the same way, recognize cities as patterns. The sensual effect of different façades and building types in a city create within us, over time, a sense of a pattern which is certainly not retained in our memory in the form of a film or an image. Designing façades by creating patterns is, of course, a classical way of creating cities.

Any highly targeted element of communication can also, naturally, be integrated at any time into this flowing pictorial structure – whether this is a piece of advertising or some other communication from an investor in a media façade. This is equally true even when the façade has been principally conceived in line with an artistic concept – because the capital of the media façade is time and one can project different batches of content at different times of the day. The decisive factor here is that hard edges between different batches of content should be avoided.

Combination and Self-generating Processes

The principle of process-based combination offers the possibility of an approach to projection which appears highly organic. Yet this is still not a truly process-based unfolding of a library of images, because the images are still held within a closed system. Through the act of combination the media modules appear in a sequence which will never be repeated but these sequences will have visual and thematic similarities.

In 2009 I started to sound out the design possibilities presented by different ways of combining images by the creation of a platform which would also offer the potential of creating self-generating processes. The name of the platform – "SITE-character" – makes, with the word SITE, a very clear reference to the idea of the website while the word site is also a very clear reference to architecture. The use of the word "character" shows that the idea of the platform is precisely to manage the character of a façade. The idea being that a façade can have different characters in the morning and the evening or appear differently during the week and at the weekend.

I used this platform to create my own website and, in doing so, created my own mini façades for demonstration purposes.

"SITE-character" uses Flash. This makes it possible to create interfaces at will using the web (and even if Flash is replaced by html the principle remains the same). It makes sense to see the content of a media façade not exclusively as specific material for the creation of a specific aesthetic in a specific location but rather, from the very beginning, as a strategic part of a website – or of several websites (in which case either a timeline or some other rule can ensure how relevant websites are used). This notion has a conceptual effect on the process-based development of the contents. "SITE-character" manages the intelligent combination of individual media modules. This is a method of combining which operates according to rules. It cannot generate something new – such as activate new images - of its own accord. However, this can happen if "SITE-character" draws data from a website and connects this with data from somewhere else. Such data is changing constantly and the content of a website seldom stays the same for long - and this is particularly true when a website is part of a social network. The result is a media façade whose content is truly (partially) self-generating in real time and this mixture of combination and self generation is important, because a purely self-generating structure is very hard to control. A fully free development of content may appear to offer fascinating possibilities - but it could also lead to direct conflict with a client who expects his media façade to have some communicative capacity. An individually determined set of rules can now combine – and seamlessly integrate – fully freely evolved and concretely designed content.

Here is a specific example: "SITE-character" draws images from a specially organized platform on Flickr. The platform to be used here has a certain pictorial theme, but access is not controlled (with the exception of a filter which blocks pornographic and violent images etc.). Such a theme could be, for example, images which express a brand. At the same time a fixed pool of images related to the brand is also created. Then, images from the pool are combined with the images from Flickr by means of some sort of on-line switching modus which is the responsibility of the designer and which must differentiate between the world of the brand images and the organic content from Flickr. The result is a process which constantly generates something new while, at the same time, guaranteeing some form of control in the interests of communication.

The transformation of façades in the public realm into self generating systems is, for me, a fascinating objective. It could lead to a completely new relationship with architecture in that people will suddenly be able to participate in and contribute directly to the appearance of a building.

I am well aware of the resistance to such notions. After all, it appears that I am questioning the very creative basis of architectural design. But this is an unnecessary – if unavoidable – misunderstanding: no new discipline has ever replaced an old one; everything is a question of evolution. The modern expectation that architecture must have both aesthetic and organizational quality remains untouched.

Of course the reality is that such unified and intensive projections are only an issue for locations with an intensive public character. Examples could include a project to improve a neglected public space - or the combination of architectural highlights, as a means of creating a positive image for a city.

BAYER **WORLD SCANNER**

The task of the Bayer Media Sculpture is to re-charge the Bayer brand: to show a Bayer which is innovative and communicative. Of course this message is not just meant for a regional audience but should have a global reach. Hence it was clear that the project would be intensively accompanied by the company's communications department. At the same time it was promised that a separate website would be created for the project and I saw this as an opportunity to not only publicize the project and ensure that it became well known but also to combine the process-based aspects of the project directly with the website. Initial problems with the implementation of the project then arose from internal work on the Bayer websites and problems with firewalls.

My opinion is that this project is a perfect example of how a media sculpture can be truly cybernetically created and of how this can not only fascinate but also open up whole new opportunities for corporate communications strategies.

The manner in which the IMPP-organized content of the media sculpture continuously emphasizes different Bayer themes can be individually controlled in real time. This permits the creation of any number of interfaces for the interactive management of the content.

This basic arrangement allows many people to manage the contents via the internet. Which obviously raises the questions of why - and what does this bring?

The answer can be found in regarding the media sculpture as a "seismograph". This seismograph reacts to the interest which the world shows to Bayer's scientific developments. The "Bayer World Scanner" web tool was created to make this process visible. The tool is integrated into the Bayer website and has the immediate task of measuring visitor traffic to the key Bayer themes. This visitor behavior then determines the extent to which the themes are presented on the Bayer Science Tower.

But who notices what?

The questioned is justified. The driver passing in front of the tower is hardly able to differentiate between the amounts of time dedicated to each theme. The key is the story which is told as a result of this combination of media sculpture and internet presence. The plot in one sentence:

Bayer interacts with its clients across the globe and takes their reactions to Bayer's scientific developments very seriously

In order to allow as many people as possible to be part of the story, the web tool must be created in such a way that visitors to the Bayer website have something to gain from using it.

These benefits are to be found in the user-friendly access to knowledge. The Bayer World Scanner gives the visitor both an overview and simple introduction to current themes. Contents are automatically linked to the existing Bayer website meaning that there is no need to create new content.

Rendering © Kronhagel Mediatecture

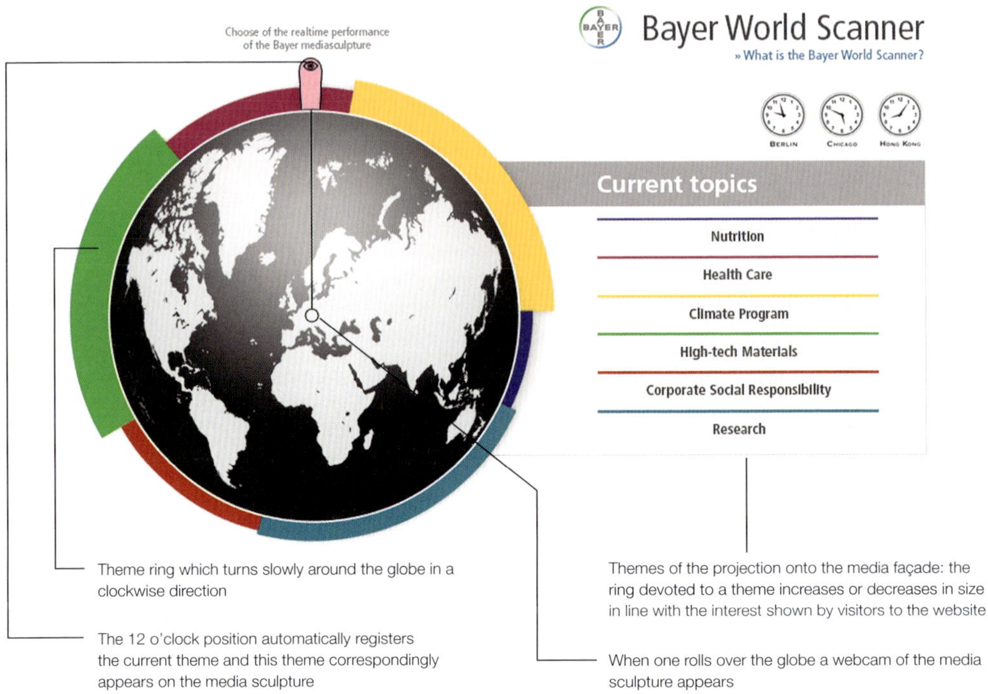

The behavior of the users is portrayed on the Bayer World Scanner by the representation of the themes on the website in the daily cloud. The more often a theme is viewed, the larger its name appears in the cloud. This record of interest then influences the themes offered by the tool and, indirectly, the behavior of the media sculpture. And this can be observed across the globe by means of a webcam.

The Bayer communications department regularly places new themes in this cloud – with middle-sized lettering. If visitors rarely click on the theme, interest is obviously not huge and the theme is removed again.

The network mentality of this holistic concept requires that the user has three experiences:

I am part of the whole...
... Bayer is open for me ...
... Bayer is modern and
innovatively organized.

And, in this way, Bayer creates a contemporary image for itself as a company which lives from science and very dynamically and efficiently networks its scientific activities

The Bayer Media Sculpture is becoming a symbol of a company which directs its activities in the interests of people: Science for a better life.

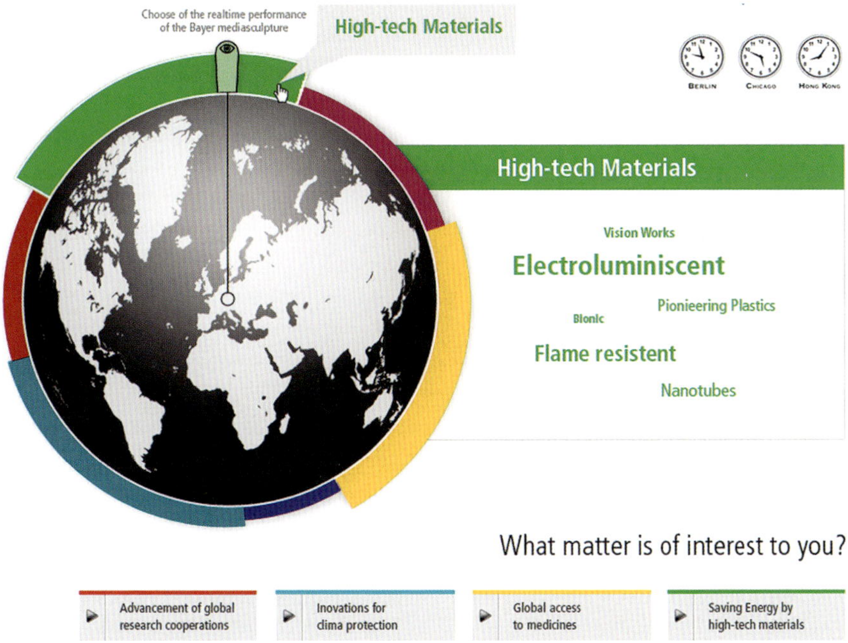

Grafik © Kronhagel Mediatecture

Design

The Bayer World Scanner can be managed with acceptable effort. The key is that the Bayer World Scanner is not a closed website but rather, principally, a technical instrument which is added to the existing Bayer website.

The reason for giving the tool a circular form arises from two factors: Firstly it makes sense to create a circular design because this is a reference to the globe. A small graphic device enables us to display the entire earth on the surface and, fortunately, one can lightly shift this graphic in such a way that Leverkusen (the home of Bayer) is precisely in the center. From this center radar-like trails oscillate out around the world.

The circular form also naturally refers to the Bayer Logo. The ring of the logo is reinterpreted as the content ring. Segments of various colors represent different themes and these colors are taken up by the words in the daily cloud. The relative sizes of the rings can change. The result is that one can sense which themes are, at any one time, provoking most interest among the global society. These themes are then correspondingly arranged on the media sculpture. At the same time, the ring is constantly moving in a clockwise direction. A "marker" at twelve o'clock measures each segment and the content management tool of the media sculpture ensures that the theme is highlighted for a corresponding period.

At the same time, these segments are a tool which, upon being rolled over, shows the relevant theme – which can then be selected with a click.

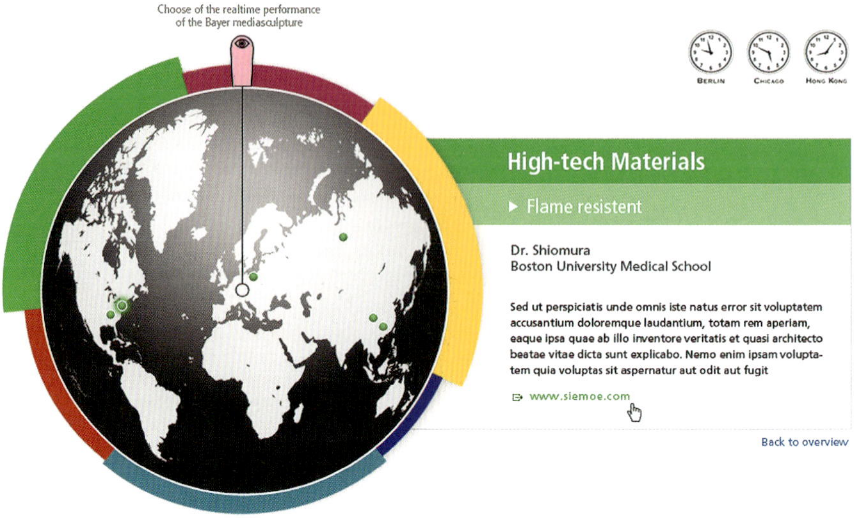

The tool also offers a true networking of spaces. When the visitor clicks on a term in the daily cloud a list appears showing the scientists who are working on that particular project. The scanner also shows where he or she is and, if one rolls over this location a pop-up then appears offering more information about the scientists and, if desired, their internet addresses. This allows visitors to the web tool to network in a location-specific way with the scientific world of Bayer and this is a capability which, incidentally, can also be of immense use to the scientists. Who always knows who is working where and on what? (The parallels with the ETH World project are not a complete coincidence because the network integration of spaces usually leads to the creation of a community).

All facts and figures are fully drawn from the existing Bayer website, as a result of which no new editorial work was required.

Wolfram Lusche proposed the idea of offering a catalog of questions which gives users the opportunity to more clearly articulate their interest. This would make it even easier for Bayer to place the relevant theme on their daily cloud.

The only administrative work will involve the daily cloud and the catalog of questions – but this is work which will provide Bayer with valuable information about the behavior of its customers.

Strategy versus Design

Such willingness to invest in a built object such as the Bayer Media Sculpture which offers no tangible financial returns is unusual. Such landmarks usually offer visitors at least a restaurant or a visitor platform.

But what is the relationship between the investment in the project and the brand benefits which it will bring? Naturally there is the so-called "media value", which will arise from the extensive publication of the project in all possible media. But this will not communicate the quality of the content. Media offer relatively short-term impressions and the only image which sticks is the fact that this huge built volume now carries images and that the effect is both impressive - and pleasing. Once again we are left with McLuhan's: "The medium is the message."

It is, incidentally, almost impossible to avoid the installation being somehow celebrated as the world's largest television or advertising column. There is also always the danger that the communicative effect of such a project is very superficial and that it is unable to serve its sustainable purpose. This happens when such a project relies on visual effects alone – when design is everything.

For this reason it is important to strategically network such built objects with the communicative processes of the new space. This clearly represents a paradigm change – this is the principle of the extension of space. And this means that such a project is never finished – which is a nightmare for managers in industry. Each proper construction has a structure which is complete and, above all, which is controllable – and this is very difficult with communicative processes. And this is precisely where the subject is both so difficult and so exciting:

When Bayer promises that science will bring us a better future then this can only mean from today's perspective that science will be used for "Climate Saving". The intercommunicative processes of the Bayer World Scanner have constantly put this subject at the center of things precisely because it is the public that influences the choice of themes. The extent of this danger can be seen in the first reactions to the installation when Bayer placed images of a test on Youtube. The interest and the enthusiasm were huge. But at the same time there were voices who asked why such a project was being implemented by a company such as Bayer. These voices had a problem with enthusing about something of this nature when it was being produced by what they saw as a typical environmentally unfriendly corporation.

But anyone who wants to win over a community must engage with that community. And this also means that the communications work of a company cannot only be driven by design but, above all, by strategic processes. And the benefits from doing this can be enormous:

Nothing better protects a brand today than a stable community.

Wolf Lieser *has been working as an art consultant since 1990. 1994 Opening of the first gallery in Wiesbaden (till 2002). 1998 Foundation of the organization Digital Art Museum [DAM] – Online museum since 2000. 1999-2002 Colville Place Gallery, London – the first gallery for digital art. 2003 Relocation and opening of the first [DAM] gallery in Berlin. 2005 [DAM] presents the d.velop digital art award [ddaa] for the first time, a biennial distinction for the lifework of a digital art pioneer.*

DIGITAL ART FOR MEDIA FAÇADES

Digital art has existed since the 1960s; media façades, on the other hand, are much older – albeit they were not yet named as such in the beginning of the 20th century. However, everybody knows the illuminated advertising signs on Times Square in New York, which can be regarded as the predecessors of today's media façades. But it was the development of the light emitting diode and, above all, the data processing power of computers in the late 20th century that made this medium also more attractive for artists.

In the past, artists won a forum in public space mainly with large steel or stone sculptures. The creation of such sculptures often required great effort, and as a consequence, their material was selected for the longest possible resistance against weathering effects. Static and designed for eternity – this is how the majority of art in public space has fashioned itself up to now. This might change with media façades; since they can function as a medium for artistic content, they will likely be the future of art in public space. There is no lack in diversity of artistic content: digital art has grown up; it has gained in quality and audience, and in the meanwhile it has become an active participant in international art discourse. The times of fast effects – whether a 3D visualization of an interior or an interactive installation with a simple stimulus-reaction scheme – are over. Artists like Manfred Mohr, Gazira Babeli, Mark Napier or Jodi deal with new media in a conscious and critical way and arrive at conclusive, convincing artistic positions.

However, digital art is not just a new medium, as was the case with the introduction of photography and video art; rather a paradigm shift is taking place along with it regarding forms of distribution and marketing. Clearly defined procedures or rules in art – such as the production of an original by the artist, uniqueness and the clear attribution of the manually produced work, as well as the differentiation between art, craftsmanship or design – are noticeably blurred or even simply ignored by some artists working with digital media. This does not only apply to digital media: Since Andy Warhol's Factory, it is common and accepted for a successful artist to run an own production company, one that might easily have 50 employees. Here, the staff designs and realizes works in accordance with the ideas of the artist. What was more of an exception in the times of Rembrandt is today a rule in the established art system.

This development also benefited from contemporary art's move away from the personal signature or unique style in painting – artists like Donald Judd or Sol LeWitt set the stage. This led to entire series of "original paintings" such as the "dot paintings" by Damien Hirst, which he did not paint himself. Here, the notion of the original can only be understood as a marketing strategy. This became particularly obvious when Damien Hirst proclaimed in 2009 that he himself was painting again! The absence of a personal signature, which in the beginning was declared the flaw of digital art, is now regarded as an especially honest form of artistic creation. The computer image, the design, is not consequently transferred onto the canvas in order to provide the buyer with an "original"; it is the digital file that becomes the artwork.

The invention of photography already heralded this shift: The "original", namely the slide or the negative, is not the actual artwork. Only the enlarged prints of the "original" make it possible to properly perceive the work. In the beginning, the later so-called vintage prints were not yet limited. However, at least since the rise of art photography in the 1990s, prints have been precisely numbered and limited. Even though the video art market has not reached a comparable scale, similar principles in the relationship between original and artwork apply. Nevertheless, these two art forms still have a certain material basis in common: the videotape and the above-mentioned slide.

Digital art, however, can be traced back to zeros and ones – electric impulse or no impulse – pure information that can only be perceived in its manifestation through so-called "hardware", meaning computers and presentation media such as printer, screen or projector. Such an artwork can be easily distributed as a file and travels around the globe in seconds via the Internet. Intellectually unfamiliar and abstract, this fluid condition of the medium has for a long time been a reason for many art lovers to only skeptically approach this genre. Thus, it is no coincidence that the first collectors primarily concentrated on the early plotter drawings, which – at least as a product of ink drawn on paper – were still embedded in a traditional context.

For the gallery, curator or conservator of digital media, new challenges arise from the above-mentioned conditions in order to do justice to the medium. Not only to convince clients but also to preserve the artworks so that they still can be experienced in 100 years demands a change in thinking and the development of new competencies. The museum conservator of the future must be skilled in programming and in the translation of older software environments into contemporary settings.

New questions also emerge with regard to marketing. The buyer of a software work receives a CD or other storage medium, such as an external hard drive, which contains a file. They can easily copy it by themselves and present it in different ways. The limited edition, and thereby the proof of legitimate acquisition of the work, exists solely in the documentation material or the certificate accompanying the storage medium. Following the viral dissemination of music through its digitalization, there will probably also be a similar development in the field of visual, digital art. Perhaps it is no coincidence that in turn the performance as live act undergoes a revival in contemporary art. Here, everything is "genuine" and "real" – "live" art, just like the many musicians who currently generate their main income through concerts.

Now we arrive at the media façade, which equally implies the performance of an artwork. Once installed, it only offers a setting for the presentation of a visual or audiovisual experience. The artistic creation is presented as a digital film, an active software program or interactive installation. One of the first media façades in Germany, mounted onto the façade of the Zeilgalerie in Frankfurt/Main in 1992, was still confined to the visualization of information about wind, rain and temperature through differently colored lights. The large LED walls that have appeared in our cities in the last 20 years or so have gained more attention but are ultimately used as big TV screens. News alternates with advertising. Artistic works are rather an exception.

Over the last years, artists like LAb[au] from Brussels and realities:united from Berlin have become recognized for dealing with the theme more comprehensively than just mounting an oversized monitor onto a house wall. In 2006/07, LAb[au] used a self-developed interactive software in the staging on an entire highrise building. realities:united have realized several projects in this field with specifically designed façades. In most cases, special light modules were developed for an outer wall, which – besides being creative elements themselves – could be purposefully controlled by a computer to create a range of different lighting situations (see page ...). The emerging potentials for commercial application also stimulated companies such as ag4 media facade GmbH to search for novel solutions. Their product MediaMesh enables a façade to be covered with an additional transparent skin made of stainless steel mesh fabric with interwoven LED profiles, which can be activated as desired. In this case, the actual outer façade of a building remains almost completely preserved and visible.

The fascination of digital art, especially of software art, resides in the possibility for a developed work, a completed software, to be displayed both on a household screen – starting with a small telephone display – and on a huge LED wall. Such a creation can be employed in very flexible and time specific ways, potentially generating additional income for the artist via licenses. In general, the idea of time limited usage rights for a software is a concept that seems to be ideally suited for this medium. Art is thus subject to a wide diffusion but is not meant for permanent possession; in essence, it becomes more of a performance. The steel sculpture stays put, but the software work, on the other hand, is fleeting and dependent on the given surrounding conditions, meaning the hardware environment. Hence, it is appropriate to our time and capable of processing topical information, and thereby able to react to the environment. Our lifestyle, which in general has become much more nomadic – due also to today's labor market – finds its counterpart as well: an art we can carry with us as a file and that can be performed at any time and place.

Artists like C.E.B. Reas can readily transfer their artworks onto a media façade: Their abstract compositions, based primarily on minimalist concepts, are ideally suited for presentation on large-scale surfaces. It is even suited for narrative concepts such as the virtual life programs by boredomreseach, which make the evolution of simple "life forms" perceivable in a poetic manner. These creatures form ever-new constellations over the years through the simulation of life cycles. Such narrative works seem predestined to emotionalize the urban sphere.

Activities such as the first Urban Screens Conference 2005 in Amsterdam and the Media Facades Festival 2008 in Berlin are evidence of an enhanced discussion around the specific potentials offered by artists, curators and theorists. They illustrate that a new branch of art in public space is emerging. One of the early projects to utilize a commercial LED screen for an artistic purpose was, by the way, our series "Digital Move", which we curated for the LED screen at the Sony Center Berlin in 2005 and 2006. Digital artworks were presented there three times per day, each with a selection from three to four artists.

One can only hope that beside this medium's commercial usage also possibilities will be created in the urban sphere where artists can realize specific projects. In any case, they are certainly able to.

Media art by C.E.B. Reas

AD

ADVERTISING?

Seen through the eyes of the real estate industry, media façades have no direct user profile. They can be financed either indirectly through the added value that they bring to the architecture or directly through the integration of advertising. This chapter shows that standard forms of advertising are not suitable for media façades and seeks out new concepts for the strategically and architecturally logical communication of brand values in urban space.

Dr. Christoph Köller *is a management consultant (Görgen & Köller GmbH). He is also involved in the field of external advertising. In 2009 he organised the "Digital out of Home Business Conference" in Amsterdam, in parallel with the exhibition: "Integrated Systems Europe – DOOH".*

THE SEVEN SINS –
of Digital Out-Of-Home Advertising

In principle, media façades have nothing to do with classical advertising. They are employed to give a building and, very often, the company behind a building an unmistakable public face.[1] However the experience of real projects shows that, as soon as the use of electronic display media on the outside of buildings is discussed, attention soon turns to the "advertising millions" which they can apparently generate. "As soon as we hang up these things, the advertisers will be knocking at the door", is a phrase which, lightly modified, can be heard in almost every project meeting.

And it cannot be denied: every country has its well-known examples of media façades which are used for advertising purposes and are, at the same time, very successful. Prominent examples include, in the USA, Times Square in New York and, in Europe, Piccadilly Circus in London. The conditions which have led to the success of such examples will be discussed below.

More recently, the use of electronic displays – both outdoors and indoors – has been given the name "Digital Signage" (DS). A range of studies has shown that the market for such signage harbors huge potential for growth. One such study has estimated that the number of digital signage displays is set to reach 2.3 million by 2012 while another has shown that the advertising revenues generated by DS facilities in Europe in the same year will exceed[2] 600 million.

The reality however is that many revenue sources which are ascribed to "Digital Signage" often have little to do with such technology and, in many cases, have been around for a long time. And while there is no end to the labels created for such revenue sources - Street TV, Mall TV, Digital Out-Of-Home and Infoterminals are just a few examples – the truth is that while some of these labels have something to do with advertising many do not.

In order to introduce some structure to these many labels – as well as to better understand this "advertising" use - it is interesting to assess the business models of operators of such technology. These can be separated into four basic use configurations:

TDK
Budweiser

"SIMPLY INFORMATION"

means that the operator of the display is fulfilling an obligation (or a wish) to provide information – such as timetable or departure information to travelers. Such systems will generally be provided and operated by the owner or operator of a building and the investment is made in order to reduce process costs and improve customer relationships.

"MERCHANDISING"

is driven by the desire to use focused information and entertainment (in, for example, a shop) as a means of pursuing specific turnover targets. The operators of such systems are often the owners of the brands (such as banks) but can equally well be third party operators. In this case Digital Signage is a marketing tool whose objective reaches from increasing brand awareness to increasing sales. Third party advertising can also be included but is clearly not the core content.

"CUSTOMER EXPERIENCE"

The principal target of "Customer Experience" is to entertain or inform visitors to larger, not necessarily branded, environments (such as shopping malls). The key content here is about the location and what it offers. Such equipment is often (but not always) operated by specialist service providers. Yet, even though the improvement of the "customer experience" is a declared objective of this approach, it is not far from a pure third party advertising model.

"ADVERTISING REVENUE"

is the only business model which envisages advertising as the sole direct source of financing for an electronic display (although indirect and unexpected sources can of course never be fully discounted). The idea here is to use a network of electronic screens to attract a slice of the advertising cake.

Naturally there are areas of overlap between these four business models: a "customer experience" model can easily, for example, become an "advertising revenue" model. And, while they are clearly differentiated by the distinct ways in which operators aim to create revenue, there is one key similarity – electronic displays will only be deployed when their business model promises a return!

And in which of these business models is "Digital Signage" to be found? In principle in all four; although only the latter can truly be called "Digital Out-Of-Home" (DOOH). The idea is that advertising is displayed outside the home - hence this is classical outdoor advertising – but it is based exclusively on the use of electronic media.

And, incidentally, DOOH can naturally also be located inside buildings – in both private and public locations. I will however below concentrate on the DOOH approach as it is employed externally – or "Open Air DOOH".

I mentioned above that DOOH is said to have a golden future, and it is certainly true that many new operators of electronic advertising media are coming to market: ECE flatmedia in Germany, Adspace in the USA and the globally active NeoAdvertising are just a few typical examples. In addition to this, established operators are enlarging their infrastructure: Infoscreen, for example, is equipping German railway stations and CBS Outdoor and JC Decaux are doing likewise in metro stations and airports. And much is also happening outdoors, with Clear Channel Outdoor and JC Decaux replacing some of their traditional advertising hoardings in London with LED-Screens and Ocean Outdoor creating new locations in the same city.

And yet, despite such activity, one does not gain the impression that classic outdoor advertising is drowning under – or even being threatened by – a flood of electronic displays. The impression is rather of a very uneven market which heavily focuses on the placement of screens in certain sorts of locations in certain cities in certain countries. Is this really much more than the ongoing "simple collection" of advertising revenue?

In order to answer this question I have drawn up a list of the seven sins of digital outdoor advertising which could slow down – if not completely impede – the spread of DOOH.

It is a DOOH sin if one,

has no network,

attracts no attention,

can deliver no proof of effectiveness,

is too focused on technology,

shows too little electronic creativity,

operates outside the world of the traditional agencies,

has no permit.

We shall investigate these sins in detail below. Before starting however, it is important to state that there are DOOH operators who do not (or, at least, hardly) commit any of these sins and that these operators are really (economically) successful. And in addition to this there are sins which should not always be regarded as such.

HAVING NO NETWORK

Advertising is often campaign-based: A company wishes to invest a certain budget at a certain moment – or over a certain pre-defined period – in order to meet a certain target, such as creating the highest possible level of awareness for a new brand. Outdoor advertising is, in this context, just one of a number of potential channels and is, to a certain extent, in competition with television, radio, print and, increasingly, online advertising. Though some advertisers may wish to do away with it, outdoor advertising still represents an average of 4% of campaign budgets. (This is the figure for Germany: in other countries it can be higher). In Germany there is good demand for locations in cities of over 500,000 inhabitants. In order to increase the number of image views there is a tendency to show the same image in several locations across any one city and, of course, in a number of cities. There is usually nothing to be gained from advertising in just one location or just one city (or shopping mall).

An extreme example for a comprehensive outdoor advertising campaign was Toyota's introduction of the Auris in 2007. Toyota and its agencies booked virtually every relevant hoarding in Germany.

Such is the predictability of advertising industry decision-makers that, in Germany, this same demand for "national coverage" is also made of the digital form of outdoor advertising. The result is that each operator of a DOOH network has to invest in a comprehensive national (or at least regional) infrastructure. Assuming that enough locations are available, the creation of such a network inside buildings is already a financial challenge. Outdoors – where LED-Displays or cholesteric LCDs are the only available display technologies for large-scale use[2] – the costs are prohibitive. It is virtually impossible for an investment calculation to show a two-year return on investment (as is often demanded in the outdoor advertising industry). The consequence is that there has been virtually no progress towards the creation of a network of outdoor electronic screens – and progress indoors has been very limited.

Yet there are a number of ways of circumventing the need for a network. In America, on the one hand, both the established mechanisms and client interests of the traditional outdoor advertising industry understand that single local locations can indeed be interesting whereas in Europe, on the other hand, there is a trend towards identifying extremely attractive single locations: And this brings us to the second sin.

ATTRACTING NO ATTENTION

Outdoor advertising obeys the old rule that there are just three important factors: location, location and location. Advertising on Times Square or Piccadilly Circus "works" because these have been the busiest locations for centuries. And the value of a location in terms of outdoor advertising increases with the quality of the location and with the likelihood that the image displayed there will be seen.

An example of this phenomenon is the company Royalscreen which, a few years ago, sought to market a Germany-wide network of LED screens. In April 2007 the Tagesspiegel newspaper in Berlin reported that around 1,000 $9m^2$ LED screens were to be erected across Europe – including 45 in Berlin alone. In reality nothing came of the European and German dreams and just seven screens were eventually set up in less sought-after locations in Berlin. The project was not an economic success and, meanwhile, even the website has now disappeared.

Lanxess Arena parking house, Cologne

LED screen: defective and highly situated, Vienna

From the point of view of the advertising agencies, however, this question of attention is not only a question of location. Once a building has been identified, the positioning of the screen on that building is also critical. If it is too high or too far from the pavement, agencies are quick to question whether, the theoretical number of "opportunities to see" will be matched by the reality.

A key to grabbing the attention of viewers is the situation in which they see the image. According to Stefan Kuhlow, CEO of Infoscreen GmbH, there is a fundamental difference between the viewer in "waiting" and "walk through" locations. The former tends to have a higher attention level for screen-based information whereas the latter tends to be distracted. This is a difference, incidentally, which applies as much to indoor as to outdoor locations.

Too much attention, however, can also be negative. In outdoor locations, for example, the themes of road safety and "light pollution" are highly relevant. And, even though studies in the USA commissioned by the OAAA appear to demonstrate that the existence of LED displays has no effect on the rate of traffic accidents, this is outweighed in Germany by a tendency towards "assumed guilt". In the words of the ADAC, "everything which could distract the driver in the act of driving should be avoided in the interests of road safety": a statement which applies, needless to say, to all forms of large-scale outdoor advertising. And yet despite this fear, large LED screens are still often approved for busy junctions in large cities – with the LED display on the Lanxess Arena parking garage in Cologne being a good example.

The visibility of electronic displays at night also raises issues of "light pollution". As a reaction to perceived harmful effects on health, active citizen's campaigns have developed – in particularly in the USA – against LED displays close to residential districts. In Los Angeles, for example, there is currently a bitter fight between the City Council and the outdoor advertisers CBS Outdoor and Clear Channel regarding the possible removal of up to 101 (already approved) LED billboards. The conflict was sparked off by the subject of "light pollution". We will return below (Sin 7) to the subject of "approval" which is central to this conflict.

Of course if – as discussed below – operators fail to prove the effectiveness of a screen, then it will not remain in place long enough to become a long-term problem for local residents.

From LED screen to "green technology", Berlin

NO PROOF OF EFFECTIVENESS

Television, radio and print have spent years investing in systems designed to judge the effectiveness of advertising. Such market researchers as GfK and Nielsen record, almost to the second, information about which viewers with which socio-demographic characteristics have seen which advertising. A particular result of such information is that the qualitative reach and, correspondingly, the per-thousand-contact price of the advertising can be calculated. The fact that these information-gathering bodies are independent is naturally of great importance.

Outdoor advertising has, both nationally and internationally, similar mechanisms. Frequency studies and MediaPlakat both led to a better documentation of the value of existing outdoor advertising locations. The "older" DOOH operators (principally Infoscreen Deutschland and Infoscreen Österreich) recognized early the need for good documentation and subjected themselves to independent, voluntary evaluation.

The accuracy of effectiveness statistics is naturally limited - especially in comparison with online media in the case of which virtually each contact can be documented. Experiments are being carried out with, for example, "face recognition" technology, but reliable applications are not yet available. Most recently both Microsoft and Intel have announced their entries into the DS market: "digital billboards" equipped with cameras and special software should note the age, gender and size of the audience and allow statistics to be drawn up showing the intensiveness with which products and images are observed. Alongside technical problems however, such solutions also raise issues of privacy and personal rights.

The advertising industry is hence still faced with questions regarding the justification of viewing figures for electronic media – especially in comparison with "classic" outdoor advertising. And, in addition to this, the fact that a display is electronic is, in itself, of little interest.

TOO FOCUSED ON TECHNOLOGY

At first glance, DOOH is a problem which requires a technical solution. Content is to be distributed across a network and displayed on screens and, for this to work, electronic screens, networks and content management systems are required.

As far as can be seen, today's networks and content management systems are able to meet all the demands of DOOH. But the screens, on the other hand, are not: at least in surroundings which are too well illuminated and – in particular – outdoors. In both these cases there are problems of either legibility or power consumption – or both. The relatively low distance from the viewer means that, in particular, electronic displays measuring between 9 and, approximately, 18m² (in Europe: 16mm Pixel Pitch or better) require high definition. Particularly in cases of intense sunlight enormous energy consumption is required if such images are to remain visible.

There are technical alternatives – such as the cholesteric LCD displays developed and produced by the German manufacturer AEG MIS in Ulm, which are notable for their reflectivity and bistability. Reflectivity refers to the fact that part of the sunlight is reflected: the more sun, the more legible the image. Bistability is a characteristic of images which enables them to be presented without the presentation consuming any energy. The downsides of this are that, until now, bistable images suffer losses in contrast and that the projection of moving pictures is, if at all possible, qualitatively very limited.

Here it is necessary to point out again that, for the advertisers, the choice of technology is, today at least, virtually irrelevant. This means that, before making an investment in a DOOH network (or even in a single screen), decisions should be based not upon technical feasibility but upon what makes sense from the point of view of the advertiser. And, above all: what is to be projected and how? This is how electronic creativity is challenged.

ELECTRONIC CREATIVITY

The presentation of image and text on an electronic display can, unfortunately, not be equated with the presentation of the same material on paper. Even the presentation of "the color white" is a very different process on an electronic display, whatever the chosen technology, than on a printer. And content and media must be adjusted to fit.

The same applies to the choice of the content. Attempts are often made, for example, to show minimally adapted television advertisements on "public displays". In technical terms this is naturally no problem but, in dramatic terms, it can be catastrophic. The average attention span for outdoor advertising is two seconds – in the case of waiting areas slightly longer. But no one has the patience to follow a 10 to 30 second television advertisement.

Furthermore, it is necessary to consider whether and how DOOH could be combined with other media. The buzzword here is "crossmedia integration", as exemplified by the current trend towards the integration of posters and radio (see the activation-combi from Ströer). Clear Channel has also attempted to combine Bluetooth keys with LED displays in order to greet individual Mini drivers. There is definitely work to be done in this direction.

For this reason, successful DOOH companies have established special departments dedicated to creating content for their own screens. This is a complicated process to which there is so far no alternative. The creative agencies which should theoretically be responsible for such content appear imprisoned in their core television and print domains and moving into other media would generate extra costs which it is not clear that future clients would cover. This leads to a fatal short-circuit:

Creative agencies find it easier to simply transfer the methods that they use in print and television advertising over to external advertising without attempting to adapt these to the requirements of the new medium.

Media façades are seldom conceived as pure advertising facilities because they are generally also bound up with cultural objectives or issues of architectural quality. But, despite this, many investors still expect a return on their investment.

And, as outdoor advertising is in any case such a small part of the market, agencies naturally question the logic of investing in the area.

With the results that very few creative agencies have experience of electronic displays and that the skills required in order to master the technology are very thinly spread. Much will have to be learnt in order to develop this market, not only as a means of addressing the above-mentioned lack of effectiveness but also in order to "Teach the Creatives". The key will be concerted action by the operators

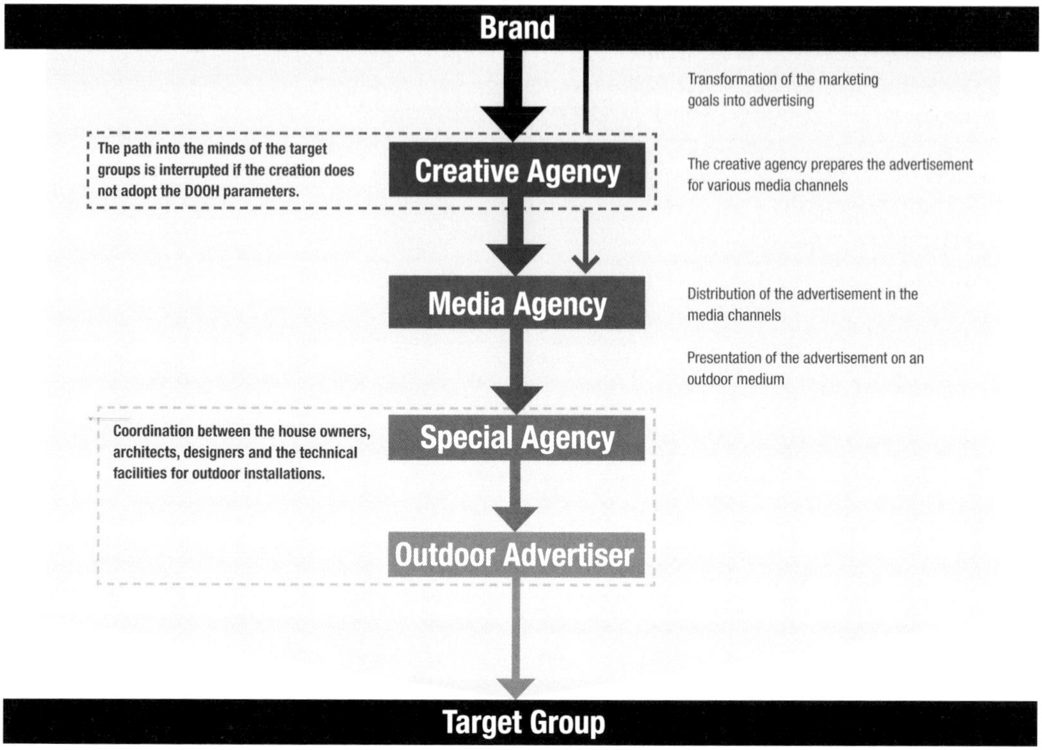

OPERATING OUTSIDE THE WORLD OF THE TRADITIONAL AGENCIES

The sins listed above show how important it is that one moves within the world of the traditional agencies. In the USA, organizations such as OVAB and OAAA have already moved in this direction. In Europe activities still tend to be one-off actions, although such players as Infoscreen have already gained a position of respect in the agency world. But what does the world of the agencies mean for DOOH?

Illustration 2 shows a simplified form of the process (although there are many exceptions): The company behind the advertising (such as a brand producer) wishes to meet certain objectives with relationship to its target customers. These could be objectives based on brand recognition levels or turnover for a particular product. Communications measures such as advertising are developed in order to achieve these targets. The producer allocates a budget for these measures to a media agency which is then responsible for identifying the right mix of media channels as well as developing the content of the advertising. The agency then reserves the required amount of television, radio and print space and, if outdoor advertising is also planned, special agencies are often involved in order to ensure that in, for example, the 13 German cities with more than 500,000 inhabitants, the most ideal locations of the outdoor advertisers Ströer, JCDecaux and Wall (or others) are reserved. In this sense the agencies take over the function of bundling. This system is more or less fully established in outdoor advertising.

If the decision is finally taken to use electronic outdoor advertising it is essential to ensure that the media agencies both understand and accept the potential of this form of advertising (remember the above sin of unproven effectiveness).

And on the other side these agencies should show good will and have the ability to design campaigns for this form of advertising. This is not yet always the case – and this will not change of its own accord. As mentioned above, it is thus necessary to "educate the market" – as some operators are already starting to do.

But even when the market has been educated, a missing permit can halt the process.

NO PERMIT

Of course individual buildings (such as shopping malls) are also required to obtain a wide range of permits – from the building permit to the fire certificate. But when the design process involves the public space, the problem is magnified. In addition to satisfying the police, city planners, historic buildings committees and transport authorities, all such relevant political bodies as the planning committee and local authorities have to be consulted. Under such circumstances the permitting process becomes an equation with many variables.

One decisive factor here is the success with which political decision makers can be convinced that an electronic display can benefit an area for which they are responsible, whether this is the city as a whole or a particular city district.

The integration of the screens with the architecture is also of special significance here, especially in the case of urban locations and larger screens. There are examples of large German cities where the relevant authorities are quite open to the idea of more screens being located in the city, as long as these are truly integrated into the architectural infrastructure and not simply grafted on.

This brings us to the end of the list of seven sins which are hindering the market breakthrough of DOOH. If these issues can be addressed then the proper establishment of digital outdoor advertising will be much closer.

And finally it is important to remember that these sins apply to media façades which are to be exclusively financed by advertising revenue. If other revenue sources are involved, then their relevance is naturally much reduced – but this is a matter for other chapters of this book.

Dr. Christoph Köller
Hürth, 2010/01/15

1 To be fair, it would be more correct to talk here about "Self Promotion": at least in cases in which the investment in a media façade has come from a company's own communications budget and in which the campaign has to do with the image of the company.

2 The LCD industry is working hard on the development of displays which can be read in sunlight, but is still limited by size (82" is the current limit) and situations of highly intense sunlight (from approx. 100,000 Lux).

MARKETING STRATEGIES
in Public Spaces

Advertising in public spaces is a delicate subject because all our experience shows that it often pays no attention whatsoever to its urban context. And advertising in public spaces becomes even more difficult as soon as electronic media are involved because moving images represent a very specific set of challenges to our perceptive abilities. Advertising is generally a cliché for aggressive communication and this cliché inhibits our ability to address the theme of advertising in the context of urban design. I will attempt to show, however, that it is precisely the creative integration of advertising into the public space which can deal with this cliché. This will become particularly true in the future if we move on to talking about marketing strategies rather than just advertising

I have two motives for seeing marketing strategies as one possible way of defining the contents of media façades:

In addition to seeing media façades as a gain for our culture it is clear to me that this new medium requires a financial element in order to flourish. Success cannot be assured by the idealism of just a few clients and investors. Such idealism is driven by the investor's objective of creating a unique positioning for himself and, over time, as many more such projects are realized, this objective becomes less attractive and the media façade is no longer a unique selling point. It would make sense to develop business models which underline the cost-effectiveness of media façades. This is true in all areas of culture. Even the wildest architectural extravagance must promise its investor some sort of return – whatever form this return might take.

TIMES ARE CHANGING!

We can seize the opportunity presented by the seismic changes going on in the advertising industry. Direct communication with the unlimited objective of selling aggressively is on the way out. Much more sensitive methods of active communication and the creation and promotion of communities are becoming ever more successful. This transformation is symbolic of a society which is undergoing deep behavioral changes brought on by the internet. Knowledge is no longer a vertical collection of information filtered by experts but rather it is something available to everyone in unlimited quantities. This opens up unimaginable opportunities for the contents of media façades. It would be simply stupid not to want to make use of and to shape these opportunities!

"Webguerillas", a social media agency from Munich, has described the new tendencies on the advertising market in a manifesto that articulates these changes in society extremely well:

1. Communities versus Classic Content Websites: Social communities are currently much more popular than classic media, content sites and portals. Traditional advertising strategies and the tried-and-tested media approaches are becoming less valuable.

2. Advertising Pull rather than Advertising Pressure: A large number of advertising contacts is no parameter for success. In the future it will be essential to animate and activate consumers and draw their own personal network into a campaign.

3. Fan Communities rather than Target Groups: the classic target group no longer exists. One must develop ways of identifying the fans of a brand regardless of the milieu in which they operate.

4. Activity Loss rather than Leakage: the term "leakage" is outdated. The basis of future advertising must be the individual dialog with consumers.

5. Thousand Conversation Price rather than Thousand Contact Price: the TCP says nothing about the potential for activating consumers. A better parameter for measuring active dialog in future will be the Thousand Conversation Price.

6. Gross Involvement rather than Gross Rating Point: the measuring of average advertising pressure is not enough. The age of Web 2.0 enables us to measure the actual effect of advertising.

7. Always On rather than Audience Rates: younger target groups watch television much more often (and in a much less linear fashion) via the internet. At the same time, they network together in communities and read the latest news. We need, however, a way of measuring this new "always on" modus

8. Involvement rather than Recipes: the active media consumer in lean-forward modus is increasingly replacing the couch potato.

9. Brand Content rather than Media Content: the classic media landscape is disappearing: brands and bloggers compete with editorial content sites for the attention of users.

10. Real-time Monitoring rather than Surveys of Coverage: current surveys of coverage are an anachronism in the age of Web 2.0. They lead to an unrealistic representation of the use of media and must be fully replaced by real-time monitoring.

11. Global Village rather than Nielsen Regions: The division of the map of Germany up into Nielsen regions harks back to the age of the postbus. Divisions and units of measurement are required which correspond with a cosmopolitan view of the world

12. Brand Messengers versus Advertising Vehicles: the advertising messages broadcast in the classical media are becoming increasingly less accepted. These are being replaced by net-based multiplicators reporting about brand preferences.

THE POTENTIAL OF ELECTRONIC MEDIA

In the meantime, the use of digital media in public space has become an international phenomenon and a "scene" has emerged which could hardly be more diverse. There are many different approaches to the use and integration of moving images in cities. Here is an overview of these various uses and of their relationship with advertising:

1. Urban Screens

A range of technically very different forms of electronic media combine into platforms on which advertising clips and posters are projected under the control of content management systems. This area of business is known as DOOH – digital out of home.

Media agencies, whose principle role is the physical organization of the campaigns of advertising agencies in public areas, are generally responsible for the management of these "monitors".

2. Public Viewing

Relatively large (and usually temporary) LED screens are constructed in public spaces in which large crowds can gather. These tend to show sports events. The events are financed by the advertising banners which are combined with the live pictures. (BBC Big Screen has developed itself strongly in this area as a result of which this is a very widespread approach in Great Britain).

3. Light+Art Projects

Many architects and investors commission media artists to enhance their façades with art. In most cases this art is abstract and highly suitable for the integration of intelligent interactivity into the visual imagery.

The art wishes to remain autonomous but it places itself at the disposal of the client who wishes to use art to create a positive image for himself.

The result is a transformation of architecture which, thanks to the dynamic identity imparted by the images, establishes a connection with the medial dynamism of society. The result is that the architecture becomes an advocate of trust in social change and appears to offer the client the perspective of being successful on the global scale.

Technologically, this transformation is achieved by means of individual façade constructions with integrated light and media technology and video projections.

4. The Façade as Medium

Electronic media are integrated into architecture in such a way that architecture becomes an interface with the virtual world. This process is often used as a means of supporting and communicating the corporate identity of the client. The borderline with light+art projects is fluid. Furthermore, the façade as medium tends to work with a media technology which is also able to show pictures of up to perfect video quality as a means of fulfilling its role of visually supporting the communications work of the client.

This technical potential leads investors to assume that investment costs can be compensated by advertising income. So far this has been shown not to be the case, although here one has to clearly differentiate between the media façade and the urban screen: a media façade is not a monitor but part of the architecture. In all corners of the world it is difficult to win approval for a project as soon as architecture is seen as being a purely advertising tool and, hence media façades tend to be designed to have a form other than a simple screen. The result is that the advertising clips produced by advertising agencies for screens tend not to be suitable for projection on media façades.

CONCEPTUAL APPROACHES TO MARKETING WITH MEDIA FAÇADES

The task is now to establish the use of media façades as a component of marketing strategies. The differentiation from classical advertising results in a different basic approach. Advertising always has an aggressive character, because its aim is to directly encourage the act of purchasing. A marketing strategy on the other hand contains a series of communications measures and our role is to find the right place for media façades in this strategy.

A further motive for the switch from advertising to marketing strategies is that a media façade should never be fed with simple series of advertising clips. Not only because this makes approval more difficult but also because such advertising appears little more than a monoculture. The façade has to create a clear role for itself in the public realm and its content must create a positive impression. Advertising which is linearly projected merges quickly into the general background noise of the city and reduces the presence of the architecture which it should be enhancing. This is extremely counterproductive. The value of the architecture should have a positive effect on brand communication.

The chain of communications of a marketing strategy can free a media façade from appearing to be a closed entity. The media façade becomes a module of the chain of communication which then continues over into other media or, better still, emerges from complex marketing structures which also incorporate other media. Marketing efforts are currently intensively shifting to online services. Marketing clearly leads to information strategies and clients trust brands which give them the best information. But no media in the public space can afford this information work because an information strategy must always lead to a way of communicating information which allows the target to focus on his computer or on other individual tools.

The Galeria Kaufhof on Alexanderplatz in Berlin was fully refurbished in 2005. The architects Kleihues+Kleihues opened up the façade with a huge window which was to be used as a media façade. Recordings made on the internal escalators during opening hours were to be projected, time-delayed, to the outside. Models would present the actual fashion from Kaufhof's brands, etc.

The space is ideal for public viewing projects.

The project failed due to the limited willingness of the media agency and Kaufhof management to agree to an individual "advertising revenue" model.

Concept by ag4 mediatecture company

www.mediatecture.info/29

As an important part of marketing work now takes place on the internet, the marketing strategy must envisage how clients can be attracted to a particular website. And here, media façades can play a valuable role. They can create emotional content which seeks to build thematic ties with the marketing work of a brand on the internet and, at the ideal moment, show a link to the relevant website. This can be connected with, for example, a QR code which can be directly read on smartphones and is correspondingly linked. The decisive thing is that the pictorial language on the media façade does not have to be (or, indeed, cannot be) based on stimulating the act of purchasing! The more magical the content of the media façade, the more attractive the connection with actual marketing work can be. But this magic is not based on storytelling. It does not have to be linear - which explains why it works in public spaces (which are also not linear) much better than clips which, by definition, are always seeking to sell something.

One can also work with greatly reduced associations. A highly refreshing but non-linear pictorial language can be created which, for example, releases a very positive mood in public spaces. A huge sailing ship with green sails comes into view: everyone in Germany immediately understands the connection with "Beck's" beer. Nothing more is necessary – one doesn't even need to show the Beck's logo. Of course, this only works when a brand has already established a strong brand identity via other media platforms.

LINEARITY – THE ENEMY OF INTERACTIVITY

Each form of interactive projection requires a modular, process-based organization of the content. This content can, by means of individual input, develop into multiple combinations. And now interactive interfaces can have an effect on the act of combination itself. In my interview with Bazon Brock, the artist hints at surprising ways of doing this through his positive interpretation of the principle of mimic. Advertising provokes responses from the public and these responses are then, by means of intelligent interfaces, fed back into the stream of contents, creating a dialog between the advertising and the viewer. The idea, however, has yet to be implemented and this is no easy task.

In the same way, content management can be linked to websites in such a way that user behavior has an influence on the character of the contents.

My platform "SITE character" begins to model the contents of media façades directly in line with the individual characteristics of a particular situation. In the early morning, for example, the contents can have a very fresh character. Appropriate brand messages can then be integrated into this mood. These messages can then be generated or controlled in real time via, for example, connections with the corresponding facesite or via smartphones.

This instrument offers many possibilities for creative design, whose communicative development could be part of a marketing strategy. In this way, advertising can spread its very subtle magic into the urban space.

The construction of the media technology shifts from the glass façade to the interior space so that the car is in front of the media display.

In 2004, Helmut Jahn realized an office building for a German investor on the Mittlerer Ring in Munich. The building was also to incorporate an Audi showroom. The transparent media façade played with the presentation of real objects superimposed with virtual images. The media design was kept deliberately simple and opened windows onto the cars.

The project failed due to the inability to obtain the agreement of the City of Munich which, at this time, was opposed in principle to granting approval to the use of interchangeable media. This led Audi to decide against moving into the space.

www.mediatecture.info/30

Concept by ag4 mediatecture company

In 2005 an open construction was created for the new circulation space of a shopping center in Oslo. The medialization was to be projected by an irregular structure of LED sections. The idea was that the use of consciously striking brand images would create a suggestive atmosphere in the shopping centre.

The concept was developed in purely architectural terms. The necessary support of neither the brands nor the shops was forthcoming. The result was that it was not possible to draw up a financial model which could offer the investor a means of covering the costs of the installation.

Concept by ag4 mediatecture company

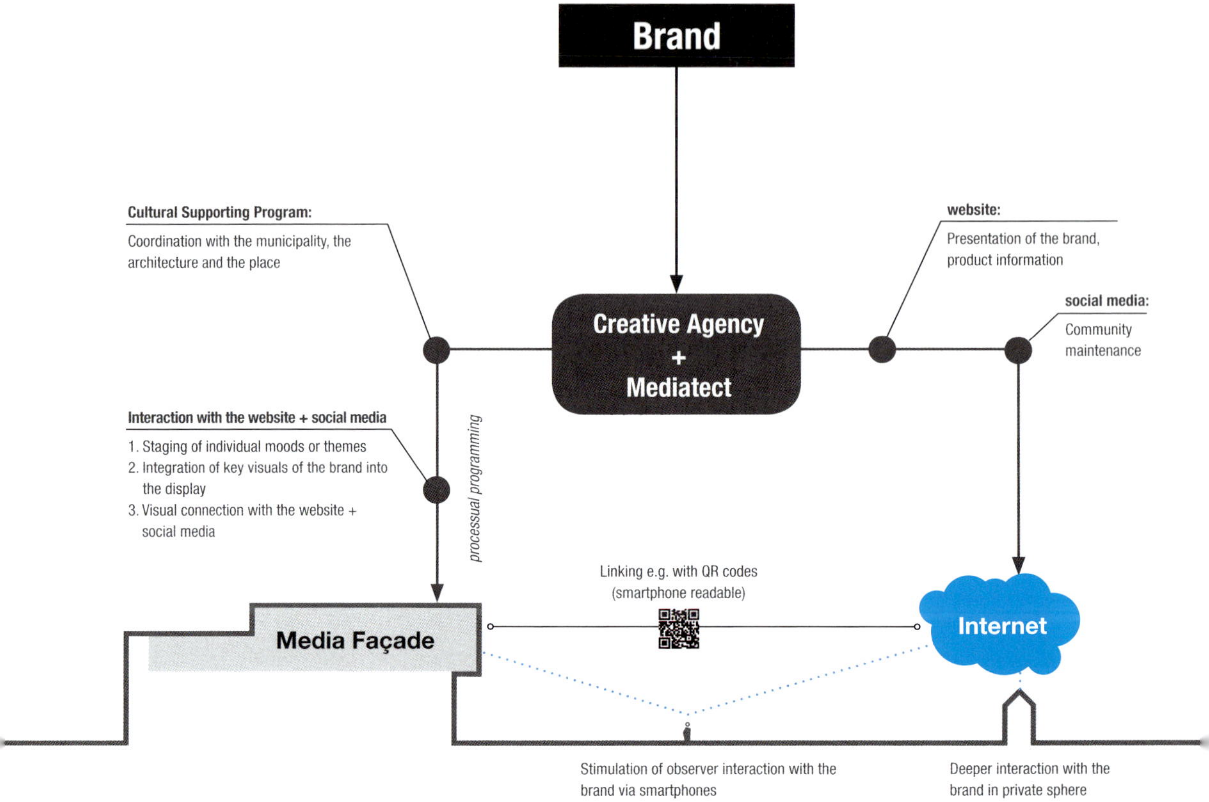

THE CONNECTION OF MARKETING AND PLACE

The standard form of content management of advertising in public places simply allows all forms of advertising to crash into each other. The important thing is that money is made. But the result is often that the advertising cannibalizes itself. How can advertising for cleaning materials build a relationship with advertising for a Jaguar? In contrast with marketing platforms, advertising in public places where many sources of advertising are close together has a simultaneous quality, whereas advertising in television and magazines is linear. And when this advertising no longer has the form of a poster but is driven by dynamic, electrical movement this feeling of simultaneous dissonance simply increases. The proximity of such different advertising content provokes a sense of irritation which, at best, is tempered by humor.

The reasons for this are two-fold. Firstly there is the sense that the types of advertising used for cleaning materials and for a brand like Jaguar could not be more different and, secondly, the feeling that advertising for, for example, a toilet cleaner is out of place on an expensive façade. The suggestion is, hence, that not every sort of marketing is appropriate for media façades.

One possibility that could be addressed is the idea of creating a profile for the advertising content which is considered appropriate for any location - taking into account the project, the client, the user and the nature of the surrounding public space. Analysis can then reveal which marketing themes can be appropriately linked with the location.

Example: *Networked presentation of the theme "sustainability" through a media façade*

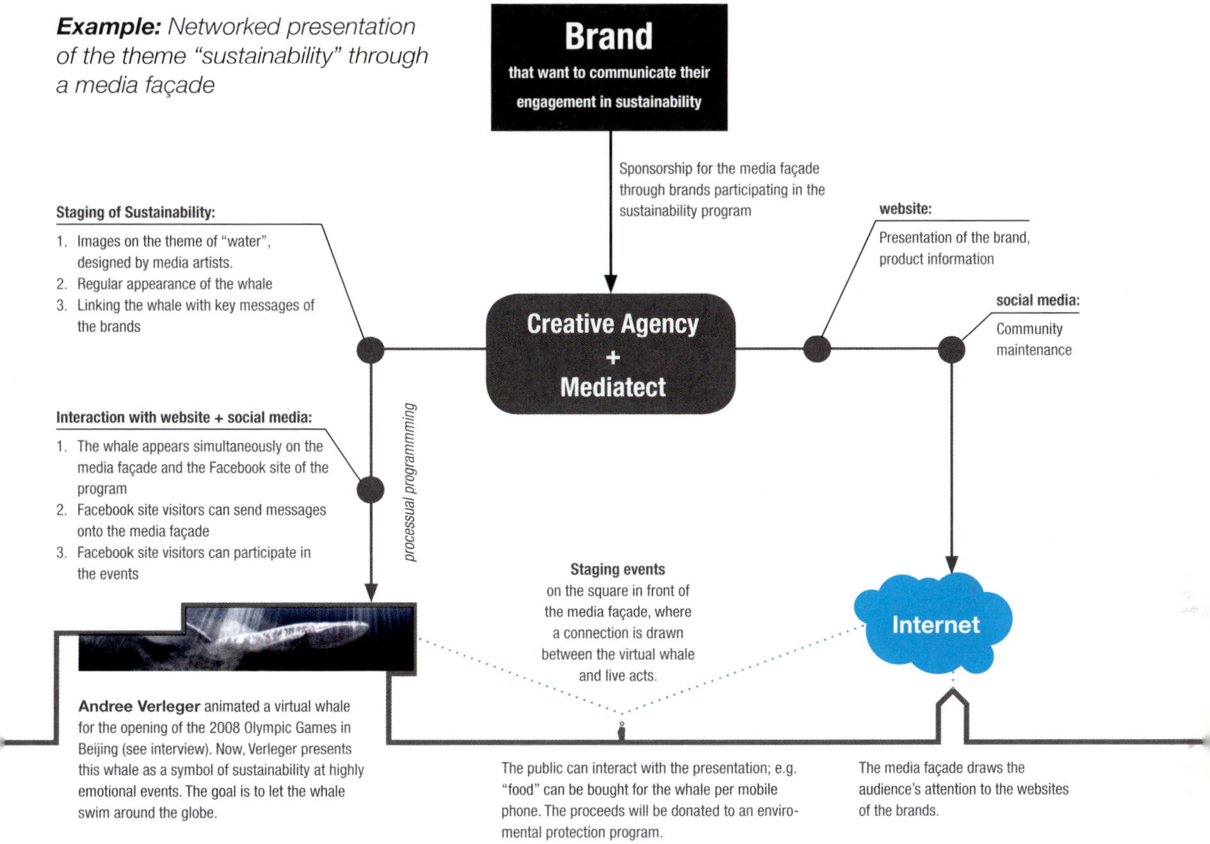

The result could be an exclusive relationship between a media façade and the brands which correspond with the defined profile.

These brands will be offered the opportunity to use the media façade as a marketing medium for a certain length of time and in line with a certain set of rules. Experience shows that a period of two to five years would be both sensible and realistic.

For media agencies operating in today's climate this idea may appear unrealistic because such agencies tend to prefer to have complete freedom in how they organize their advertising.

However, scarcity always leads to good business and, if a city were to agree to such a system as defined above, the exclusivity which it offers could become highly valued.

Such scarcity would surely be accompanied by an improvement in the visual quality of the advertising, not only because the notion of exclusivity is one which is bound up with good design but also because of the requirement to harmonically coordinate the marketing strategy with the nature of the surroundings. This could also encourage the city to permit advertising in locations which, as a result of their urban importance, are usually advertising-free and this, in turn, would offer marketing agencies the opportunity to occupy locations which allow their material to work more effectively then when it is simply displayed anywhere.

In the best case scenario this could lead to some form of image transfer between architecture and design. It could suddenly be of great benefit to a brand to be connected with particularly attractively designed architecture. Furthermore, the architect and the client will naturally only accept the advertising if it is conceived to form an aesthetic unity with the architecture – and the design work involved means that this is naturally only really thinkable if the brand – and, really, only this one brand - has some sort of exclusivity arrangement with the location.

That this can be successful can be seen in the established relationships between sporting arenas and brands.

And what is the problem when a building built by an investor adopts a brand identity? The tenants in such a building can vary widely and they often do not remain long enough to allow a sense of identity to develop. And the emotional values of a brand can certainly help in the process of identification of a workplace or living place. Such a transfer of identity is exemplified by the Sony Center on Berlin's Potsdamer Platz. The unique and self confident atmosphere of the complex has helped to increase rental levels, which suggests that such a comprehensive concept can bring dividends for an investor. At the same time, however, one should not forget that the development of the Sony Center – and the creation of this association between brand identity and building – was extremely expensive for Sony due to the huge work involved in assuring the all-important architectural quality of the development. Bearing this in mind, the creation of identity using a media façade is considerably more economical. And this approach is also relatively variable which, in an age in which brands can change quickly, is perhaps no bad thing. What happens, for example, when the architecture and the brand are no longer compatible?

THE RESPONSIBILITY OF CITIES

Marketing with media façades is complex. This can only work when a bundle of measures are coordinated with each other. But such a new form of advertising will probably not be pioneered by the advertising industry but rather by cities. For when urban decision makers really begin to address the issue of how to make their cities both more attractive and economically viable, they will soon realize that resources are scarce and that new creative approaches are essential.

In his archeological analysis of media, Erkki Huhtamo describes how the possibility of copying posters led to advertisers anarchically occupying every available surface in the city. Only with the introduction of rules and of tenants' rights was it possible to bring this process under control and this led to the development of a serious advertising industry with a professional organization.

The current emergence of the possibilities of electronic media is comparable with this historical situation from the 19th century. The sensual effect of moving images is so unbelievably strong that they very quickly dominate a space. The coordination of these sensual effects and of the effect they can have on a space is essential. But it can hardly be expected that those who produce advertising today are any more likely to voluntarily regulate their activities than were their 19th century forebears in the example described above. And it would hardly make sense for a single advertiser to limit his own activities, even if he was acting with the best motivations, because in this case his advertising would be sensually overwhelmed by its neighbors.

In the chapter about the Network (R)evolution I describe mediatects as creative developers of rules. In the context of advertising and media façades this role is very clear. The mediatect will not reinvent advertising but, much more; discover systemic connections between architecture, urban space and the challenges of the psychology of perception. And he will abstract all these aspects into a platform with a multiplicity of means of adjustment which he, depending on the context, employs sensitively to create a harmonious entity. The result could be to create rules which help the city to develop a modern method of city management in such a way that advertisers can work with architecture and the urban space in order to create magical effects.

It is hardly realistic to expect that planning authorities will, in the near future, be so convinced by these arguments that they draw up completely new rules for entire cities. When one considers such issues as long-term contracts etc, one can see that the economic implications of such a policy could, for instance, be far too complex. It would, however, be quite possible to establish sets of rules which guarantee design coherence in individual, closed locations. And this would be a way of establishing nuclei of good practice whose evident success could lead to something bigger.

RÉSUMÉ

Advertising in public places will always have both its friends and its enemies. Erkki Huhtamo's analysis shows that this argument is as old as advertising itself. And it is good that this question is controversial because this very controversy offers an opportunity for aesthetic and cultural development. And yet I am personally surprised by how ideological this discussion often is. Ideological demands for the removal of advertising from public spaces in order to protect architecture simply miss the point because good architecture itself has a promotional character. It is hard to understand why the principle of advertising is often seen as so negative, because it is a principle which, in the evolutionary sense, is part of life itself: plants produce blossom in order to attract bees to collect pollen. If there was no advertising, a whole range of vital processes would not take place. An urban space in which advertising is not permitted becomes unstable because it simply denies a key element of our social processes.

It makes much more sense to fight about the details of advertising design. There is dumb, ugly and arrogant advertising - and the same can be said about architecture!

Larry Weber, *an internationally renowned expert in communications and marketing, is passionate about the convergence of technology, the web and communications. He is the Chairman of the W2 Group, an ecosystem of digitally driven marketing services companies including A2a Media, a leader in outdoor digital display.*

INTERVIEW WITH LARRY WEBER

Dec. 2009, Boston

Having the opportunity to interview you about the issue of media façades and advertising is for me a special situation. On the one hand you make a very convincing case for the argument that traditional advertising - with its simple aim of direct selling - has little future. You are a supporter of the social web, in which consumers are directly involved in the development of content. And yet on the other hand you are a strategic partner of A2aMedia, a company whose area of operations is the marketing of urban screens. And urban screens usually show a linear series of advertising clips the role of which is to aggressively encourage the passer-by to make a purchase. This is doubtlessly not the aim of A2aMedia, but it is the reality, because the closed system of the media agencies permits no other concept.

I could of course question you here about your vision of advertising with electronic media in the public realm. But my fear is that your response would be very cautious because media façades are certainly not the focus of your work. For at the end of the day the success of the social web is not dependent upon medialized architecture, which is relatively cumbersome in comparison with the World Wide Web: not only because architecture cannot alter its physical position, but also because the technology behind media façades is complex – which differentiates them very clearly from web hardware.

But if we assume that a media façade is something different from an urban screen and, furthermore, that the content of media façades should offer investors some sort of financial return, then we can certainly discuss how media façades can support the creation of social networks. Do you have some ideas?

Lots! I believe that marketing in the future will be fully aligned with the search for experiences. The task will be to get people to experience something as a means of encouraging them to do something. It is this encouragement that I describe as "emotive", meaning that the experience must be increasingly emotional. This sensual experience can even go as far as using the sense of smell. Hence it is important that the media façade frees itself from being a form of one-way communication and becomes, instead, a bridge. Media agencies have believed up to this point that media façades are places for selling impressions but my view is that the media façade will become much more an experience in its own right which must be used as an interactive bridge. The media façade can be used to carry paid messages – but not in the way that this has happened for the past 75 years. This means that firms who operate both digital and real time strategies must now synchronize the various building blocks of these communication strategies – in just the same way that they must synchronize the buildings and other locations of their clients.

But the observation that social media are growing so fast - faster than any other form of media has ever grown before - is much more important. One only has to think that Facebook alone brings 400 million people together. Just imagine what it would mean if media façades could carry Facebook–updates and render such communication between people visible. Just imagine if the themes being discussed on Facebook could also appear on media façades.

But who will pay for this?

That is a fair question. Facebook presents itself as a "concept call global conversation" and that it also what we (w2group) are talking to Facebook about. Say that we go to Facebook together with, for example, General Electric with the idea of starting such a concept call global conversation. General Electric wishes to position itself as a Green Company. We guarantee that 100 million people will witness this communication and that 50 million people will actively take part. We could arrange that, for example, the CEO of GE has a conversation with Al Gore and then (for – say - 500,000 dollars) we could broadcast this communication on our media façades. In such a way media façades become part of a global strategy in which communication becomes visible in the public realm with people using their iPhones and Blackberrys in order to spontaneously make their own contribution.

The result is that, rather than being driven by advertising, content is driven by sponsoring: which, in turn, can naturally be accompanied by logos.

I also believe that media façades can work in the interests of "social goodness" by supporting communities of local groups and highlighting, for example, local heroes who are notable for their sporting achievements or social engagement.

In addition to this I feel that brands must develop their presence in all dimensions. An example: if I buy a book from Amazon I can look at a video of the author, read something about the book and even write my own criticism. That is to say that I remain for some time in a specific virtual context – and this is something which a media façade can, in some reduced way, also enable. More than anything, a media façade can be part of this comprehensive brand presence.

Away from the world of brands it is necessary that a framework emerges which permits media façades to develop an organic structure. In the resulting data-flow an editor could build in – just as an example – an image of an artwork such as the Mona Lisa and program it to appear 30 times a year as a means of establishing a certain atmosphere. This would simply be one basic element in a collection of planned content. This content would be organized - but not simply, as if by some random generator. Rather the content must make sense and be structured according to a specific set of rules which guarantees output which entertains and informs as required - as well as encouraging consumption.

I like this organic description of content design. Indeed it is a theme which I would like to address in more detail – albeit from another direction. You support the creation of communities. A community is something very alive and you describe in your books that it makes no sense to want to control communities. But as an architect or, better, a mediatect I have more affinity with symbolism, as a result of which I am interested in the creation of identity. I interviewed a historian and we agreed that the usual principles for creating identity – such as the meaning of home - are less effective today than they used to be. And I was surprised when my interviewee identified our consumption patterns as a key element of our creation of identity. Being a fan of a particular brand codes me in a way that other people can understand. But the meaning of this brand may for me be very abstract in that it has no face and I cannot communicate with it.

This is surely precisely your theme. And you pay for example a great compliment to the apple brands because they are brands which know what it means to present themselves with great transparency and concreteness. I would like to add that these are also brands which understand how to use flagship stores (which for me are the best examples of mediatecture) as a means of creating a face for their brands. But not every brand can present itself through such flagship stores. Maybe brands can gain a face through such media façades in that these become a part of the website of the brand. Maybe media façades can play a role in the social web in that they become the face of a brand with which one can interact. Here I am referring to a face in the sense of an organic symbol. The term that I use for this is "facesite". What do you think of this?

I like the term facesite. I think that architecture can change the face of a community. I was once lucky enough to have dinner with Frank O. Gehry and he explained to me what a terrible and dirty city Barcelona used to be and how the identity of the citizens had changed as the architecture became better: Or about the influence of Sydney Opera House and how proud people now are to belong to such a city. And the feeling that I get at the moment at which I realize, 'hey, this is all a part of me,' is a feeling which can also be strongly reinforced by digital façades. Brands will be heavily influenced over the next 30 years by the consistency of their communication strategies. They must be consistent in the way in which they steer the discussion and represent their ideas and façades or - as you say - facesites can be a very important part of this because they reach people every day.

But I would like to suggest a difference in the way one evaluates the architecture of star architects. A building by Frank O. Gehry could stand anywhere in the world because its design has, in general, little to do with the regional culture. It is acquired as a means of enhancing a place and when such architecture becomes famous then the local people are certainly very proud. This is doubtlessly an identification factor. But I also feel impelled to criticize such a situation in the sense that such architecture only fulfils its own identity and does little to enhance any particular social system. Hence my hope that, even in the purely commercial sense of media façades, brands can also be convincingly represented as part of our processes of social communication and identification. Not via the humdrum playing of advertising clips but rather via the tie-up between the media façade and the web-presence of the brand. But the question for me is how?

Well, media façades can establish direct contact with people via the processes of the social web. This is the social web as an information carrier. The media façade can actively ask people on the street what they think about a certain issue – and they can then use a smartphone to answer. I think that one can encourage a lot of communication.

But one should consider that the possibilities of text-based communication in public spaces are limited simply because one cannot concentrate in the same way as in front of the computer at home. Car drivers - in particular – cannot react at all.

It is not necessary to directly copy standard forms of conversation – there are many ways of inviting people to respond. One can present a subject and request that people use Myspace or Facebook to respond as and when they see fit. Those who have the time and the means can do so immediately with their smartphone while others can do so later from home.

I can imagine that the presentation of a brand on a media façade is in reality a sort of introduction – as already happens now on the internet – and that this introduction invites the viewer to check something special on the website.

Exactly – it's merely a question of establishing contact.

And this is of course perfectly in line with the definition and role of mediatecture.

My personal experience is, incidentally, that the creation of such connections is generally blocked by advertising and media agencies. I have worked on numerous proposals for media façades where the architect or investor has made the assumption that the advertising income would finance the project. But media agencies only seem open to projects which use the media façade as a traditional television, with a perfect image and a classical format. But such a form of billboard is of course – for reasons which I can completely understand - difficult to establish in cities all around the world. Yet on the other hand the market for communications in public areas is growing. Why do advertising agencies seem so indifferent to this problem and what is the source of this contradiction?

It's all a question of timing. We have had 75 years of one-way communication and that is something which you cannot reverse overnight. I once had a teacher who explained to me that it is not as if humankind went to bed one night in the renaissance and awoke in the baroque age.

On the other hand one must naturally expect that a certain form of media-buying will remain relevant.

But this form seems so dumb and inefficient to me!

I agree with you – but things are changing. Many brands are remodeling their internal communication and agencies are springing up everywhere which are much more like systems integrators. The problem with media agencies is also that the key figures tend to be people with somewhat aggressive natures who will push their strategy through, even when it makes little sense for their clients. This is a personality structure with which the world of social networks has very little in common.

Marius Watz (NO) *works with software processes as an aesthetic material. His work is concerned with hard-edged rule-based geometric abstraction, exploring the expressive potential of generative systems. He works with a range of output formats, from realtime software works and public projections to physical objects created using digital fabrication.*

Watz has exhibited at venues like Victoria & Albert Museum (London), Todaysart (The Hague), Künstlerhaus (Vienna), Museumsquartier (Vienna), ITAU Cultural (São Paulo) and Club Transmediale (Berlin). He also does audiovisual performances with musician Alexander Rishaug.

Watz is the founder of Generator.x, a curatorial platform for generative art and design. He is a lecturer at the Oslo School of Architecture and at the Oslo National Acad- emy of the Arts, Department of design. He has given workshops and lectures on computational aesthetics, live cinema and digital fabrication.

Marius Watz is represented by [DAM]Berlin. He is currently based in Oslo and New York.

MEDIA ARCHITECTURE
and data art

Architecture is generally understood as the practice of shaping space. Obviously, this covers the building of houses, but it also implies extended types of spatial intervention that does not necessarily include material structures. Landscape architecture manipulates natural spaces through terra-forming and biological processes. Urban planning deals with the macro-parameters of cities, shaping their population patterns and commuter flows to optimize productivity and quality of life.

Similarly, modern buildings are more than a collection of walls and spaces. They constitute complex infrastructures, the majority of their features hidden from view. Without ventilation shafts, cable chutes, electrical circuits, computer networks, and security systems, the physical space would be of little use to their inhabitants beyond providing basic shelter. More often than not, this invisible infrastructure is of greater importance than the physical walls it runs through.

In this immaterial conception of architecture, the extension of physical space through electronic displays is a logical next step. The emergence of the field of media architecture might be the direct result of advances in lighting technologies, but its invention is an inevitable consequence of new functions architecture is expected to fulfil. As the nature of human activity increasingly becomes intertwined with the intangible world of electronic communication, the need to manifest this invisible sphere becomes ever more pressing.

Media architecture presents a possible response by enabling buildings to become carriers of information, whether explicitly by providing information displays or more obliquely as a signifier of architectural identity. In the latter case, media facades adopt a function historically fulfilled by ornamentation, signifying the function of the building and establishing its social standing. But as an architectural gesture media facades can go far beyond simple ornament, serving to externalize the internal status of the building. Ambient environmental conditions or data points relevant to resident corporations can be made explicitly visible, revealing an internal logic or intelligence both of the building and its inhabitants.

An early example of this strategy is Toyo Ito's Tower of Wind (1986). By cladding an exhaust tower with computer-controlled lighting, Ito transformed a utilitarian structure into an ambient display of local environmental conditions. Lab[au]'s chrono.tower design for the Dexia Tower facade in Brussels maps time to light by translating hours and minutes into RGB values. Yet another solution is found in The Source, a kinetic sculpture for the London Stock Exchange by artist group Greyworld. A grid of spheres suspended from 32 meter high cables rise and fall in response to changes in the market.

In my own work as an artist I deal with abstractions based on computational processes. I create images by defining the rules and algorithms describing every aspect of their composition, from relationships between visual elements to their colour and movement through the virtual spaces they inhabit. While my work is often concerned with formalist aesthetics, my inspiration comes from self-organizing phenomena and my work often explores forms latent within such systems.

As part of my practice I occasionally take on commissions to create visualizations for corporations based on data sets central to their business. My goal in these projects is not to create a utilitarian representation lending new insights into its semantic meaning (I am an artist, not a scientist). Rather I am looking to convey aesthetic and conceptual aspects of the information, as well as the intrinsic nature of the signal flow. An essential part of this process is conducting interviews with users whose work is closely linked to it. This is then followed by a material exploration of the data set, looking for hidden patterns and identifying parameters most suitable for visualization.

STOCKSPACE FOR KNIGHT CAPITAL GROUP

An example: "Stockspace" was commissioned by Knight Capital Group, a company providing electronic trading services for capital markets. Their activities include trade execution aided by advanced algorithms, access to multiple markets, and fully automated trading based on quantitative models. Knight's business is based on massive amounts of information measured in milliseconds. Their systems monitor thousands of stock symbols across multiple international exchanges, executing more trades than the New York Stock Exchange over its network. Its traders make decisions in real time based on miniscule fluctuations in stock prices and financial indicators, including non-numerical sources like news stories.

Asked to create a series of visualizations of this complex world of information flows, I decided to focus on two parameters: The sheer size of the data set and its temporal qualities. Time was investigated on two levels: A macro-level tracking changes of stock price over the space of several years and on the micro-level of intraday trading. Price fluctuations were mapped to spatial structures, revealing the oscillating nature of the input signal. By juxtaposing structures generated from a large number of stock symbols a landscape of data emerges, a symbolic landscape that Knight's traders and analysts rely on for their work.

New media theorists like Lev Manovich have posited the notion of "data art", in which information flows become the focus for aesthetic investigation. This idea is supported by popular response to data visualization, which despite its dry scientific connotations has been embraced with enthusiasm by lay audiences. Perhaps this is a response to the saturation of everyday life by electronic data sources, from always-on smart phones to online banking and augmented reality services. If the role of art is to address the human condition, then it must also address the intermingling of physical reality with intangible networks of information.

Understood in this context, media architecture has the potential to combine the traditional function of the facade as cultural signifier with a new role as ambient information displays. By communicating data points of significance to the function of building and its inhabitants, the facade becomes a dynamic mirror of its activities. Virtual layers of information could thus be manifested in physical space, responding to an extended understanding of the site that includes its intangible infrastructure.

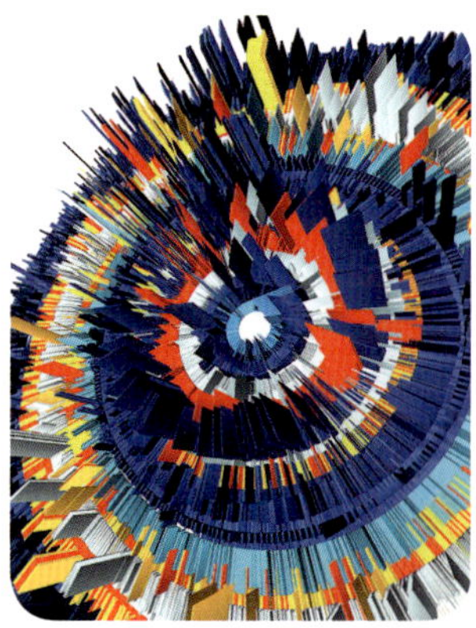

Stockspace

COMMENT

This work by Marius Watz brings the book back to its beginning. Bruno Schindler's essay "Schauplätze der Macht" (Sites of Power) challenged architecture to be a tool for sensual orientation. I began the book with the belief that our contemporary architecture could hardly fulfill this role anymore. However, Marius Watz's illustration of electronic trade on the capital markets clearly illustrates where the real power resides today. The influence these financial transactions have on the lives of all of us is tremendous, and it is only now becoming apparent. This power exists in total concealment, and it is clear that exactly this fact is extremely dangerous. But what kind of architecture could ever sensually describe such a power, a power whose potential lies in constant change? I believe I needn't further explain how this power can be made perceivable with mediatecture.

One might counter that this power very much likes to remain in concealment. On the other hand, we also have to admit that hitherto hardly anyone was interested in these complex processes. Nowadays, this is changing, and everyone is talking about new rules. This provides me with the opportunity to emphasize the "(R)evolution of Networks" and my opinion that it is an important task for mediatects to create sets of rules. It would seem that sets of rules could be extremely beneficial for our culture.

Today, Bruno Schindler's call for a sensual orientation takes stage at a new pace. A world that no longer understands itself anymore is at risk to succumb to a calamity. Soon or later, the actionists in the finance industry will be confronted with the fear of the incomprehensible in the form of aggression. They will have to go door to door for trust. Thus, transparency gains an enhanced meaning, one that can hardly be attained with full glass façades.

The work of Marius Watz completes another arc of the book. There isn't a more vivid example for a spatial extension than the expansion of the stock market via electronic media. I can't shake loose the suspicion that the "paradoxical spatial perception" will cause complications. The construction of development spaces can innocently appear to the actors as the creation of "spaces of possibilities". Technology, in the sense of modernity, enables variable action. But the virtual rooms of electronic media turn this game into uncompromising constellations, which could have very concrete consequences for people all over the world. Could it be that we, influenced by the thinking of modernity, are more or less blind in this eye? And that it is so utterly difficult for us now to set really restrictive rules as this would contradict our notion of modernity?

Best Practice advertising with mediatecture: the Apple entrance pavilion in Manhattan, Photo © Trey Ratcliff

SECURITY

Perspectives for Mediatecture

Designs where the technical potential of electronic media is integrated into architecture can already be seen in numerous facets of our living environment. There is little reason for further discussion about whether media have relevance in architecture.

Arriving at the end of this book, the questions arise of what level of maturity mediatecture has attained to date and what we can expect from this new discipline in the future.

Maturity

Many mediatecture projects live off the fact that the design is, above all, informed by the technical potentials of electronic media. Critically seen, it is the application of technology itself that becomes the concept, which naturally becomes problematic when we speak of a content-based orientation.

This also took place in music when Synthpop emerged with the appearance of the synthesizer. The mere availability of affordable synthesizers on the market influenced the music style. However, the majority of the bands, such as Kraftwerk or Depeche Mode, used the same synthesizer, and thus, at the end of the 80s, one had to admit that the sounds of the different bands were very similar. But this does not mean that electronic technologies lost any importance – on the contrary. They were subsequently used quite naturally and combined with classical methods of music production. In the music industry, the real business with electronic media took off when they were no longer seen as something special. They were employed and further developed for situations where they simply made sense creatively or reduced production costs. The synthesizers of the 80s were both unhandy and unreliable. The Synthpop scene alone was certainly not the only incentive for the industry to further develop this technology to the current state where almost any sound can be produced electronically. But Synthpop did play an important role: It destabilized entrenched structures in music, therewith opening new paths for innovations. And we are well aware that entirely new forms of social conduct evolved in the process; the idea of what one can and wants to understand as music changed radically (naturally, without putting an end to previous ideas of music). However, without the combination of new creative opportunities and profit maximization through reduced production costs, this development likely would not have happened. Today, a complex recording can be produced with considerably less effort. And as this is the case, young creative spirits, who previously wouldn't have had such a chance, can also experiment along new paths. It is another example of how market and culture intertwine with one another.

It is now apparent that mediatectural work has primarily established itself in two fields: first, in the scenographic charging of interior spaces such as trade fairs and exhibitions or forums and lobbies; and secondly, in the hype of media façades. However, it seems to be valid in both cases that the technological seduction is often in the foreground and the sustainable usage remains to be seen. Often the reason for this is that mediatecture and architecture do not trust one another – each discipline wants to outdo the respective other when they meet because of the wishes of a client[1]. The media installation attempts to capture the realm of sensory perception for itself, and then architecture strikes back with particularly intense spatial conditions.

Nevertheless, it is nothing new that emergent new media are always pursued for their technological aspects in the beginning and only later do suitable applications evolve. This can especially be observed in the context of the media façade. In effect, it is culturally at the stage of Screensaver 2.0, says Sven Miebach, founder of ecue, a company with a focus on complex light and media control systems[2]. He is not entirely wrong. Most media façades gain their allure through the novel forms of expression enabled by innovations in media technology.

What would have been the response if one foretold that the protagonists of Synthpop would be the pioneers of a vast industrial development! They just wanted to be different and break down the decrepit traditional structures in music culture. The motivations of media façade designers couldn't be more similar! The supposed main motive for the medialization of façades is to overcome their boredom and monotony.

It can be expected that also building culture will be more and more profoundly effected by the digitalization of our living environment and will change fundamentally; a process likely, but not solely, triggered by the desire to break through the boredom and the globally visible self-similarity of real estate around the world. Indeed, the current forms of media façades have what it takes to destabilize the societally anchored and disseminated understanding of what the role of architecture is[3]. It will be exciting to see what becomes of it!

The media façade will only reach its cultural prime when it can also promise economic return. There are still numerous technical challenges that need to be solved, for example, a sensible energy balance or overcoming so-called "light pollution". The industry will address these tasks when it becomes a profitable venture. The question is whether one should leave this development and the possible benefits that emerge from it to the industry and the market, or whether it is possible to build in ideas and aspirations that evoke a cultural impetus.

In this book, I have attempted to provide some ideas on this subject. And I would be thrilled when these thoughts have the faculty to stimulate discussion. They all rotate around the observation that the Internet is causing fundamental changes in our societal structures.

However, I am of the opinion that any cultural claim in this context only has a chance of success if the mediatecture building culture becomes a business model that bears advantages for real estate investors. The following is one aspect: In building projects where emphasis is placed on enhancing the identity of a place in the sense of the Bilbao effect, the costs of medialization would definitely be comparable with the costs of endowing a building with the desired attractivity necessary for it to be perceived around the world. And it would likely be cheaper to abandon the technical efforts of trying to make buildings defy the laws of gravity. One way or the other, it is about the "media value" – the value that is generated by reports about the project on display. This media value is the business model by which cities and companies around the world can afford exceptionally prestigious buildings. When architecture and mediatecture successfully engage with one another at this point, magnificent results can be expected – and quite possibly at a lower cost. This vision naturally implies that the intention for a medialization of architecture is not simply accomplished with the application of a media skin, rather that architecture itself must become a medium. Employing electronic media can play a role, but it is not absolutely necessary. At the forefront should be the will to shape architecture as a medium that connects with people and their ideas instead of a competitive exploration of the physical limitations of buildings.

The Future

Our society has difficulties with the changes that have been triggered by the digital revolution. A vivid illustration of this is the call for the reconstruction of the baroque Berlin City Palace. A quick summary: Prussian rulers ordered the construction of the palace on the Spree Island in the 16th century. In 1950, GDR leaders had the palace demolished and built the "Palast der Republik" (Palace of the Republic) in its place. Upon German reunification in 1990, there was fierce dispute whether this palace was worth preserving. At the same time, the idea arose to reconstruct the baroque palace, which ultimately prevailed. An international competition selected the plan by Italian architect Francesco Stella: a new building functionally divided into one third for the municipal museum's collection of non-European art and culture, one third for the Central and Regional Library, and one third for the collections of the Humboldt University. This incredibly expensive project reveals a considerable bewilderment, at least in the minds of the German people. The longing for old glory is fed by the promise of winning back something substantial and valuable for our living environment. And simultaneously, people would receive exactly the opposite. In the magazine "Zeit" from June 2010, Florian Illies compared this project with Andy Warhol's "Brillo Box": "It has only been decided that there should be a historical façade, and the box will be built with the cubature of the former palace. Without noticing it, we have once again fallen into the trap of Andy Warhol and Pop Art: A shell is a shell is a shell."

One can also draw a wonderful comparison with René Magritte's picture that doesn't show a pipe. With that, we approach the image's mode of effect: "We humans need images about a thing only so long as we don't understand this thing, but we also always need these images when we want to understand a thing anew. And we use them when we do not want to understand something at all or in an 'art trade' where it should have nothing to do with an understanding," explain Klaus Wassermann and Vera Bühlmann[4]. How we deal with images is incredibly complex – and in no case does it redeem what the iconoclasts of the city palace had hoped for: stability and reliability.

If this is the case, then all the pursuit of clarified relationships is an ill-fated romance, which prevents us from arriving in our time with all of its complexities.

Would it be any surprise if I append this with the pledge that a mediatectural approach would provide perspectives of how to find a way out of this dilemma?

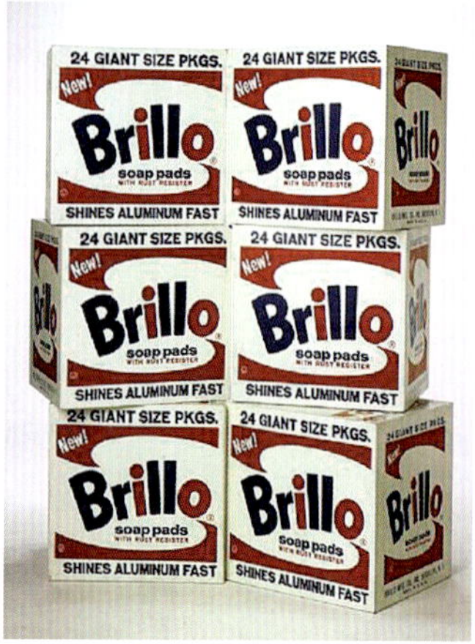

A media solution for the Spree Island building site could indeed attend to the very different reflexes and desires, and would never run into the quandary of wanting to be anything more than a shell. The group that wanted to keep the "Palast der Republik" clearly expressed the need for a medium that preserved a link with a societal reality, which still has not really been processed in people's minds.

Whether one palace or the other, it is all about a virtual space! In the middle of Berlin, in the middle of Germany (if not Europe), people are craving for a connection with their projections, desires and hopes. But it is simply impossible to unite people with one symbol – unless one chooses a method that doesn't require any content-related determination. And by inviting interaction, it could at once become a medium for the people – a medium that, on the side, could even be set up within the framework of a sustainable business model in which the marketing interests of the financial backers are taken into consideration. It was, by the way, exactly the lack of a business model that ultimately caused the city palace project to fail. It makes little sense economically to construct a new platform for the content (the three institutions and their collections). In the end, one has to pay and one needs a good reason to do it.

Mediatecture not only provokes a change in mindset in individual disciplines, its challenge lies in the connection of very different realms. Such a change will not take place over night, but it belongs to a cultural and economic dynamic characterized by dissolving borders.

It is imaginable that the current interim solution for the Schlossplatz square in the form of a big field will become cult because it represents in a wonderful way the dissolution of confined spaces and simultaneously guarantees real encounters between people.

1. The reader might sense my tension as I feel connected to both mediatecture and architecture.
2. ecue designed the media control systems for the Yas Hotel in Abu Dhabi.
3. Naturally, this also applies to all influences of digitalization on other processes in building culture. See "The Network (R)evolution", in this volume.
4. Streaming Spaces, Space + Media

It is also a form of mediatecture:

Photo © Julia Zimmermann

... the space is the medium,
and the grass with its bridges is the tecture ...

Author and Editor:

Christoph Kronhagel, born in Wolfsburg in 1958, studied architecture at the Technical University of Aachen. During his studies he learnt about the architectural profession from his experience in a series of large offices. In 1991, three years after completing his diploma, he founded ag4 (Arbeitsgemeinschaft für 4-dimensionales Bauen – working group for four-dimensional building) in Cologne. In the 1990s this interdisciplinary group became one of the first teams worldwide which attempted to actively bring together architecture and media. It was soon clear that this convergence was leading to a new discipline: mediatecture.

ag4 became a collection of specialists from highly diverse disciplines. The result was that the group was soon able to offer not just architectural design but also media production and the integration of media technology into architecture. Since 1997, ag4 has been realizing projects in the role of general contractor with full responsibility for costs. This has enabled the completion of integrated projects in which ag4 has used mediatectonic content to enable large corporations to raise awareness of both their identity and their corporate objectives.

The first ideas for media façades can be traced back to 1992 but it was not until 2004 that LED technology was far enough advanced to enable the realization of the first transparent media façade for T-Mobile in Bonn. Since then, Christoph Kronhagel has supported internationally famous architectural offices and investors in their projects for media façades.

ag4 develops its own technology and products for the variable integration of media technology into façades. Following the successful launch of these products, Christoph Kronhagel sold his share of ag4 to his partners in order to allow himself to continue concentrating, free and unattached, on mediatectonic projects, medial façades and the challenge of defining content. This book marks the start of this process.